INTRODUCTORY MATHEMATICS
A Prelude to Calculus

INTRODUCTORY MATHEMATICS
A Prelude to Calculus

Robert C. Fisher
Florida International University

John Riner
The Ohio State University

Jerry Silver
The Ohio State University

Bert K. Waits
The Ohio State University

Charles E. Merrill Publishing Company
A Bell & Howell Company
Columbus, Ohio

Published by
Charles E. Merrill Publishing Company
A Bell & Howell Company
Columbus, Ohio 43216

International Standard Book Number: 0–675–08750–3

Library of Congress Catalog Card Number: 74–16988

2 3 4 5 6 7 8 9 10 — 80 79 78 77 76 75

Printed in the United States of America

To
Arnold E. Ross

PREFACE

This text is designed to be used by students with various backgrounds and abilities. The main body of the text is a thorough, yet concise, presentation of the essential precalculus topics. The supplements after each section contain solved and partially solved examples and additional problems for the students to work. Students who need to study more examples and work more problems have that opportunity. The advanced students can skip many or all of these supplements without loss of continuity.

The text has been classroom-tested for five years at The Ohio State University. The precalculus classes normally cover the entire book in one term, but the text can be covered successfully at a more relaxed pace in two terms. A spiral approach is used in treating difficult concepts such as absolute value and inverse functions, and geometrical interpretations are liberally interspersed. In addition, an increasing level of sophistication is used in developing the topics as the students gain experience and confidence with the material.

The Q-questions scattered throughout the text check the students' understanding of the immediately preceding material and actively involve them in the learning process. The students also are actively involved in solving problems in the specially designed supplements. An achievement test at the end of each section serves as an additional check on the students' progress. Altogether, about 600 solved examples and over 1100 problems with answers are provided.

Chapter 1 begins with a review of algebra by introducing numbers via sets and the number line. The rules for algebraic manipulations are reviewed as the axioms for real numbers are mentioned briefly. (A more axiomatic treatment appears in Appendix A.) The remainder of the chapter is devoted to overcoming the common stumbling blocks of (fractional) exponents, inequalities, and absolute value.

Chapter 2 introduces the concept of functions and deals primarily with polynomial functions and their graphs. Translations are introduced in the chapter, although the graphs, not the axes, are moved. In addition, other "mappings" are introduced to facilitate graphing. This unique geometric approach has been very effective in extending students' abilities to sketch graphs quickly. These techniques, used throughout the remainder of the text in graphing transcendental functions, have proved quite useful during and after the study of calculus.

Chapter 3 begins with the study of inverse functions, and immediately uses these concepts to develop exponential and logarithmic functions. Properties of logarithmic functions are reinforced by computational work, which could be omitted, considering the increasing popularity of electronic calculators. The last section of the chapter (also optional) deals with summation notations, geometric series, and problems involving compound interest.

Chapters 4 and 5 provide a concise, yet thorough, coverage of trigonometry. The concepts of functions and inverse functions as well as graphing techniques are reinforced here. Chapter 4 introduces the trigonometric functions, covers the usual identities, and includes applications to right triangles. Chapter 5, which could be omitted all or in part, develops trigonometry further, and, in addition, introduces complex numbers. Complex numbers are introduced to provide an introduction to polar coordinates and to further reinforce certain facts about the trigonometric functions.

This book has served as the text for two precalculus courses at The Ohio State University. Administrative science majors cover Chapters 1 through 3, while engineering, physical science, and mathematics majors also cover Chapters 4 and 5.

Finally, we wish to thank all the individuals who have contributed to this book. Literally thousands of students and many teachers have contributed greatly to its present form. Our colleagues Larry Elbrink, Dave Mader, Nick Moore, and Jim Schultz have been especially helpful. But certainly our highest distinguished service award must go to Dodie Huffman, who faithfully deciphered horrendous scribbling in order to type and retype the manuscript in its several versions.

CONTENTS

INTRODUCTION

Before you begin to use this book, you should be aware of a number of special features incorporated in it. Interspersed in the usual text material are a number of Q-questions which appear in boxes. As soon as you encounter a Q-question, you should answer it and write down the answer. The Q-questions are designed to test your understanding of the immediately preceding material. The answers appear at the top of the next page you *turn*. If you cannot answer a Q-question correctly, you should reread the preceding paragraph or two.

Q0: Where do you find the answer to Q0?

At the end of each section is a problem set containing two types of problems. The early problems are basic and direct applications of the material in the section, while those following the double line are more difficult. You should work all the basic problems and check your answers with those provided immediately after the problem set. Additional basic problems are found in the supplement, a section which immediately follows each text section and which provides more work in that section, if you feel that you need it. The supplement consists of solved problems of the

> A0: Here!

type encountered in the immediately preceding section, followed by partially solved problems to which you are to contribute. There are drill problems for each major topic in the supplement, and, finally, a set of supplementary problems. After completing the supplement, you should rework the problems that gave you difficulty at the end of the text section. Of course, you can use the supplement to obtain additional practice and to improve understanding, even if you have successfully completed a problem set.

The process of learning mathematics is one of internalizing it. While a good teacher can provide guidelines to make the process easier, in the end it is the work and effort of the student that accomplishes the internalization. It is our hope that the format and style of this text will help you to learn in a way that is more personal and independent.

1 REAL NUMBERS

1.0 INTRODUCTION

Real numbers are the most useful mathematical objects that man has created. In this first chapter, some basic properties of the system of real numbers will be considered, including some review of exponential and radical notation. We begin with some work involving sets to develop our ability to express ourselves clearly and concisely. We introduce the number line in order to picture sets of real numbers and to study the concepts of order and absolute value.

While it is not our intention to provide a rigorous development of the real number system, we do want to point out that a few basic rules and their logical consequences govern the algebra of real numbers. It is important that you be able to use this algebra easily and confidently in your study of mathematics and its applications.

1.1 SETS, NUMBER LINES, AND INTERVALS

Rationals = Q = p/q where p +q are integers.

The language of sets and set notation is used widely in mathematics and its applications. The word *set* is used to denote a collection of objects. This usage is consistent with the use of the word set in some common phrases such as "a set of dishes" or "a tool set."

It is possible to describe a set by listing its members or elements; for example, {1, 2, 3, 6} is the set whose members are the numbers 1, 2, 3, and 6. Frequently, braces are used as above in set notation. At other times a capital letter may be used to denote a set. As a matter of convenience, in this text we will consistently designate certain sets of numbers by particular letters. Among the conventions we will use are the following:

N is the set of natural numbers {1, 2, 3, ...}.
Z is the set of integers {..., −2, −1, 0, 1, 2, ...}.
R is the set of real numbers.

p/q where p + q are integers.

If x is a member of set A, we write

$$x \in A$$

and read this statement as "x belongs to A," or "x is an element of A," or "x is a member of A." For example, 2/3 is a real number so we can write $2/3 \in R$. On the other hand, 2/3 is not a natural number, indicated by $2/3 \notin N$.

Q1: Which of the following statements are true:

$$0 \notin N, \quad \sqrt{3} \in R, \quad \sqrt{4} \in Z, \quad \sqrt{-2} \notin R, \quad \sqrt{9} \in N, \quad -\frac{4}{3} \in R$$

Another method of designating a set is to state a property that enables one to tell what objects belong to a set. For example, the set {1, 2, 3, 6} can be described as the set of natural numbers that are divisors of 6. This fact is written in *set builder* notation as follows:

$$\{1, 2, 3, 6\} = \{n : n \in N, n \text{ divides } 6\}$$

We read the right side of this equation as "the set of all n such that n is a natural number and n divides 6." Notice that braces are used again to indicate a set. To the left of the colon is a letter denoting an element of the set; to the right of the colon is a statement expressing the condition required for membership in the set.

Because most of our work will be concerned with real numbers, we will agree that, unless otherwise specified, all our sets contain real numbers although we do not explicitly mention that fact. For example,

$$\{x : x^2 = 4\} \quad \text{means} \quad \{x : x^2 = 4 \quad \text{and} \quad x \in R\}$$

and, of course, this set is $\{-2, 2\}$.

When conditions are used to define a set, the conditions may be such that no element belongs to the set. For example, since the square of every real number is positive, the set $\{x : x^2 = -4 \text{ and } x \in R\}$ has no elements in it. This set is the *empty set*, or *null set*, denoted by the symbol $\emptyset$.

If every member of a set A is a member of a set B, then A is a *subset* of B, or A is *contained in B*, written as $A \subseteq B$. For example, $\{1, 3\} \subseteq \{1, 2, 3\}$. This concept is stated formally in the following definition.

**Definition
1.1.1**

$A \subseteq B$ (read "A is a subset of B") if and only if $x \in A$ implies $x \in B$.

Notice that the relations $A \subseteq A$, and $\emptyset \subseteq A$ are true for every set A.

Q2: Let $A = \{n : n \in N, n \text{ divides } 6\}$ and $B = \{k : k \in N, k \text{ divides } 12\}$. Which of the following statements is true: (a) $A \subseteq B$, or (b) $B \subseteq A$?

A1: All are true. (Note: $\sqrt{4} = 2$, $\sqrt{4} \neq -2$. The symbol $\sqrt{a}$ represents the *nonnegative number* whose square is a.)

A2: (a) is true.

Two sets C and D may be described by different conditions, but each may contain exactly the same elements. In that case the two sets are equal, written $C = D$. To prove that two sets are equal, we show that each set is a subset of the other. In symbols, equality of sets is defined as follows.

**Definition
1.1.2**

Let C and D be sets. $C = D$ if and only if $C \subseteq D$ and $D \subseteq C$.

Q3: Is $N = Z$? Explain.

As we have said, much of the time in elementary mathematics we deal with subsets of the set R, the set of real numbers. Thus when we say "number," we mean "real number." A real number can be pictured as a point on a *number line*. We construct a number line, a geometric representation of R, by choosing two points on a line and labeling one O and the other U (see Figure 1.1.1). These two points are called the *origin* and the *unit point*, respectively. We take the length $\overline{OU}$ to be 1 and assume that the length of every line segment is a number. If a point P of a number line lies on the same side of the origin as the unit point U, then P represents a positive number x, called the *coordinate of P*. This positive number x is the length $\overline{OP}$. The points O and U have coordinates zero and one. A point S which lies on the side of the origin opposite to that of U represents a negative number y. The positive number $-y$ is the length $\overline{OS}$ (see Figure 1.1.1). Conversely, it is true that for any real number z there is a point R of the number line whose coordinate is z.

Figure 1.1.1

In summary, each real number corresponds to exactly one point on the number line. As a matter of convenience we speak of "the point 2" instead of "the point whose coordinate is 2," and so on. In particular, we talk about the origin 0 and the unit point 1.

A number line can be pictured as a horizontal line with the unit point to the right of the origin, as shown in Figure 1.1.2. When we compare the "sizes" of any two numbers, we obtain the *order relation* in R. This order relation can be described graphically. For example, the number x is *less than* the number y if x is pictured *to the left of y*, as shown in Figure 1.1.2. We write this fact using an inequality sign ($<$) as follows: $x < y$. For example, $2 < 3$ means that 2 is less than 3 and that 2 lies to the left of 3 on the number line, as in Figure 1.1.2.

Figure 1.1.2

Q4: Which of the following statements is true:

$$11 < 7, \quad -2 < -3, \quad -4 < -1$$

Sometimes it is convenient to reverse the inequality symbol. The statement $y > x$, read "y is greater than x," means that y is to the right of x on the number line. Thus, $y > x$ means exactly the same thing as $x < y$.

A real number x is *positive* if $x > 0$. Thus, positive real numbers are pictured as points on the number line to the right of the origin (see Figure 1.1.3).

Figure 1.1.3

Q5: List the numbers in the set $A = \{n : n \in N, n < 4\}$. Let $B = \{x : x \in Z, x > 0 \text{ and } x < 4\}$. How are the sets A and B related?

If $z < 0$, then z is a *negative* number and is pictured as a point to the left of the origin (see Figure 1.1.4). The number zero is neither positive nor negative.

Figure 1.1.4

A3: $N \neq Z$. $N \subseteq Z$ but $Z \not\subseteq N$.

A4: Only $-4 < -1$ is true. -2 *is not* to the left of -3, but -4 *is* to the left of -1.

A5: $A = \{1, 2, 3\}, B = \{1, 2, 3\}, A = B$.

Q6: Which of the following numbers are negative: (a) $-n$, where $n \in N$; (b) x, where $x > -3$; or (c) $-r$, where $r \in R$?

If a number x simultaneously satisfies the two statements $a < x$ and $x < b$, we write $a < x < b$. For example, the statement $-2 < x < 4$ means that x is *both* greater than -2 *and* less than 4. The set $I = \{x : -2 < x < 4\}$ is the set of all real numbers between -2 and 4 (see Figure 1.1.5). However, according to this usage, the set $A = \{x : 4 < x < -2\}$ is empty, since there are *no* real numbers that are both less than -2 *and* greater than 4. This set A should not be confused with the set $B = \{x : x < -2$ or $x > 4\}$ pictured in Figure 1.1.6.

Figure 1.1.5

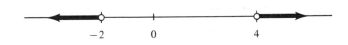

Figure 1.1.6

The set $I = \{x : -2 < x < 4\}$ is called an *open interval* and is denoted $I = (-2, 4)$. Notice that $-2 \notin I$ and $4 \notin I$. Our notation for open intervals is explained by the following definition.

Definition 1.1.3

If $a < b$, the *open interval* $(a, b) = \{x : a < x < b\}$.

Notice that (a, b) is a subset of R. In later sections we will consider intervals in more detail.

Q7: Sketch each of the following on a number scale:

$$A = (1, 4), \quad B = \{x : x > 4 \text{ or } x < 1\},$$
$$C = \{x : x > 4 \text{ and } x < 1\}, \quad D = \{x : 4 < x < 1\}$$

We shall use two operations to combine sets, union and intersection. The *union* of two sets A and B is the set whose elements belong either to A or to B (or to both), designated $A \cup B$. Thus, we have the following definition.

Definition 1.1.4

The *union* of two sets A and B is the set

$$A \cup B = \{x : x \in A \text{ or } x \in B\}$$

For example, $\{1, 2, 3\} \cup \{2, 4, 6\} = \{1, 2, 3, 4, 6\}$. The shaded region in Figure 1.1.7 (a *Venn diagram*) graphically illustrates Definition 1.1.4.

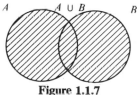

Figure 1.1.7

Q8: $Z \cup R =$ _____ , $A \cup \emptyset =$ _____

As the answer to Q8 shows, it is possible that the union of two sets may be equal to one of the sets, and it is always true for any set A that $A \cup A = A$. If $A \subseteq B$, then $A \cup B = B$ for any sets A and B.

The *intersection* of two sets A and B is the set whose elements belong to both A and B, designated $A \cap B$. Thus, we have Definition 1.1.5.

Definition 1.1.5

The *intersection* of two sets A and B is the set

$$A \cap B = \{x : x \in A \text{ and } x \in B\}$$

See Figure 1.1.8. For example, $\{1, 2, 3\} \cap \{2, 4, 6\} = \{2\}$.

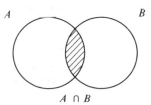

$A \cap B$

Figure 1.1.8

A6: (a) If $n \in N$, then n is to the right of the origin. Hence, $-n$ is to the left of the origin, so $-n$ is negative.
(b) x may be positive or negative.
(c) $-r$ may be either positive or negative. If r is positive, $-r$ is negative. But if r is negative, $-r$ is positive. If $r = 0$, $-r = 0$ and $-r$ is neither positive nor negative.

A7: *A*:

    B:

Sets *C* and *D* are empty.

A8: $Z \cup R = R$. Since integers are also real numbers, $Z \subseteq R$. $A \cup \varnothing = A$.

Q9: $Z \cap R = $ _____, $A \cap \varnothing = $ _____, $A \cap A = $ _____

Unions and intersections of sets of real numbers can be pictured as sets of points on a number scale. In particular, the unions and intersections of open intervals can be displayed graphically on a number line. For example, suppose that

$$I = \{x : -2 < x < 4\} = (-2, 4)$$

and

$$J = \{x : -1 < x < 5\} = (-1, 5)$$

From the pictures of *I* and *J* in Figure 1.1.9, it is easy to see that

$$I \cup J = \{x : -2 < x < 5\} = (-2, 5)$$

and

$$I \cap J = \{x : -1 < x < 4\} = (-1, 4)$$

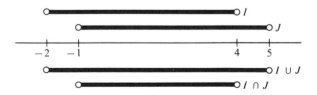

Figure 1.1.9

Of course, the union or intersection of two intervals may not be an interval, as the following examples show:

$$\{x : -1 < x < 0\} \cup \{x : 0 < x < 1\}$$
$$\{x : -1 < x < 0\} \cap \{x : 0 < x < 1\}$$

Q10: Picture the preceding sets on a number scale.

We will use the language and notation of sets developed in this section throughout our work, but you should not believe that we are studying set theory! In fact, we are using set language and notation only to clarify statements about real numbers. We conclude this section with two examples.

Example 1.1.1

Let $A = \{a : a \in N \text{ and } a < 6\}$, $B = \{b : b \in A \text{ and } b > 3\}$, $C = \{c : c \in Z \text{ and } c < 1\}$. Compute $A \cap B$ and $(A \cup B) \cap C$.

Solution

A is the set of all natural numbers less than 6. Thus $A = \{1, 2, 3, 4, 5\}$. B is the set of all elements in A greater than 3. Thus $B = \{4, 5\}$. So $A \cap B = B$. Also, $C = \{0, -1, -2, -3, \ldots\}$, and $A \cup B = A$, so $(A \cup B) \cap C = A \cap C = \varnothing$.

Example 1.1.2

Picture the sets $D \cup E$ and $D \cap E$ on a number scale if $D = \{x : x \in R, -1 < x < 4\}$ and $E = \{x : x \in R, x > 0\}$.

Solution

D is the set of all real numbers between -1 and 4 (see Figure 1.1.10). E is the set of positive numbers. A number belongs to D or E (or to both) if it is greater than -1. A number belongs to both D and E only if it is a positive number less than 4. Thus, the sets $D \cup E$ and $D \cap E$ can be pictured as shown in Figure 1.1.10.

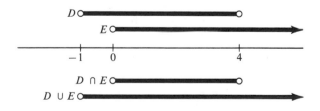

Figure 1.1.10

PROBLEMS 1.1
1. True or false? (a) $5 \in N$, (b) $2/3 \notin Z$, (c) $\sqrt{3} \in Z$, (d) $\pi \in R$, (e) $-10 \in Z$, (f) $0 \in N$

A9: $Z \cap R = Z$, $A \cap \emptyset = \emptyset$, $A \cap A = A$.

A10: $\{x : -1 < x < 0\} \cup \{x : 0 < x < 1\}$ is shown below.
$\{x : -1 < x < 0\} \cap \{x : 0 < x < 1\} = \emptyset$.

2. Let $A = \{n : n \in N$ and $n < 5\}$, $B = \{x : x \in Z$ and $x^2 < 9\}$, and $C = \{d : d \in N$ and d is a divisor of $9\}$. List the elements in each of the following sets: (a) A, (b) B, (c) C, (d) $A \cap C$, (e) $A \cup C$, (f) $(A \cap B) \cup C$

3. Let $D = \{n : n \in N$ and $1 < n < 2\}$ and $E = \{x : x > 0\}$.
(a) $D \cup E = $ _____ (b) $D \cap E = $ _____ (c) Is $D \subseteq E$?

4. Use set builder notation to describe the set of all even natural numbers.

5. Let $a \in A$, $c \in C$, $A \subseteq C$, $B \subseteq C$.
(a) Does $a \in C$?
(b) Does $c \in A$?
(c) Can there be an element in C which is in $A \cap B$?
(d) Can there be an element in C which is not in $A \cup B$?
(e) Is $A \cap C = C$?
(f) Is $B \cup C = C$?

6. Let $A = \{a, b, c\}$. There are 8 subsets of A. List them.

7. Let $M = \{x : x > 2\}$, $N = (-3, 1)$, and $P = \{x : x < 4\}$. Sketch each of the following on a number line: (a) $M \cap P$, (b) $M \cup N$, (c) $M \cup P$, (d) $M \cap N$

8. True or false? (a) $-3 > -1$, (b) $(-4)(-2) > 8$, (c) $-4/3 < -1.3$, (d) $-x < 0$, for all $x \in R$

9. If possible, write each of the following sets as a single interval. Sketch a picture of each set. (a) $(-3, 4) \cup (1, 2)$, (b) $(-5, 1) \cap (-1, 8)$, (c) $(-1/2, 1) \cup (1, 4)$, (d) $\{x : x < 5\} \cap (0, 1)$

10. Let $A = \{n : n \in N$ and n is a divisor of $12\}$, and let $B = \{n : n \in N$ and n is a divisor of $30\}$. Is $A \subseteq B$?

11. (a) If $A \cap B = \emptyset$, does $A = \emptyset$? (b) If $A \cup B = \emptyset$, does $A = \emptyset$?

12. Let m be a fixed positive integer, $A = \{n : n \in N$ and n is a divisor of $m\}$, $B = \{n : n \in N$ and n is a divisor of km, $k \in Z\}$. Prove that $A \subseteq B$.

13. If $a, b \in N$, then there are integers q and r such that $a = bq + r$, where $0 < r < b$ or $r = 0$. For example, if $a = 16$ and $b = 3$, then $16 = 3 \cdot 5 + 1$, and $0 < 1 < 3$. Let $P = \{n : n$ is a divisor of both a and $b\}$. Let $Q = \{n : n$ is a divisor of both b and $r\}$. Prove that $P = Q$.

14. Using Problem 13, prove that the greatest common divisor of a and b is also the greatest common divisor of b and r.

ANSWERS
1.1

1. (a) True, (b) True, (c) False, (d) True, (e) True, (f) False

2. (a) {1, 2, 3, 4}, (b) {−2, −1, 0, 1, 2}, (c) {1, 3, 9}, (d) {1, 3}, (e) {1, 2, 3, 4, 9}, (f) {1, 2, 3, 9}

3. (a) E, (b) ∅, (c) Yes

4. {x : x = 2k, k ∈ N}

5. (a) Yes, (b) Not necessarily, (c) Yes, (d) Yes, (e) No; A ∩ C = A, (f) Yes

6. {a}, {b}, {c}, {a, b}, {a, c}, {b, c}, A, and ∅

7. (a) See Figure 1.1.11.

2   4

Figure 1.1.11

(b) See Figure 1.1.12.

−3   1 2

Figure 1.1.12

(c) The entire line,
(d) ∅

8. (a) False, (b) False, (c) True, (d) False

9. (a) (−3, 4). See Figure 1.1.13.

−3   4

Figure 1.1.13

(b) (−1, 1). See Figure 1.1.14.

−1   1

Figure 1.1.14

(c) See Figure 1.1.15.

−1/2   1   4

Figure 1.1.15

(d) (0, 1). See Figure 1.1.16.

Figure 1.1.16

10. No; 12 ∉ *B*.

11. (a) No, (b) Yes

Supplement 1.1

1. Sets — Language and Notation

Problem S1.   Let $A = \{n : n \in N \text{ and } n < 3\}$. List the elements in A.

Solution: The phrase "$n \in N$ and $n < 3$" describes the conditions for membership in set A. It means that in order to belong to A, an object must be a natural or whole number ($n \in N$) and also must be less than 3 ($n < 3$). Remember that N, the set of natural numbers, is the set $\{1, 2, 3, 4, \ldots\}$, where the "dots" . . . mean "continue consecutively." The natural numbers less than 3 are 1 and 2. Hence, $A = \{1, 2\}$.

Problem S2.   Let $B = \{n : n \in Z \text{ and } -3 < n < 2\}$. List the elements in B.

Solution: First recall that $n \in Z$ means that n is an integer. The set Z of integers consists of the whole numbers, the negative whole numbers, and zero. $Z = \{0, \pm1, \pm2, \pm3, \ldots\}$. The statement "$-3 < n < 2$" means that n is between -3 and 2. The integers between -3 and 2 are $-2, -1, 0,$ and 1. Hence, $B = \{-2, -1, 0, 1\}$.

Problem S3.   List the elements of C, where $C = \{n : n \in N \text{ and } n^2 < 6\}$.

Solution: The elements of C are all the _____ numbers whose _____ are less than ____. These are the numbers _____. Therefore, $C = $ _____.	natural squares, 6 1 and 2 $\{1, 2\}$

Problem S4.   Let $D = \{n : n \in N, n \text{ is a divisor of } 16\}$. List the elements of D.

Solution: The elements of D are all the _____ _____ which	natural numbers

_____ 16. To find the divisors of 16, factor 16. $16 = 8 \cdot$ ____ $=$ $4 \cdot$ ____ $\cdot$ ____ $= 2 \cdot$ ____ $\cdot$ ____ $\cdot$ ____. Hence, the divisors of 16 are 1, 2, $2 \cdot 2 = 4$, 8, and ____. Therefore, $D =$ _____.	divide 2 $2 \cdot 2$, $2 \cdot 2 \cdot 2$ 16, $\{1, 2, 4, 8, 16\}$

DRILL PROBLEMS

S5. List the elements in M if $M = \{n : n \in Z \text{ and } -4.4 < n < 3\}$.
S6. Let $T = \{t : t \in Z \text{ and } t^2 < 16\}$. Let $S = \{s : s \in T \text{ and } s > -2\}$. List the elements in T. List the elements in S.
S7. Let $D = \{n : n \in N \text{ and } n \text{ is a divisor of } 36\}$. List the elements in D.

Answers

S5. $M = \{-4, -3, -2, -1, 0, 1, 2\}$
S6. $T = \{\pm 3, \pm 2, \pm 1, 0\}$, $S = \{-1, 0, 1, 2, 3\}$
S7. $D = \{1, 2, 3, 4, 6, 9, 12, 18, 36\}$

II. Unions and Intersections

Problem S8. Let $A = \{-1, 1, 3, 5, 7\}$ and let $B = \{0, 1, 2, 3\}$. Find (a) $A \cup B$, and (b) $A \cap B$.

Solution: (a) $A \cup B$ is the set of all elements which are in A *or* in B (see Definition 1.1.4). Each number in the list $-1, 0, 1, 2, 3, 5, 7$ is in either A or B. Hence, $A \cup B = \{-1, 0, 1, 2, 3, 5, 7\}$.
   (b) $A \cap B$ is the set of all elements which are in *both* A and B (see Definition 1.1.5). The numbers 1 and 3 are the only numbers which are in both sets. Therefore, $A \cap B = \{1, 3\}$.

Problem S9. Write $(-2, 3) \cap (1, 5)$ as a single interval, if possible.

Solution: First picture the intervals $(-2, 3)$ and $(1, 5)$ as sets of points on the number line, as in Figure 1.1.17. Remember that the symbol $(-2, 3)$ denotes *all* real numbers between -2 and 3. $(-2, 3)$ is not to be confused with $\{-2, 3\}$, which is the set consisting of the two *integers* -2 and 3. From Figure 1.1.17 it is

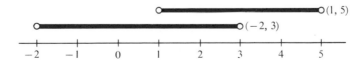

Figure 1.1.17

clear that the numbers which are in *both* (−2, 3) and (1, 5) are precisely those numbers less than 3 and greater than 1. This set is written in interval notation as (1, 3). Therefore, (−2, 3) ∩ (1, 5) = (1, 3). (Pictures such as Figure 1.1.17 are often convenient when one is graphing different intervals in the same problem.)

Problem S10.   Write (−2, 3) ∪ (1, 5) as a single interval, if possible.

Solution: Using the graphs from the preceding explanation, we observe that the numbers which are in (−2, 3) *or* in (1, 5) are precisely the numbers less than 5 and greater than −2. In interval notation this set is written (−2, 5). Thus, (−2, 3) ∪ (1, 5) = (−2, 5).

Problem S11.   Let $A = \{x : x \in R \text{ and } x > 3\}$, and let $B = \{x : x \in R \text{ and } x < 5\}$. Find $A \cup B$ and $A \cap B$.

Solution: On the number line, set A consists of those points to the _____ of 3, and B consists of those points to the _____ of 5. A picture of the sets A and B on a number line is _____. It is evident

| right |
| left |
| See Figure 1.11.8 |

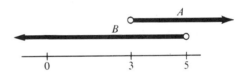

Figure 1.1.18

that $A \cap B$ is the set of all points between ____ and ____. In set builder notation, we write this as $A \cap B = \{x : x \in R \text{ and } _____\}$. In interval notation, $A \cap B = \_\_\_\_\_$. Furthermore, $A \cup B = \_\_\_\_$, since every $x \in R$ is either in ____ or in ____.

| 3, 5 |
| 3 < x < 5 |
| (3, 5) |
| R |
| A, B |

Problem S12.   Let $S = (−3, 2)$ and $T = \{−3, 2\}$. Find $S \cap T$.

Solution: S is the set of _____ real numbers between ____ and ____. Set T consists of the two _____ −3 and 2. Since neither −3 nor 2 is between −3 and 2, we conclude that $S \cap T = \_\_\_\_$.

| all |
| −3, 2 |
| integers |
| ∅ |

DRILL PROBLEMS

S13. If $A = \{1, 2\}$, $B = \{0, 1, 2, 3, 4\}$, and $C = \{n : n \in N \text{ and } n > 2\}$, then
(a) $A \cup B = \_\_\_\_\_$ (b) $A \cap B = \_\_\_\_\_$

(c) $B \cup C =$ _____ (d) $B \cap C =$ _____
(e) $A \cup C =$ _____ (f) $A \cap C =$ _____
S14. If $D = (2, 5)$ and $E = \{3, 4\}$, then $D \cap E =$ ____.
S15. Let $A = \{x : x \in R$ and $x > -1\}$, $B = \{x : x \in R$ and $0 < x < 2\}$, $C = (1, 4)$. Then $(A \cap B) \cup C =$ ____.

Answers

S13. (a) $\{0, 1, 2, 3, 4\}$, (b) $\{1, 2\}$, (c) $\{0, 1, 2, \ldots\}$, (d) $\{3, 4\}$, (e) N, (f) $\varnothing$
S14. $\{3, 4\}$
S15. $(0, 4)$

III. Subsets

Problem S16. Let $A = \{n : n \in N$ and $n < 3\}$, $B = \{x : x \in Z$ and $x < 3\}$, and $C = (1, 2)$. Show that $A \subseteq B$, $C \nsubseteq A$, and $A \neq B$.

Solution: The natural numbers which are less than 3 are 1 and 2. Hence, $A = \{1, 2\}$. To show that $A \subseteq B$, we must show that every number in A is also in B (see Definition 1.1.1). Since 1 and 2 are integers each less than 3, they are both elements of B. Hence, $A \subseteq B$.

To demonstrate that $C \nsubseteq A$, it is sufficient to find one number in C that is not in A. Since C consists of all real numbers between 1 and 2, we simply choose a number in that interval which is not an integer, for example, $3/2$. Then $3/2 \in C = (1, 2)$, but $3/2 \notin A$. Hence, $C \nsubseteq A$.

Finally $A \neq B$ because $B \nsubseteq A$. For example, $0 \in B$ but $0 \notin A$.

Problem S17. Let $A = (0, 2)$ and $B = (-1, 4)$. Is $A \subseteq B$? Is $B \subseteq A$?

Solution: If $x \in A$, then x is between ____ and ____. But a number between 0 and 2 is also between -1 and 4. Hence, $x \in$ ____, and thus _____. However, $B \nsubseteq A$, as we can demonstrate by finding one number in B which _____. For example, $0 \in$ ____ but 0 ____ A. Thus, _____.

$0, \quad 2$
B
$A \subseteq B$
is not in A
$B, \quad \notin$
$A \nsubseteq B$

Problem S18. Show that $(1, 4) \subseteq (0, 3) \cup (2, 5)$.

Solution: Written as a single interval, $(0, 3) \cup (2, 5) = (0, 5)$. Thus, we wish to show that $(1, 4) \subseteq (0, 5)$. But this simply says that all real numbers between 1 and 4 are also between 0 and 5. This is clearly true since $0 < 1$ and $4 < 5$. That is, $1 < x < 4$ implies $0 < 1 < x < 4 < 5$, or $0 < x < 5$.

DRILL PROBLEMS

S19. Let $E = \{-1, 0, 2\}$. Is $E \subseteq N$? Is $E \subseteq Z$? Why?
S20. Is $T = \{n : n \in N$ and $n^2 < 2\} \subseteq (0, 2)$?

Answers

S19. No; -1 and 0 are not natural numbers. Yes.
S20. Yes. $T = \{1\}$ and $1 \in (0, 2)$, so $\{1\} \subseteq (0, 2)$.

SUPPLEMENTARY PROBLEMS

S21. Let $A = \{n : n \in N$ and n is a divisor of 12$\}$, and let $B = \{n : n \in N$ and n is a divisor of 36$\}$. Show that $A \subseteq B$ and that $B \nsubseteq A$.
S22. If $P = \{n : n \in N$ and $n < 5\}$, $S = \{n : n \in Z$ and $-3 < n < 3\}$, and $T = (1, 2)$, then find the following sets: (a) $S \cap P$, (b) $(S \cap P) \cap T$, (c) $(S \cap P) \cup T$
S23. Describe $A \cap B$ if (a) $A = B$, (b) $A \subseteq B$, (c) $B \subseteq A$.
S24. Describe $A \cup B$ if (a) $A = B$, (b) $A \subseteq B$, (c) $B \subseteq A$.
S25. Let $A = \{x : x^2 < 4\}$ and $B = \{x : 0 < x < 10\}$. Sketch each of the following sets on a number line: (a) A, (b) B, (c) $A \cap B$, (d) $A \cup B$, (e) $((-1, 1) \cup (1, 5)) \cap (0, 2)$, (f) $(-4, 4) \cap Z$

Answers

S21. $A = \{1, 2, 3, 4, 6, 12\}$, $B = \{1, 2, 3, 4, 6, 9, 12, 18, 36\}$
S22. (a) $\{1, 2\}$,   (b) $\varnothing$,   (c) All real numbers between and including 1 and 2
S23. (a) A or B,   (b) A,   (c) B
S24. (a) A or B,   (b) B,   (c) A
S25. (a) See Figure 1.1.19.   (b) See Figure 1.1.20.

-2	2

Figure 1.1.19

0	10

Figure 1.1.20

(c) See Figure 1.1.21.   (d) See Figure 1.1.22.

0	2

Figure 1.1.21

-2	10

Figure 1.1.22

(e) See Figure 1.1.23.   (f) See Figure 1.1.24.

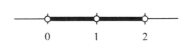

0	1	2

Figure 1.1.23

$-4 \quad -3 \quad -2 \quad -1 \quad 0 \quad 1 \quad 2 \quad 3 \quad 4$

Figure 1.1.24

1.2 BASIC PROPERTIES OF REAL NUMBERS

Geometrically, we can use the number line introduced in Section 1.1 to describe R, the set of real numbers. Each real number corresponds to a point of the number line, and, conversely, each point on the number line corresponds to a real number. Thus, we can think of the set of real numbers as the set of points on a number line.

Given the points 0 and 1, it is easy to visualize the natural numbers N and the integers Z on a number line, as indicated in Figure 1.2.1. Of course, there exist many real numbers other than the integers. For example, we can designate the real number 2/3 on a number scale by dividing the segment from 0 to 1 into 3 equal lengths. One way to do this is to use the method illustrated. You should be aware that it really is not "fair" to use a ruler to subdivide this segment because use of the ruler assumes that we already know exactly where 2/3 is located. The method we now describe uses only a straightedge (not scaled) and a compass.

Given the points 0 and 1 of the number line, we draw another line, obliquely intersecting the number line at 0 as in Figure 1.2.2.

Figure 1.2.1

Starting at zero on this second line, we mark a convenient length 3 times in succession, indicated *A, B, C.* We connect *C,* the third mark (corresponding to the denominator 3 of 2/3), with 1 on the number line. Parallel to the line *C*1 is placed another line through *B,* the second mark (corresponding to the numerator 2 of 2/3). This line then will intersect the number scale at a point corresponding to the real number 2/3. In a similar manner we can designate a point on the number line which corresponds to any real number of the form *p/q* where *p* and *q* are integers.

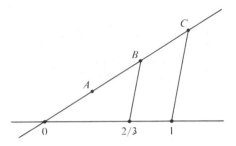

Figure 1.2.2

Q1: Construct the point $-2/5$ on the number line.

Those real numbers described above are commonly called *fractions* or *rational numbers,* defined formally as follows.

**Definition
1.2.1**

The set of rational numbers, denoted *Q,* is defined by

$$Q = \left\{ x : x = \frac{p}{q}, p \in Z, q \in Z, \quad \text{and} \quad q \neq 0 \right\}$$

Clearly, from Definition 1.2.1, $Q \subseteq R$; but $Q \neq R$. That is, there exist real numbers which are not rational (called *irrational*). In Appendix A to this section, we sketch a proof of the fact that $\sqrt{2}$ is an irrational number. Nevertheless, $\sqrt{2}$ corresponds to a point of the number line as illustrated in Figure 1.2.3. The right triangle shown has two sides of length 1 and hence, by the Pythagorean theorem, a hypotenuse of length $\sqrt{2}$.

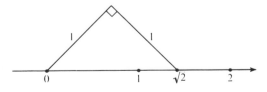

Figure 1.2.3

To summarize, every real number corresponds to a point on the number line, and conversely. Furthermore, $N \subseteq Z \subseteq Q \subseteq R$, but $N \neq Z \neq Q \neq R$ (see Figure 1.2.4).

Figure 1.2.4

A formal definition of the set of real numbers would consist of a collection of axioms that characterize the set of real numbers. We will present some of these axioms here. From these axioms, we will develop certain other basic facts in order to illustrate the method of proof that is used in mathematics. Some of the more important facts are labeled *theorems*. The proof of a theorem consists of a number of logical steps in which each step is justified by an axiom, definition, or a previously proved fact. The familiar algebraic rules or laws that apply to the real numbers are, in fact, either axioms or logical consequences of axioms for the set of real numbers. Appendix A deals more completely with the axiomatic development of the real number system.

In the set R of real numbers there are defined two operations or rules of combination, called *addition* ($+$) and *multiplication* ($\cdot$). The real numbers and these operations satisfy the following axioms.

Closure. If a, $b \in R$, then $a + b \in R$ and $a \cdot b \in R$. (This axiom says that the sum and the product of any two real numbers are real numbers. The number $a \cdot b$ is also written ab.)

The commutative axioms. If a, $b \in R$, then $a + b = b + a$ and $a \cdot b = b \cdot a$. (This axiom says that the order in which we add or multiply real numbers does not change the result.)

The associative axioms. If a, b, $c \in R$, then $(a + b) + c = a + (b + c)$ and $(a \cdot b) \cdot c = a \cdot (b \cdot c)$. (This axiom says that results in addition or multiplication are not changed by the way in which we group the numbers.)

The identity axioms. There exist two distinct real numbers, zero and one, written 0 and 1, such that if $a \in R$, then $a + 0 = 0 + a = a$, and $a \cdot 1 = 1 \cdot a = a$. (It can be proved that the numbers 0 and 1 are unique. The real number zero is called the *identity element for addition*, while the number one is called the *identity element for multiplication*.)

The inverse axioms. If $a \in R$, then there exists another real number, written $-a$ (read "negative a"), such that $a + (-a) = (-a) + a = 0$. Also, if $a \neq 0$, there exists another real number, written $1/a$ or a^{-1} (read "the reciprocal of a"), such that $a \cdot (1/a) =$

A1:

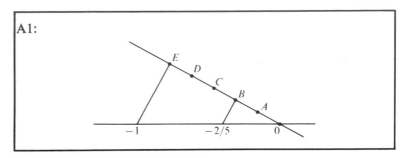

(1/a)·a = 1. (It can be proved that −a and a⁻¹ are unique. The
number −a is called the *additive inverse* of a, and 1/a is called the
multiplicative inverse or *reciprocal* of a.)

The distributive axioms. If a, b, c ∈ R, then a·(b + c) = ab +
ac, and (a + b)·c = ac + bc.

Q2: Which axioms are used to justify the following statements?
    (a)   (a + x) + b = b + (a + x)
    (b)   x + (y + (−y)) = x
    (c)   (a + c) + b = c + (a + b)
    (d)   −a + 0 = −a
    (e)   1·1 = 1

These axioms form the basis for developing the familiar "rules
of algebra" for numbers. (Remember, when we use the word
number, we mean *real number*.) Some of these rules are established
in Appendix A. For example, there we prove that if $a \in R$, then
$-(-a) = a$, and $a \cdot 0 = 0$. This second fact tells us why the number
zero cannot have a reciprocal, because no matter what number
would be assigned to the symbol 1/0, the product of that number
times zero would be zero (by the above) and not one as required
by the inverse axiom.

Another important consequence of the real number axioms is
the following theorem.

**Theorem
1.2.1**

If $a \cdot b = 0$, then $a = 0$ or $b = 0$.

This theorem, proved in Appendix A, is very useful in finding
solutions to certain equations. If we are given an equation and a
permissible set of numbers which may be replaced for the variable,
then any number from this set which makes the equation true is
called a *solution*. The set of all such solutions from the permissible set is called the *solution set*. For example, if the permissible
set is Z, then the equation 2x = 5 has no solution. That is,

$\{x : 2x = 5, \ x \in Z\} = \emptyset$. However, the solution set for $2x = 5$ in R is $\{5/2\}$. Unless the set of permissible solutions is explicitly mentioned, we shall assume it to be R.

**Example
1.2.1**

Find the solution set for $3x^2 - 8x = 0$, and justify each step in the computation.

Solution

One way is to factor the left-hand side:

$$3x^2 - 8x = x(3x - 8) \qquad \text{Distributive Law}$$

Now, $x(3x - 8) = 0$, if and only if $x = 0$ or $3x - 8 = 0$ (Theorem 1.2.1). Solve $3x - 8 = 0$:

$$(3x - 8) + 8 = 0 + 8 \qquad \text{Property of Equality}$$
$$3x + (-8 + 8) = 0 + 8 \qquad \text{Associative}$$
$$3x + 0 = 0 + 8 \qquad \text{Inverse}$$
$$3x = 8 \qquad \text{Identity}$$
$$\tfrac{1}{3}(3x) = \tfrac{1}{3}(8) = \tfrac{8}{3} \qquad \text{Inverse and Arithmetic}$$
$$\left(\tfrac{1}{3} \cdot 3\right)x = \tfrac{8}{3} \qquad \text{Associative}$$
$$1x = \tfrac{8}{3} \qquad \text{Inverse}$$
$$x = \tfrac{8}{3} \qquad \text{Identity}$$

Thus, the solution set is $\{0, 8/3\}$.

Example 1.2.1 illustrates the detailed use of the properties of real numbers to solve a simple equation. Normally we will not make such explicit mention of the properties of real numbers used when we solve a simple equation, but it is important to realize that there is a valid reason for each step in the solution of an equation.

In Appendix A, subtraction is defined in terms of addition, specifically, the addition of the additive inverse. Division of real numbers is defined in terms of multiplication by the reciprocal. Since the reciprocal of zero does not exist, division by zero is left undefined.

If $b \neq 0$, the number a/b is called the quotient of the number a divided by the number b and is also written $\frac{a}{b}$, or $a \div b$. The number a is the *numerator* of the quotient, and the number b is the *denominator* of the quotient. *Remember*, a quotient is undefined if the denominator is zero. If the denominator is not zero, the quotient is zero, if and only if the numerator is zero.

A2: (a)   Commutative law for addition
 (b)   Inverse and identity laws for addition
 (c)   Commutative and associative laws for addition
 (d)   Identity law for addition
 (e)   Identity law for multiplication

Q3: Find the set $\{x : A$ is undefined$\}$ if A is the quotient:

(a) $\dfrac{4}{x-1}$ (b) $\dfrac{-5}{x^2}$ (c) $\dfrac{x-3}{x-3}$ (d) $\dfrac{\sqrt{3}}{x^2+1}$

A quotient a/b is also called a *fraction*, and the rules for operating with fractions are consequences of the definition of division and the axioms for real numbers. You should be familiar with the following rules:

$$\frac{a}{1} = a \tag{1}$$

$$\frac{a}{b} = \frac{c}{d} \quad \text{if and only if } ad = bc \tag{2}$$

$$\frac{a}{c} + \frac{b}{c} = \frac{a+b}{c} \tag{3}$$

$$\frac{a}{b} \cdot \frac{c}{d} = \frac{ac}{bd} \tag{4}$$

$$\frac{a}{b} \div \frac{c}{d} = \frac{ad}{bc} \tag{5}$$

$$\frac{ca}{cb} = \frac{a}{b} \quad \text{for } c \neq 0 \tag{6}$$

The proofs of these properties are not difficult and are left for the exercises.

The following rule for adding or subtracting fractions follows from Equations (3) and (6):

$$\frac{a}{c} \pm \frac{b}{d} = \frac{ad \pm bc}{cd} \tag{7}$$

Q4: Prove that Equation (7) is true.

Let us conclude this section with several more examples.

**Example
1.2.2**

Find $\{x : x^2 - 4 = 0\}$.

Solution   Since $x^2 - 4 = (x - 2)(x + 2)$, we have $(x - 2)(x + 2) = 0$. Theorem 1.2.1 tells us that this product can be zero if and only if $x - 2 = 0$ or $x + 2 = 0$. Thus, $x = 2$ or $x = -2$, and the solution set is $\{-2, 2\}$.

Example 1.2.3   Solve $x^2 - x - 6 = 0$.

Solution   Factoring, we have

$$x^2 - x - 6 = (x - 3)(x + 2) = 0$$

But this can be true only when $x - 3 = 0$ or $x + 2 = 0$. Thus, the solution set is $\{-2, 3\}$.

Example 1.2.4   Solve $\dfrac{1}{1 + (1/x)} = \dfrac{1}{2}$.

Solution   First, note that $x \neq 0$ because division by zero is undefined. Next, simplify the left side:

$$\frac{1}{1 + \dfrac{1}{x}} = \frac{1}{\left(\dfrac{x + 1}{x}\right)} = \frac{x}{x + 1}$$

Here we see also that $x \neq -1$, because the fraction $\dfrac{x}{x + 1}$ is not defined for $x = -1$. So for $x \neq 0$ and $x \neq -1$, multiply both sides of the equation $\dfrac{x}{x + 1} = \dfrac{1}{2}$ by $2(x + 1)$, obtaining

$$2(x + 1)\frac{x}{x + 1} = 2(x + 1)\frac{1}{2} \quad \text{or} \quad 2x = x + 1$$

Adding $-x$ to both sides yields $x = 1$. Thus, the solution set is $\{1\}$.

Example 1.2.5   Solve $\dfrac{1}{1 + (1/x)} = 0$.

Solution   Remember that $a/b = 0$ only when $a = 0$ and $b \neq 0$. Thus, the solution set is the empty set $\varnothing$.

Example 1.2.6   Solve $\dfrac{x}{x - 1} - \dfrac{2x - 1}{x + 1} = 1$.

Solution   Notice first that $x \neq 1$ and $x \neq -1$. Subtracting the fractions [using Equation (7)], we obtain

$$\frac{x(x + 1) - (2x - 1)(x - 1)}{(x - 1)(x + 1)} = \frac{x^2 + x - 2x^2 + 3x - 1}{x^2 - 1} = 1$$

A3: (a) $\{1\}$ (b) $\{0\}$ (c) $\{3\}$ (d) $\emptyset$

A4: By Equation (6), $\dfrac{a}{c} = \dfrac{ad}{cd}$ and $\dfrac{b}{d} = \dfrac{cb}{cd}$. Thus $\dfrac{a}{c} + \dfrac{b}{d} = \dfrac{ad + bc}{cd}$

by Equation (3). Similarly, $\dfrac{a}{c} - \dfrac{b}{d} = \dfrac{ad - bc}{cd}$.

Simplify and multiply both sides by $x^2 - 1$ to obtain

$$-x^2 + 4x - 1 = x^2 - 1$$

or
$$0 = 2x^2 - 4x$$
$$0 = 2x(x - 2)$$

Thus, $x = 0$ or $x = 2$. You should verify by substitution that $\{0, 2\}$ is the desired solution set.

Of course, it is not always so easy to find solution sets to equations. Indeed, there are many equations which do not permit explicit solutions. Later we shall investigate techniques which help us approximate solutions of such equations.

One class of equations, however, which we encounter often and which easily lends itself to explicit solutions is the *quadratic equation*, $ax^2 + bx + c = 0$, $a \neq 0$. In Examples 1.2.2 and 1.2.3 we observed that such expressions often can be factored. But even if we cannot (or do not) factor the expression, the real number solution set is given by the quadratic formula which is discussed in Section 2.3.

PROBLEMS
1.2

1. State properties of the real number system which are illustrated in each of the following:
(a) $(x + y) + z = x + (z + y)$ (b) If $c + y = y + d$, then $c = d$.

2. For what value of x, if any, is each of the following expressions undefined:

(a) $\dfrac{1}{x^2}$

(b) $\dfrac{4x - 1}{5}$

(c) $\dfrac{(x + 1)^2}{(x^2 - 1)(x + 3)}$

(d) $\dfrac{1 + 3x}{x^2 + 7x - 8}$

3. For what value of x, if any, does each of the expressions in the preceding problem equal zero?

4. Show, by example, that subtraction and division are neither commutative nor associative.

5. Find the additive and the multiplicative inverses of each of the following numbers: (a) 2, (b) $-\dfrac{1}{3}$, (c) $\dfrac{14}{11}$, (d) 0

6. Find the solution set for each of the following equations:

(a) $13x - 6 = 19x + 42$

(b) $\dfrac{a}{4} + \dfrac{1}{2} = 2a - \dfrac{1}{4}$

(c) $(x - 1)(x + 7) = 0$

(d) $\dfrac{4}{x - 1} = 0$

(e) $\dfrac{x^2}{x - 1} = 0$

(f) $(x - 4)(x - 3) = 42$

(g) $\dfrac{x - 4}{3x + 1} = \dfrac{x - 5}{3x}$

(h) $\dfrac{5}{x + 4} = \dfrac{x}{3x + 12} - \dfrac{20}{3}$

(i) $2x^2 - 7x + 5 = 0$

(j) $\dfrac{x}{x - 1} + \dfrac{x + 1}{3} = \dfrac{1}{3}$

(k) $\dfrac{x^2 - 1}{x - 1} - 1 = 0$

(l) $x^2 + 3 = 0$

(m) $\dfrac{3x + 3}{x^2 - 4} + \dfrac{x - 3}{x^2 - x - 2} = \dfrac{-2}{x + 1}$

(n) $\dfrac{3x + 4}{x^2 - 4} + \dfrac{4x + 1}{x^2 - x - 2} = \dfrac{1}{x + 1}$

(o) $\dfrac{2x + 4}{x^2 - 4} + \dfrac{x - 8}{x^2 - x - 2} = \dfrac{3}{x + 1}$

---

7. Prove the so-called cancellation property for addition. That is, prove that for any $x \in R$, $a + x = b + x$ implies $a = b$.

8. Prove the cancellation law for multiplication; namely,

$$\text{if} \quad a \cdot x = b \cdot x, \quad \text{and} \quad x \neq 0, \quad \text{then} \quad a = b$$

If $x = 0$, can we conclude $a = b$? Why?

9. Prove that the numbers 0 and 1 are unique. (Hint: Assume that there is another number $0'$ such that $a + 0' = a$ for *all* $a \in R$. Then show that $0' = 0$.)

10. For $x \in R$, prove that $-x$ is the only real number with the property that $x + (-x) = 0$.

11. For $x \in R$ and $x \neq 0$, prove that $1/x$ is the only real number with the property that $x \cdot (1/x) = 1$.

In Problems 12 through 16, the following explanation is relevant. A 2-by-2 *matrix* is a rectangular array of the form

$$\begin{pmatrix} a & b \\ c & d \end{pmatrix}$$

where a, b, c, and d are real numbers. M will denote the set of all 2-by-2 matrices. We define "addition" in M by

$$\begin{pmatrix} a & b \\ c & d \end{pmatrix} + \begin{pmatrix} e & f \\ g & h \end{pmatrix} = \begin{pmatrix} a + e & b + f \\ c + g & d + h \end{pmatrix}$$

and "multiplication" in M by the equation

$$\begin{pmatrix} a & b \\ c & d \end{pmatrix} \cdot \begin{pmatrix} e & f \\ g & h \end{pmatrix} = \begin{pmatrix} ae + bg & af + bh \\ ce + dg & cf + dh \end{pmatrix}$$

12. Compute (a) $\begin{pmatrix} 1 & 2 \\ 3 & 1 \end{pmatrix} + \begin{pmatrix} 3 & -10 \\ -1 & 3 \end{pmatrix}$, and (b) $\begin{pmatrix} 1 & 2 \\ 3 & 1 \end{pmatrix} \cdot \begin{pmatrix} 2 & 0 \\ 1 & -1 \end{pmatrix}$.

13. Do the operations of matrix addition and multiplication satisfy a closure property?

14. (a) Prove that matrix addition is associative and commutative.
(b) Is there a matrix that is an additive identity?
(c) Does every matrix have an additive inverse?

15. (a) Prove that matrix multiplication is associative.
(b) Show, by example, that matrix multiplication is *not* commutative.
(c) Is there a matrix that is a multiplicative identity?
(d) Find a matrix with a, b, c, and d not all zero which does not have a multiplication inverse.

16. What condition is necessary to insure that the matrix $\begin{pmatrix} a & b \\ c & d \end{pmatrix}$ will have a multiplicative inverse? Prove your answer.

ANSWERS
1.2

1. (a) The commutative and associative properties, (b) the commutative, associative, identity, and inverse properties

2. (a) 0, (b) none, (c) 1, -1, -3, (d) 1, -8

3. (a) none, (b) 1/4, (c) none, (d) $-1/3$,

4. For example: $1 - 2 = -1$ but $2 - 1 = 1$; $1 \div 2 = 1/2$ but $2 \div 1 = 2$; $3 - (2 - 1) = 2$ but $(3 - 2) - 1 = 0$; $(4/2)/2 = 1$ but $4/(2/2) = 4$,

5. Additive: (a) -2, (b) 1/3, (c) $-14/11$, (d) 0; multiplicative: (a) 1/2. (b) -3, (c) 11/14, (d) none

6. (a) $\{-8\}$, (b) $\{3/7\}$, (c) $\{1, -7\}$, (d) $\varnothing$, (e) $\{0\}$, (f) $\{-3, 10\}$ (g) $\{-5/2\}$, (h) $\{-5\}$, (i) $\{1, 5/2\}$, (j) $\{-2, 0\}$, (k) $\{0\}$, (l) $\varnothing$, (m) $\{1, -11/6\}$, (n) $\{-5/3\}$, (o) $\{x \in R : x \neq 2, -2, -1\}$

Supplement 1.2

I. Properties of the Real Number System

Problem S1. State the axiom, definition, and/or theorem which justifies each of the following steps:

$$\underset{\text{(a)}}{(495 + 674) - 495} = \underset{\text{(b)}}{(495 + 674) + (-495)} = 495 + (674 + (-495))$$

$$\underset{\text{(c)}}{= 495 + (-495 + 674)} = \underset{\text{(d)}}{(495 + (-495)) + 674}$$

$$\underset{\text{(e)}}{= 0 + 674} = \underset{\text{(f)}}{674 + 0} = \underset{\text{(g)}}{674}$$

Solution:
(a) The definition of subtraction.
(b) The associative property. Notice that the order of the numbers was unchanged and only the "grouping" was changed.
(c) The commutative property. Notice that the only change was in the order of addition.
(d) The associative property.
(e) The inverse property.
(f) The commutative property.
(g) The identity property.
   Notice that $(495 + 674) - 495 = 674$ can be computed easily and quickly, for it is not necessary to add 495 and 674 and then subtract 495.

Problem S2. Supply reasons for the steps in the following proof of the well-known result that $a \cdot 0 = 0$ for any $a \in R$:

Solution:

$$
\overset{\text{(a)}}{a \cdot 0} = \overset{\text{(b)}}{a \cdot 0 + 0} = a \cdot 0 + (a + (-a)) = \overset{\text{(c)}}{(a \cdot 0 + a) + (-a)}
$$

$$
= \overset{\text{(d)}}{(a \cdot 0 + a \cdot 1) + (-a)} = \overset{\text{(e)}}{a(0 + 1) + (-a)} = \overset{\text{(f)}}{a(1 + 0) + (-a)}
$$

$$
= \overset{\text{(g)}}{a \cdot 1 + (-a)} = \overset{\text{(h)}}{a + (-a)} = \overset{\text{(i)}}{0}
$$

Therefore, $a \cdot 0 = 0$.

(a) _____	(a) identity property
(b) _____	(b) inverse property
(c) _____	(c) associative property
(d) _____	(d) identity property
(e) _____	(e) distributive property
(f) _____	(f) commutative property
(g) _____	(g) identity property
(h) _____	(h) identity property
(i) _____	(i) inverse property

DRILL PROBLEMS

S3. Supply reasons for each step in the following:

$$
\overset{\text{(a)}}{(76 - 139) - 76} = \overset{\text{(b)}}{(76 + (-139)) + (-76)} = (-139 + 76) + (-76)
$$

$$
= \overset{\text{(c)}}{-139 + (76 + (-76))} = \overset{\text{(d)}}{-139 + 0} = \overset{\text{(e)}}{-139}
$$

S4. Justify each step in the solution of the following problem: Find the solution set for the equation $4/(x + 1) + 1 = 2$.

Solution: Multiply both sides by $x + 1$. The equation becomes

(a) $(x + 1)\left(\dfrac{4}{x + 1} + 1\right) = (x + 1)2$

$\dfrac{(x + 1)4}{x + 1} + (x + 1)1 = x(2) + 1(2)$

(b) $\dfrac{(x + 1)4}{x + 1} + x + 1 = 2x + 2$. Since $\dfrac{(x + 1)4}{x + 1} = 4$ if $x \neq -1$, the equation becomes $4 + x + 1 = 2x + 2$.

(c) This easily reduces to the equation $x = 3$. Therefore, since $\dfrac{4}{3 + 1} + 1 = 2$, the solution set is $\{3\}$.

Answers

S3. (a) Definition of subtraction,   (b) the commutative property,   (c) the associative property,   (d) the inverse property,   (e) the identity property
S4. (a) The distributive property,   (b) the identity property, the commutative property,   (c) the commutative, associative, identity, and inverse properties

---

II. Operations Involving Rational Expressions

Problem S5. Compute (a) $\dfrac{7}{2} + \dfrac{3}{2}$, (b) $\dfrac{2}{5}\left(\dfrac{2}{7} + \dfrac{3}{2}\right)$, (c) $\dfrac{4 + (2/3)}{3/4}$, and (d) $\dfrac{(4/5) - (3/2)}{(1/2) - (3/5)}$.

Solution:

(a) $\dfrac{7}{2} + \dfrac{3}{2} = \dfrac{7 + 3}{2} = \dfrac{10}{2} = 5$

(b) $\dfrac{2}{5}\left(\dfrac{2}{7} + \dfrac{3}{2}\right) = \dfrac{2}{5}\left(\dfrac{2 \cdot 2 + 3 \cdot 7}{2 \cdot 7}\right) = \dfrac{2}{5}\left(\dfrac{4 + 21}{14}\right) = \dfrac{2}{5} \cdot \dfrac{25}{14} = \dfrac{2 \cdot 5 \cdot 5}{5 \cdot 2 \cdot 7} = \dfrac{5}{7}$

(c) $\dfrac{(4/1) + (2/3)}{3/4} = \dfrac{(4 \cdot 3 + 2 \cdot 1)/1 \cdot 3}{3/4} = \dfrac{14/3}{3/4} = \dfrac{14}{3} \cdot \dfrac{4}{3} = \dfrac{56}{9}$

(d) $\dfrac{(4/5) - (3/2)}{(1/2) - (3/5)} = \dfrac{(8 - 15)/10}{(5 - 6)/10} = \dfrac{-7/10}{-1/10} = 7$

Problem S6. Simplify (a) $\dfrac{x - 1}{x + 2} + \dfrac{x}{x + 1}$ and (b) $\dfrac{1}{x - 2}\left(\dfrac{2x - 4}{3} - \dfrac{x - 2}{x + 1}\right)$.

Solution: (a) $\dfrac{x - 1}{x + 2} + \dfrac{x}{x + 1} = \dfrac{(x - 1)(x + 1) + x(x + 2)}{(x + 2)(x + 1)}$ by Equation (7) for adding fractions. Notice that no factor is common to both the numerator and the denominator. By multiplying out and combining terms, we obtain

$$\dfrac{x - 1}{x + 2} + \dfrac{x}{x + 1} = \dfrac{2x^2 + 2x - 1}{x^2 + 3x + 2}$$

Further simplification is not possible since there are no common factors.

(b) First combine the fractions inside the grouping symbols, obtaining

$$\frac{1}{x-2}\left(\frac{2x-4}{3}-\frac{x-2}{x+1}\right)=\frac{1}{x-2}\left(\frac{(2x-4)(x+1)-3(x-2)}{3(x+1)}\right)$$

Next, notice that the numerator can be factored, yielding

$$\frac{1}{x-2}\left(\frac{2x-4}{3}-\frac{x-2}{x+1}\right)=\frac{1}{x-2}\cdot\frac{(x-2)[2(x+1)-3]}{3(x+1)}$$

Thus, it follows that

$$\frac{1}{x-2}\left(\frac{2x-4}{3}-\frac{x-2}{x+1}\right)=\frac{2x-1}{3(x+1)}$$

(We assume that $x\neq 2$.) Notice that in this problem the numerator could have been factored first, simplifying the solution.

Problem S7. Simplify $\frac{5}{6x+8}\left(\frac{x}{2}+\frac{2}{3}\right)$.

Solution: The equation

$$\frac{5}{6x+8}\left(\frac{x}{2}+\frac{2}{3}\right)=\left(\frac{5}{6x+8}\right)\left(\frac{3x+4}{6}\right)$$

follows from the rule for _____. | adding fractions
Notice that $6x+8=2(\_\_\_\_)$. | $3x+4$

Thus, $\frac{5}{6x+8}\left(\frac{3x+4}{6}\right)=\frac{5(3x+4)}{12(3x+4)}=$

_____ provided that $3x+4\neq 0$ or | $5/12$
$x\neq\_\_\_$. | $-4/3$

DRILL PROBLEMS

S8. Compute (a) $2/5-7/4$ and (b) $(2/3)(1/4-3/5)$.

S9. Simplify (a) $\left(\frac{2+x}{x-2}\right)\left(\frac{x}{2}-\frac{2}{x}\right)$ and (b) $\frac{1}{x}-\frac{1}{x-1}$.

Answers

S8. (a) $\frac{-27}{20}$, (b) $\frac{-7}{30}$

S9. (a) $\frac{(x+2)^2}{2x}$, (b) $-\frac{1}{x(x-1)}=\frac{1}{x(1-x)}$

III. Finding Solution Sets

Problem S10. Find (a) $\left\{x:\frac{x-2}{x^2+6x-7}\text{ is undefined}\right\}$ and

(b) $\left\{x:\frac{x-2}{x^2+6x-7}=0\right\}$.

Solution: (a) A fraction a/b is undefined if and only if $b = 0$. Hence, $\dfrac{x - 2}{x^2 + 6x - 7}$
is undefined precisely when $x^2 + 6x - 7 = 0$. Next, factor $x^2 + 6x - 7$ to obtain
the equation $(x + 7)(x - 1) = 0$. Therefore, by Theorem 1.2.1, $x + 7 = 0$ or
$x - 1 = 0$. So $x^2 + 6x - 7 = 0$ when $x = -7$ or $x = 1$. Thus, the set of numbers
such that the fraction $\dfrac{x - 2}{x^2 + 6x - 7}$ is undefined is $\{-7, 1\}$.

(b) A fraction a/b is zero when and only when $a = 0$ and $b \neq 0$. Therefore,
$\dfrac{x - 2}{x^2 + 6x - 7} = 0$ exactly when $x - 2 = 0$ and $x^2 + 6x - 7 \neq 0$. Clearly, $x - 2 =$
0 when $x = 2$. Notice that 2 does not make the denominator zero. Thus, the set
of numbers such that the fraction $\dfrac{x - 2}{x^2 + 6x - 7}$ is zero is $\{2\}$.

Problem S11.   Find the solution set for the equation $3x - 5 = (1/2) + 7 - 4x$.

Solution: First, add $4x$ to both sides of the equation:

$$3x - 5 + 4x = \frac{1}{2} + 7 - 4x + 4x \quad \text{or} \quad 7x - 5 = \frac{1}{2} + 7$$

Add 5 to both sides of the equation:

$$7x - 5 + 5 = \frac{1}{2} + 7 + 5 \quad \text{or} \quad 7x = \frac{25}{2}$$

Next, multiply both sides of the above equation by $1/7$. The result is

$$\frac{1}{7}(7x) = \frac{1}{7}\left(\frac{25}{2}\right) \quad \text{or} \quad x = \frac{25}{14}$$

Since $3(25/14) - 5 = (1/2) + 7 - 4(25/14)$, the solution set is $\{25/14\}$.

Problem S12.   Find the solution set for the equation $(x - 3)(x - 2) = 30$.

Solution: Notice that the indicated product, $(x - 3)(x - 2)$, does *not* equal zero.
Therefore, it is *not* true that $x - 3 = 0$ or $x - 2 = 0$. First, expand the left side
of the equation to obtain

$$x^2 - 5x + 6 = 30$$

Now, subtract 30 from both sides of the equation:

$$x^2 - 5x - 24 = 0$$

Next, factor the quadratic expression on the left to obtain the equation

$$(x - 8)(x + 3) = 0$$

Now, use Theorem 1.2.1. As an immediate consequence of this theorem, $x - 8 = 0$
or $x + 3 = 0$. So $x = 8$ or $x = -3$. Since $(8 - 3)(8 - 2) = 30$ and $(-3 - 3)\cdot$
$(-3 - 2) = 30$, the solution set is $\{-3, 8\}$.

Problem S13. Find the solution set for the equation $\dfrac{x+1}{x} + \dfrac{-1-x}{x-2} = 0$.

Solution: Adding the two fractions, we obtain

$$\frac{(x+1)(x-2) + x(-1-x)}{x(x-2)} = 0$$

or, when we simplify further,

$$\frac{-2x-2}{x^2-2x} = 0$$

Since a fraction is zero precisely when its numerator is zero (and its denominator is not zero), it follows that $-2x - 2 = 0$ or that $x = -1$. Since $\dfrac{-1+1}{-1} + \dfrac{-1-(-1)}{-1-2} = 0$, the solution set is $\{-1\}$.

Problem S14. Find the solution set for the equation $\dfrac{x+1}{1} + \dfrac{2}{x+3} = 1$.

Solution: Adding the fractions on the left side of the equation yields

$$\frac{(x+3)(x+1) + 2}{x+3} = 1$$

which becomes

$$\frac{x^2 + 4x + 3 + 2}{x+3} = 1$$

Multiplying both sides by $x + 3$, we obtain the equation $x^2 + 4x + 5 = x + 3$. Combining terms yields $x^2 + 3x + 2 = 0$. When factored, it becomes

$$(x+1)(x+2) = 0$$

Therefore, $x + 1 = 0$ or $x + 2 = 0$, by Theorem 1.2.1. So $x = -1$ and $x = -2$ are the possible solutions. Since $\dfrac{-1+1}{1} + \dfrac{2}{-1+3} = 1$ and $\dfrac{-2+1}{1} + \dfrac{2}{-2+3} = 1$, the solution set is $\{-1, -2\}$.

Problem S15. Find the solution set for the equation $\dfrac{x^2}{x-4} - \dfrac{4x}{x-4} = 0$.

Solution: Notice that the denominators are the same. Hence, the equation is simply

$$\frac{x^2 - 4x}{x-4} = 0$$

Thus, $x^2 - 4x = 0$, since a fraction is zero only when its numerator is zero. Factoring, we obtain the equation $x(x-4) = 0$. So $x = 0$ or $x - 4 = 0$. Hence, 0 and

4 are possible solutions. Substituting these values into the original equation gives us

$$\frac{0^2}{0-4} - \frac{0}{0-4} = 0 \quad \text{and} \quad \frac{4^2}{4-4} - \frac{4}{4-4} \neq 0$$

($4 - 4 = 0$ in the denominators of the second equation means that the left side is undefined and thus not equal to zero.) Hence, the solution set is $\{0\}$.

Problem S16.   Find $\left\{ x : \dfrac{x+2}{x^2-4} \text{ is undefined} \right\}$.

Solution: A fraction a/b is undefined if and only if _____. Therefore, $\dfrac{x+2}{x^2-4}$ is undefined precisely when _____. Next, factor $x^2 - 4$ to obtain the equation _____. By a previous theorem, it follows that $x - 2 = 0$ or _____. Thus, the set of numbers such that $\dfrac{x+2}{x^2-4}$ is undefined is $\{$_____$\}$.

$b = 0$
$x^2 - 4 = 0$
$(x-2)(x+2) = 0$
$x + 2 = 0$
$2, -2$

Problem S17.   Find the solution set of the equation $\dfrac{1}{x+1} - 3 = 10$.

Solution: By adding _____ to both sides of the equation, we obtain _____. Multiplying both sides by $x + 1$ yields _____. So $1 = 13x + 13$. It follows that $13x =$ _____ or $x = \dfrac{-12}{13}$. Since $\dfrac{1}{(-12/13)+1} - 3 =$ _____, the solution set is $\left\{ \dfrac{-12}{13} \right\}$.

3
$\dfrac{1}{x+1} = 13$
$\dfrac{x+1}{x+1} = 13(x+1)$
-12
10

Problem S18.   Find the solution set of the equation $\dfrac{10x}{9} + \dfrac{1}{x} = \dfrac{7}{3}$.

Solution: First, _____ the fractions on the left to obtain _____. Now, multiply both sides by _____ to obtain _____. After simplification, the equation becomes $10x^2 - 21x + 9 = 0$. Factoring

add
$\dfrac{10x^2 + 9}{9x} = \dfrac{7}{3}$
$9x$
$10x^2 + 9 = 21x$

yields _____. Therefore,
$5x - 3 = 0$ or $2x - 3 = 0$. Hence,
$x = 3/5$ or $x =$ ____. Since
$\dfrac{10(3/5)}{9} + \dfrac{1}{3/5} = \dfrac{7}{3}$ and

_____, the solution set is

$\{3/2, 3/5\}$.

$(5x - 3)(2x - 3) = 0$

$3/2$

$\dfrac{10(3/2)}{9} + \dfrac{1}{3/2} = \dfrac{7}{3}$

DRILL PROBLEMS

S19. Find the set $\left\{x : \dfrac{x+2}{x^2-4} = 0\right\}$.

S20. Find the solution sets for each of the following equations:

(a) $\dfrac{x-3}{8} = 5$

(b) $2x^2 - 13x + 15 = 0$

(c) $x^2 - 3x + 1 = -1$

(d) $\dfrac{x}{x+1} + 1 = 0$

Answers

S19. $\emptyset$ (For $x = -2$, the fraction is undefined.)
S20. (a) $\{43\}$, (b) $\{3/2, 5\}$, (c) $\{1, 2\}$, (d) $\{-1/2\}$

SUPPLEMENTARY PROBLEMS

S21. For what values of x is $\dfrac{x-1}{x^2+4x+4}$ undefined? Zero?

S22. Justify the following steps:
$$(-a)(-b) + (-a)b \overset{(a)}{=} (-a)(-b+b) \overset{(b)}{=} (-a)0 = 0$$

S23. Find the solution set for each of the following equations:
(a) $16 - x + 5 = 10x - 7$ (b) $6x^2 - 17x + 12 = 0$
(c) $\dfrac{1}{x-1} + 2 = \dfrac{1}{x-1}$ (d) $\dfrac{1}{x+1} + \dfrac{1}{x-3} = 0$

Answers

S21. $-2, 1$
S22. (a) The distributive property, (b) the inverse property. Therefore,
$(-a)(-b) + (-ab) = ab + (-ab)$. So, by the *cancellation* property (see Problem 7), it follows that $(-a)(-b) = ab$.
S23. (a) $\{28/11\}$, (b) $\{3/2, 4/3\}$, (c) $\emptyset$, (d) $\{1\}$

1.3 EXPONENTS AND RADICALS

Positive integer exponents are used to indicate repeated multiplications of the same number. For example, $5 \cdot 5$ is written as 5^2, and $(x + y)^3$ represents the product $(x + y)(x + y)(x + y)$. In this section, we shall review the properties of exponential notation and extend its use to rational exponents as well.

Definition 1.3.1

If $b \in R$ and $n \in N$, then b^n represents the product of n factors, each equal to b. We write

$$b^n = \underbrace{b \cdot b \cdot \ldots \cdot b}_{n \text{ factors}}$$

In the expression b^n, the integer n is called the *exponent* and the number b is called the *base*.

This notation will help us count factors as illustrated in Example 1.3.1.

Example 1.3.1

By counting and using Definition 1.3.1, simplify each of the following: (a) $4^3 \cdot 4^2$, (b) $(4^3)^2$, (c) $4^5/4^2$, and (d) $4^3 \cdot 3^3$.

Solution

$$\text{(a) } 4^3 \cdot 4^2 = \overbrace{(4 \cdot 4 \cdot 4)}^{3} \cdot \overbrace{(4 \cdot 4)}^{2} = \overbrace{4 \cdot 4 \cdot 4 \cdot 4 \cdot 4}^{3 + 2 = 5} = 4^5.$$

38

(b) $(4^3)^2 = \overbrace{(4^3) \cdot (4^3)}^{2} = \overbrace{(4 \cdot 4 \cdot 4)}^{3} \cdot \overbrace{(4 \cdot 4 \cdot 4)}^{3} = \overbrace{4 \cdot 4 \cdot 4 \cdot 4 \cdot 4 \cdot 4}^{2 \cdot 3 = 6} = 4^6.$

(c) $\dfrac{4^5}{4^2} = \dfrac{\overbrace{4 \cdot 4 \cdot 4 \cdot 4 \cdot 4}^{5}}{\underbrace{4 \cdot 4}_{2}} = \overbrace{4 \cdot 4 \cdot 4}^{5 - 2 = 3} = 4^3$

(d) $4^3 \cdot 3^3 = \underbrace{(4 \cdot 4 \cdot 4)}_{3} \cdot \underbrace{(3 \cdot 3 \cdot 3)}_{3} = \underbrace{(4 \cdot 3) \cdot (4 \cdot 3) \cdot (4 \cdot 3)}_{3} = 12^3.$

Following the procedure in Example 1.3.1, we can establish the following equations for $b, c \in R$, and $m, n \in N$.

$$b^m \cdot b^n = b^{m+n} \tag{1}$$
$$(b^m)^n = b^{mn} \tag{2}$$
$$(bc)^n = b^n c^n \tag{3}$$

For $c \neq 0$,

$$\left(\frac{b}{c}\right)^n = \frac{b^n}{c^n} \tag{4}$$

For $b \neq 0$,

$$\frac{b^n}{b^n} = 1$$

$$\frac{b^m}{b^n} = b^{m-n} \quad \text{if } m > n$$

$$\frac{b^m}{b^n} = \frac{1}{b^{n-m}} \quad \text{if } n > m \tag{5}$$

To show that Equation (1) is true, simply count each factor b using Definition 1.3.1.

$$b^m \cdot b^n = \underbrace{(b \cdot b \cdot \ldots \cdot b)}_{m \text{ factors}} \cdot \underbrace{(b \cdot b \cdot \ldots \cdot b)}_{n \text{ factors}} = \underbrace{b \cdot b \cdot \ldots \cdot b}_{m + n \text{ factors}} = b^{m+n}$$

We ask you to verify Equations (2) through (5) in the problems at the end of this section.

Q1: Simplify each of the following:

    (a) $x^4 \cdot x^5$ (b) $2^4 + 2^5$ (c) $(2^3)^2$

    (d) $2^{(3^2)}$ (e) $(2xy^3)^2$ (f) $\dfrac{(6x^2y)^2}{(3xy^2)^3}$

We now assign a meaning to the expression b^0. If Equation (1) is to hold for $n = 0$, we must have $b^m = b^{m+0} = b^m \cdot b^0$. If $b \neq 0$, dividing by b^m yields $b^m/b^m = b^0$ or $b^0 = 1$. In the problems, you are asked to show that if the value 1 is assigned to the symbol b^0 when $b \neq 0$, then Equations (2) through (5) are valid for non-

A1: (a) x^9, (b) $2^4(1 + 2) = 3 \cdot 2^4 = 48$, (c) 64, (d) 512, (e) $4x^2y^6$, (f) $4x/3y^4$

negative integers. Thus, we can make the following definition, and the usual properties for exponents will be true.

Definition
1.3.2

If $b \neq 0$, then $b^0 = 1$.

The symbol 0^0 poses a special problem. It appears that it should be 1 since $b^0 = 1$ for $b \neq 0$, or that it should be 0 since $0^n = 0$ for $n \neq 0$. Equations (1) and (2) both hold for $n = 0$, regardless of what value is assigned to 0^0. Since there is more than one "reasonable" value which could be assigned to 0^0, we shall leave it undefined.

Let us now seek to assign a meaning to 2^{-3}. Again, for Equation (1) to hold, $1 = 2^0 = 2^{3+(-3)} = 2^3 \cdot 2^{-3}$. Dividing by 2^3 yields $1/2^3 = 2^{-3}$. Similarly, if n is a positive integer, $1 = b^0 = b^{n+(-n)} = b^n \cdot b^{-n}$. Then, if $b^{-n} = 1/b^n$ ($b \neq 0$), Equation (1) remains valid. Again, in the problems you are asked to show that Equations (2) through (5) also remain valid if we formally extend the meaning of exponents by the following definition.

Definition
1.3.3

If $b \neq 0$, $b^{-n} = \dfrac{1}{b^n}$ ($n \in N$).

Q2: Write the following without exponents:

(a) 7^{-2} (b) -2^{-4} (c) $(-2)^{-4}$ (d) $\dfrac{3a^0}{(3a)^0}$ (e) $\left(\dfrac{x}{2}\right)^0$

We have now defined b^n for both positive and negative integers. Our next task is to define $b^{1/q}$ for $q \in N$. If we want Equation (2) to hold for $m = 1/q$ and $n = q$, then we must have $(b^{1/q})^q = b^{(1/q) \cdot q} = b^1 = b$; that is, the number $b^{1/q}$ should be such that when raised to the qth power, it yields b. Such numbers are called *qth roots* of b. (We say square roots for $q = 2$ and cube roots for $q = 3$.) Some difficulties remain, however. For example, there are at least two 4th roots of 16, namely, 2 and -2, since $2^4 = 16$ and $(-2)^4 = 16$. Although we shall not prove it here, each positive number b has exactly one positive qth root called the *principal* qth root, denoted by $b^{1/q}$.

Now, if x is real, x^2 is positive or zero. Thus, the equation $x^2 = -9$ has no real number solution, and the negative number -9 has no real square roots. In fact, no negative number can have an even root. On the other hand, negative numbers can have odd roots. For example, the equation $x^3 = -27$ has at least one solution, namely, -3. Again we state without proof that if b is negative and q is an odd integer, then b has exactly one real qth root (which is negative) called the *principal qth root of b*, again denoted by $b^{1/q}$.

Q3: Find

(a) $9^{1/2}$ (b) $8^{1/3}$ (c) $(-8)^{1/3}$ (d) $(-16)^{1/4}$

The preceding discussion is summarized in the following definitions.

Definition 1.3.4

If $q \in N$ and $r^q = b$, then r is called a qth root of b.

Definition 1.3.5

If $b > 0$ and if $q \in N$, there is exactly one real positive qth root of b. If $b < 0$, and if q is an odd integer, there is exactly one real qth root of b. In each case we call this qth root the *principal qth root of b*, denoted by $b^{1/q}$. If $b < 0$ and if q is an even integer, then $b^{1/q}$ is undefined.

Let us emphasize two points about Definition 1.3.5: (1) *The symbol $b^{1/q}$ may not be defined [for example, $(-9)^{1/2}$]; but* (2) *if it is defined, it is just one number.* For example, $4^{1/2} = 2$, not -2. Notice that we need to be especially careful with "even" roots (i.e., q is an even number). Odd roots present no special difficulties. Henceforth, we require that $b^{1/q}$ exist to circumvent the "problems" of even roots.

As before, in the problems you are asked to show, using our definition of $b^{1/q}$, that Equations (1) through (5) hold when m and n are positive integers or reciprocals of positive integers.

We are now ready to define rational exponents. For Equation (2) to hold, we must have $(b^{1/q})^p = b^{(1/q)\cdot p} = b^{p/q}$. If we use this equation to define $b^{p/q}$, the properties of rational exponents are the same as the properties of integer exponents. Thus, we are led to the following formal definition.

Definition 1.3.6

Let $p \in Z$ and $q \in N$. If $b^{1/q}$ exists, then $b^{p/q} = (b^{1/q})^p$.

A2: (a) $1/49$, (b) $-1/16$, (c) $1/16$, (d) 3 if $a \neq 0$, (e) 1 if $x \neq 0$

A3: (a) 3 (b) 2 (c) -2 (d) not defined

Using Definition 1.3.6 one can show that Equations (1) through (5) are true for any two rational numbers m and n whenever all terms (i.e., b^m, b^n, c^n, b^{mn}, etc.) are defined. Usually Equation (5) is written more simply as

$$\frac{b^m}{b^n} = b^{m-n} \tag{6}$$

whenever $b \neq 0$ and all terms are defined.

In particular, if $m = 0$ and $n = r$, any rational number, Equation (6) becomes

$$\frac{1}{b^r} = b^{-r} \tag{7}$$

whenever $b \neq 0$ and b^r is defined.

As you may recall, the principal square root of a number b is denoted $\sqrt{b}$ as well as $b^{1/2}$. For an integer $q > 2$, the number $b^{1/q}$ is written in a similar manner, namely, $\sqrt[q]{b} = b^{1/q}$, consistent with the following definition.

Definition 1.3.7

$$\sqrt[q]{b^p} = b^{p/q}.$$

Thus, combining Definitions 1.3.6 and 1.3.7, we obtain, provided $b^{1/q}$ exists,

$$b^{p/q} = \sqrt[q]{b^p} = (\sqrt[q]{b})^p \tag{8}$$

Example 1.3.2

Simplify each of the following: (a) $32^{3/5}$, (b) $9^{-1/2}$, (c) $\sqrt[3]{b} \cdot (\sqrt[4]{b})^3$, (d) $\sqrt{(-3)^2}$, (e) $(\sqrt{-3})^2$, (f) $(\sqrt{x})^2$, (g) $\sqrt{x^2}$

Solution

(a) $32^{3/5} = (32^{1/5})^3 = 2^3 = 8$, and (b) $9^{-1/2} = 1/9^{1/2} = 1/3$.

(c) $\sqrt[3]{b} \cdot (\sqrt[4]{b})^3 = b^{1/3} \cdot (b^{1/4})^3 = b^{1/3} \cdot b^{3/4} = b^{1/3+3/4} = b^{13/12}$.
Notice that b must be positive or zero so that all terms in these equations are defined.

(d) $\sqrt{(-3)^2} = \sqrt{9} = 3$.

(e) $(\sqrt{-3})^2$ is undefined since $\sqrt{-3}$ is undefined.

(f) $(\sqrt{x})^2 = (x^2)^{1/2} = x^{2/2} = x$ provided that x is positive or zero. If x is negative, $(\sqrt{x})^2$ is undefined [see part (e)].

(g) $\sqrt{x^2} = (x^2)^{1/2} = x^{2/2} = x$ provided that x is positive or zero. If x is negative, $\sqrt{x^2} = (x^2)^{1/2}$ is defined but does *not* equal x. In Section 1.5 we show that $\sqrt{x^2} = -x$ when x is negative, as Part (d) above illustrates.

Q4: Simplify each of the following:

(a) $\sqrt[3]{x^{15}}$ (b) $27^{2/3}$ (c) $\left(-\dfrac{1}{8}\right)^{-(5/3)}$ (d) $8^{1/3} \cdot 16^{3/4}$

For the remainder of this book, we shall write equations involving exponents and radicals as well as divisions with the understanding that each is a statement of equality, provided that all terms in the equation are defined. Thus, we write

$$(x^2 - 1)/(x - 1) = x + 1$$

with the understanding that $x \neq 1$. Likewise, we write $(\sqrt{x})^2 = x$ with the understanding that x is positive or zero. [Recall Example 1.3.2(f).] Remember, however, that $\sqrt{x^2} \neq x$, as shown in Example 1.3.2(d).

We conclude this section with some examples which illustrate some of the properties of exponents and radicals that we have developed.

Example 1.3.3

Simplify each of the following: (a) $(b^{-1}c^{-1})/(b^{-1} - c^{-1})$, (b) $(a^{x+2y}/a^y)^{(x+y)^{-1}}$ for $a > 0$, (c) $2x^2y(\sqrt[3]{16xy^4})$, (d) $\sqrt[3]{\sqrt{64x^3}}$, (e) $\sqrt{8} + \sqrt{18}$

Solution

(a) $\dfrac{b^{-1}c^{-1}}{b^{-1} - c^{-1}} = \dfrac{\frac{1}{bc}}{\frac{1}{b} - \frac{1}{c}} = \dfrac{\frac{1}{bc}}{\frac{c - b}{bc}} = \dfrac{1}{bc} \cdot \dfrac{bc}{c - b} = \dfrac{1}{c - b}.$ Another

procedure is to multiply by $1 = \dfrac{bc}{bc}$ to obtain $\dfrac{(bc)(b^{-1}c^{-1})}{(bc)(b^{-1} - c^{-1})} = $

$\dfrac{bb^{-1}cc^{-1}}{bb^{-1}c - bcc^{-1}} = \dfrac{1}{c - b}.$

(b) $(a^{x+2y}/a^y)^{(x+y)^{-1}} = (a^{x+y})^{1/(x+y)} = a^{(x+y)/(x+y)} = a.$

(c) $2x^2y(\sqrt[3]{16xy^4}) = 2x^2y(2^4xy^4)^{1/3} = 2x^2y2^{4/3}x^{1/3}y^{4/3} = 2^{7/3}x^{7/3}y^{7/3} = (2xy)^{7/3}.$

(d) $\sqrt[3]{\sqrt{64x^3}} = ((64x^3)^{1/2})^{1/3} = (8x^{3/2})^{1/3} = 8^{1/3}x^{1/2} = 2\sqrt{x}.$

(e) $\sqrt{8} + \sqrt{18} = 2\sqrt{2} + 3\sqrt{2} = 5\sqrt{2}.$

Example 1.3.4

Find the solution set to $\sqrt{x + 2} = x$.

Solution

Squaring both sides yields

$$x + 2 = x^2 \quad \text{or} \quad x^2 - x - 2 = 0$$

A4: (a) x^5 (b) 9 (c) -32 (d) 16

Since $(x - 2)(x + 1) = 0$, $x = 2$ or $x = -1$. Substituting each of these numbers into the original equation yields $\sqrt{2 + 2} = \sqrt{4} = 2$, but $\sqrt{-1 + 2} = \sqrt{1} \neq -1$. Thus, the solution set is $\{2\}$. Notice that, as in this example, squaring an equation may introduce extraneous solutions.

PROBLEMS 1.3

1. Simplify each of the following expressions:

(a) $(a^2 \cdot a^{-3})^3 / a^4$

(b) $4^2 xy^3 / 2^3 x^2 y^2$

(c) $0^0 \cdot 9^{-1/2}$

(d) $\left(\dfrac{8}{27}\right)^{-2/3}$

(e) $-9^{1/2}$

(f) $\sqrt{12} + \sqrt{27}$

(g) $(-16)^{1/2}$

(h) $(-27)^{4/3}$

2. Write each of the following without exponents:

(a) $(-1)^{-3}$

(b) $(2^3 \cdot 3^0)4^{-2}$

(c) $(-2x^4)^{1/2}$

(d) $\left(-\dfrac{a^5}{32}\right)^{2/5}$

(e) $\left(\dfrac{8x^3 y^6}{27 y^3}\right)^{-1/3}$

(f) $\left(\dfrac{1}{3}\right)^2 \cdot \left(\dfrac{1}{3}\right)^0 \cdot \left(\dfrac{1}{3}\right)^{-4} \cdot 3^{-5}$

3. Simplify:

(a) $\dfrac{x^{1/2} y^3}{x^{-1} y^{2/3}}$

(b) $\dfrac{(x^{-2} y^5)^{-3}}{(xy)^2 x^5 y^{-3}}$

(c) $(\sqrt[3]{b^2} \cdot \sqrt{b^4})^2$

(d) $(\sqrt[3]{\sqrt{x^4}})^{1/2}$

(e) $\dfrac{a^{-1} + b^{-1}}{b^{-2} - a^{-2}}$

(f) $\left(12^0 + \left(\dfrac{27}{64}\right)^{-2/3}\right)^{-3/2}$

4. Find the solution set for each of the following:

(a) $\sqrt{2 - x} - 3 = 0$

(b) $\sqrt{x + 2} - 3 = x - 1$

(c) $2\sqrt{x + 3} = x$

5. Verify that Equations (2) through (5) hold for positive integer exponents.

6. Show, by example, that $\sqrt{a} + \sqrt{b} \neq \sqrt{a + b}$.

7. What is wrong with the following "proof" that $-1 = 1$?

$$1 = \sqrt{1} = \sqrt{(-1)^2} = -1$$

8. Show that Equations (5) are equivalent to Equation (6).

9. Does $\sqrt{ab} = \sqrt{a} \cdot \sqrt{b}$? What if $a = -4$ and $b = -9$?

10. Simplify:

(a) $(x^{1/2} + y^{1/2})(x^{1/2} - y^{1/2})$

(b) $\dfrac{x - a}{\sqrt{x} - \sqrt{a}}$

(c) $(a^4 + 2a^2b^2 + b^4)^{1/2}$

(d) $\dfrac{\sqrt{5} + \sqrt{2}}{\sqrt{5} - \sqrt{2}}$

11. Show that Equation (3) holds for nonnegative integral exponents.

12. Show that the following rules hold for any integers m and n such that $n > 2$, provided that $\sqrt[n]{a}$ and $\sqrt[n]{b}$ are defined:

(a) $(\sqrt[n]{a})^m = a^{m/n}$

(b) $\sqrt[n]{a}\sqrt[n]{b} = \sqrt[n]{ab}$

(c) $\sqrt[n]{a}/\sqrt[n]{b} = \sqrt[n]{a/b}$

13. Show that Equations (2), (4), and (5) hold for:

(a) nonnegative integral exponents,

(b) integral exponents.

14. Show that Equations (1), (2), (4), and (5) hold for:

(a) exponents that are reciprocals of natural numbers,

(b) rational exponents.

15. If $r_1 = \dfrac{-b + \sqrt{b^2 - 4ac}}{2a}$ and $r_2 = \dfrac{-b - \sqrt{b^2 - 4ac}}{2a}$, evaluate:

(a) $r_1 + r_2$

(b) $r_1 - r_2$

(c) $r_1 r_2$

16. Write as a fraction with no radicals in the denominator:

(a) $\dfrac{\sqrt{3} - 5}{\sqrt{5} - 3}$

(b) $\dfrac{\sqrt{x} - \sqrt{a}}{\sqrt{a} + \sqrt{x}}$

ANSWERS
1.3

1. (a) a^{-7}, (b) $\dfrac{2y}{x}$, (c) undefined, (d) $\dfrac{9}{4}$, (e) -3, (f) $5\sqrt{3}$, (g) undefined, (h) 81

2. (a) -1, (b) $1/2$, (c) undefined, (d) $a \cdot a/4$, (e) $3/2xy$, (f) $1/27$

3. (a) $x^{3/2}y^{7/3}$, (b) $x^{-1}y^{-14}$, (c) $b^{16/3}$, (d) $\sqrt[3]{x}$ (e) $ab/(a - b)$, (f) $27/125$

4. (a) $\{-7\}$, (b) $\{-1, -2\}$, (c) $\{6\}$

5. Equation (2):

$$(b^m)^n = \overbrace{\underbrace{(bb \ldots b)}_{m}\underbrace{(bb \ldots b)}_{m} \ldots \underbrace{(bb \ldots b)}_{m}}^{n} = \underbrace{bbb \ldots b}_{mn} = b^{mn}$$

Equation (3):

$$(bc)^n = \underbrace{(bc)(bc) \ldots (bc)}_{n} = \underbrace{(bbb \ldots b)}_{n}\underbrace{(cc \ldots c)}_{n} = b^n c^n$$

Equation (4):

$$\left(\frac{b}{c}\right)^n = \underbrace{\left(\frac{b}{c} \cdot \frac{b}{c} \cdot \frac{b}{c} \ldots \frac{b}{c}\right)}_{n} = \frac{\overbrace{bbb \ldots b}^{n}}{\underbrace{cc \ldots c}_{n}} = \frac{b^n}{c^n}$$

Equation (5):

$$(m > n) \; \frac{b^m}{b^n} = \frac{\overbrace{bb \ldots b}^{m}}{\underbrace{bb \ldots b}_{n}} = \frac{\overbrace{(bb \ldots b)}^{n}(bb \ldots b)}{\underbrace{(bb \ldots b)}_{n}} = \overbrace{(bb \ldots b)}^{m-n} = b^{m-n}$$

6. For example, $\sqrt{4} + \sqrt{9} = 2 + 3 = 5 \neq \sqrt{4+9} = \sqrt{13}$ since $5^2 \neq 13$.

7. It does not follow that $\sqrt{(-1)^2} = -1$ since the symbol $\sqrt{}$ denotes the principal (and, in our case, the *positive*) square root. Thus,

$$1 = \sqrt{1^2} = \sqrt{(-1)^2} = 1$$

8. $b^{m-n} = b^{-(n-m)} = 1/b^{n-m}$ by Definition 1.3.3, and $b^0 = 1$ by Definition 1.3.2.

9. The equation is valid if $a \geq 0$ and $b \geq 0$.

10.
(a) $x - y$ (b) $\sqrt{x} + \sqrt{a}$
(c) $a^2 + b^2$ (Notice that $a^2 + b^2$ is nonnegative.)
(d) $\dfrac{7 + 2\sqrt{10}}{3}$

Supplement 1.3

I. Simplifying Expressions with Exponents and Radicals

Problem S1.   Simplify (a) $2^3(-16^{1/2})$ and (b) $9^2 x^{-2} y^4 / 3^3 x^{-3} y^2$.

Solution: (a) Notice that $-16^{1/2} = -\sqrt{16} = -4$. (Remember that $-16^{1/2}$ is not to be confused with $(-16)^{1/2}$ which, by Definition 1.3.5, is undefined.) Thus, $2^3(-16^{1/2}) = 2 \cdot 2 \cdot 2(-4) = 8(-4) = -32$.
   (b) Apply Equation (5) to obtain $9^2 x^{-2} y^4 / 3^3 x^{-3} y^2 = (81/27) x^{-2-(-3)} y^{4-2} = 3xy^2$.

Problem S2.   Write, if possible, without using exponents (a) $(27/8)^{-2/3}$ and (b) $(-25/16)^{-3/2}$.

Solution: (a) One approach: The equations $(27/8)^{-2/3} = (8/27)^{2/3} = 8^{2/3}/27^{2/3} = (8^{1/3})^2/(27^{1/3})^2 = 2^2/3^2 = 4/9$ follow from applications of Equation (7) (let $b = 27/8$; then $1/b = 8/27$), Equation (4), Definition 1.3.6, Definition 1.3.5 ($8^{1/3} = 2$ because $2^3 = 8$ and $27^{1/3} = 3$ because $3^3 = 27$), and Definition 1.3.1, respectively.
   (b) According to Definition 1.3.6, $b^{p/q}$ is defined only if $b^{1/q}$ is defined. Now $(-25/16)^{1/2}$ is undefined since $-25/16 < 0$. Thus, $(-25/16)^{-3/2} = [(-25/16)^{1/2}]^{-3}$ is undefined.

Problem S3. Simplify (a) $2\sqrt{60} + \sqrt{135}$ and (b) $\dfrac{x^{-2}y^{1/3}}{(xy^{-1})^{-1/2}}$.

Solution: (a) $2\sqrt{60} + \sqrt{135} = 2\cdot(4\cdot15)^{1/2} + (9\cdot15)^{1/2} = 2\cdot4^{1/2}\cdot15^{1/2} + 9^{1/2}15^{1/2} = (2\cdot2 + 3)15^{1/2} = 7\sqrt{15}$ by applications of Equation (3) and the distributive law. So $2\sqrt{60} + \sqrt{135} = 7\sqrt{15}$. (Note that $\sqrt{60} = (60)^{1/2}$ because $\sqrt{60}$ is just another symbol for $(60)^{1/2}$.)

(b) $[(x^{-2}y^{1/3})/(xy^{-1})^{-1/2}] = (x^{-2}y^{1/3})/(x^{-1/2}y^{1/2}) = x^{-2+(1/2)}y^{(1/3)-(1/2)} = x^{-(3/2)}y^{-(1/6)}$ by applications of Equations (3) and (6). So

$$\frac{x^{-2}y^{1/3}}{(xy^{-1})^{-1/2}} = x^{-3/2}y^{-1/6}$$

Problem S4. Simplify (a) $-2^2(9/4)^{1/2}$ and (b) $\dfrac{3^{-1}}{2^{-1} + 4^{-1}}$.

Solution: (a) Since Equation (4) is true for $m, n \in Q$, $(9/4)^{1/2} = $ _____.
Since $9^{1/2} = $ _____, $4^{1/2} = $ _____,
and $-2^2 = $ _____, it follows that
$-2^2(9/4)^{1/2} = $ _____. (Notice
that $-2^2 \neq (-2)^2$ since $-2^2 = -(2\cdot2) = -4$ and $(-2)^2 = $ _____.)

(b) $\dfrac{3^{-1}}{2^{-1} + 4^{-1}} = \dfrac{1/3}{1/2 + 1/4} = \dfrac{1/3}{3/4} = $ _____.

$9^{1/2}/4^{1/2}$
$\sqrt{9} = 3, \quad \sqrt{4} = 2$
-4 (not 4)
$-4(3/2) = -6$

$(-2)(-2) = 4$

$\dfrac{1}{3}\cdot\dfrac{4}{3} = \dfrac{4}{9}$

Problem S5. Simplify (a) $\sqrt[3]{\sqrt{25} + \sqrt{9}}$ and (b) $\left(\dfrac{x^{n^2+1}}{x^{1-n}}\right)^{(n+1)^{-1}}$ for $x > 0$.

Solution: (a) $\sqrt[3]{\sqrt{25} + \sqrt{9}} = \sqrt[3]{5 + 3} = \sqrt[3]{8} = 2$.

(b) It follows from Equation (5) that $x^{n^2+1}/x^{1-n} = x^{n^2+1-(1-n)} = x^{n^2+n}$. Thus, $(x^{n^2+1}/x^{1-n})^{(n+1)^{-1}} = (x^{n^2+n})^{1/(n+1)} = x^{(n^2+n)/(n+1)}$ follows from Equation (2). Next, notice that $(n^2 + n)/(n + 1) = [n(n + 1)/(n + 1)] = n$ provided that $n \neq -1$. Therefore,

$$\left(\frac{x^{n^2+n}}{x^{1-n}}\right)^{(n+1)^{-1}} = x^n$$

Problem S6. Simplify (a) $\dfrac{x^{-1}\cdot x^2\cdot x^6(x^{-3})^2}{x^{-4}}$, (b) $\left[\dfrac{9^{3/2} - 4\cdot8^{-2/3}}{2\cdot b^0 + (1/4)^{-2}}\right]^{-1}$

Solution: (a) It follows from Equation (1) that $x^{-1}x^2x^6 = $ _____. It also follows that $(x^{-3})^2 = $ _____ from Equation (2). So $\dfrac{x^{-1}x^2x^6(x^{-3})^2}{x^{-4}} = $

$\dfrac{x^7x^{-6}}{x^{-4}} = \dfrac{x^{7+(-6)}}{x^{-4}} = \dfrac{x}{x^{-4}}$. Thus, from

$x^{-1+2+6} = x^7$
x^{-6}

Equation (5), $\dfrac{x}{x^{-4}} =$ _____ .

Therefore, $\dfrac{x^{-1} \cdot x^2 \cdot x^6(-x^3)^2}{x^{-4}} = x^5.$

$x^{1-(-4)} = x^5$	

(b) $9^{3/2} = (9^{1/2})^3 = ($ _____ $)^3 = 27.$

$8^{-2/3} = ($ _____ $)^{2/3} = [(1/8)^{1/3}]^2 =$

$($ _____ $)^2 = 1/4.$ Also $(1/4)^{-2} =$

$($ _____ $)^2 = 16.$ Since $b^0 = 1,$ it follows

that

$$\left[\frac{9^{3/2} - 4 \cdot 8^{-2/3}}{2 \cdot b^0 + (1/4)^{-2}}\right]^{-1} = \left[\frac{27 - 4 \cdot 1/4}{2 \cdot 1 + 16}\right]^{-1}$$

$$= \left(\frac{26}{18}\right)^{-1} = \underline{\qquad}$$

$\sqrt{9} = 3$	
$1/8$	
$\sqrt[3]{1/8} = 1/2$	
4	
$18/26 = 9/13$	

DRILL PROBLEMS

S7. Simplify: (a) $(-32)^{-3/5}$ (b) $(1/2)(2x)^0$ (c) $\sqrt{27} + \sqrt{9}$

S8. Simplify: (a) $(-1/25)^{-1/2}$ (b) $(a^2b^{-3}/a^{-3}b^2)^{-1}$ (c) $\dfrac{a^{-2/3}(a^{-1/2})^3}{(a^{1/6})^{-2}}$

Answers

S7. (a) $-1/8$, (b) $1/2$, (c) $3(\sqrt{3} + 1)$

S8. (a) Not defined, (b) $a^{-5}b^5$, (c) $a^{-11/6}$

II. Basic Properties of Exponents and Radicals

Problem S9. Show that for $a \neq 0$, $(1/a)^n = 1/a^n$ for any $n \in Z$ by using properties of whole number exponents and the basic definitions.

Solution: If $n > 0$, then $(1/a)^n = 1^n/a^n = 1/a^n$ by Equation (4). If $n = 0$, it follows from Definition 1.3.2 that $(1/a)^0 = 1$ and $1/a^0 = 1/1 = 1.$ Thus, $1/a^0 = (1/a)^0.$ If $n < 0$, then $n = -t$ where $t > 0.$ So the equations $(1/a)^n = (1/a)^{-t} = \dfrac{1}{(1/a)^t} = \dfrac{1}{1^t/a^t} = \dfrac{1}{1/a^t} = \dfrac{1}{a^{-t}} = \dfrac{1}{a^n}$ follow from applications of Definition 1.3.3 (twice) and Equation (4).

Problem S10. Show that for $a > 0$ and $b > 0$, $\sqrt{a}/\sqrt{b} = \sqrt{a/b}.$

Solution: $\sqrt{a} = a^{1/2}$ and $\sqrt{b} = b^{1/2}.$ Therefore, the equations $\sqrt{a}/\sqrt{b} = a^{1/2}/b^{1/2} = (a/b)^{1/2} = \sqrt{a/b}$ follow from Equation (4). (Remember that $\sqrt{a}$ and $a^{1/2}$ are different symbols for the same number.)

Problem S11. Show that Equation (3) holds for integral exponents.

Solution: We are to show that $(bc)^n = b^n c^n$ for any integer n. [Here we assume that Equation (3) holds for non-negative integers.] Let $n < 0$. Then $n =$ ____ for $t > 0$. Thus, the equations $(bc)^n = (bc)^{-t} = 1/(bc)^t = 1/(b^t c^t) =$ ____ $= b^{-t} c^{-t} = b^n c^n$ follow from Definition 1.3.3 and Equation (3) for nonnegative integers. Therefore, $(bc)^n = b^n c^n$ for any $n \in Z$.

$-t$

$(1/b^t) \cdot (1/c^t)$

Problem S12. Show that Equation (3) holds for exponents that are reciprocals of positive integers.

Solution: We are to show that $(bc)^{1/n} = b^{1/n} c^{1/n}$ for $n \in N$. It follows from Definitions 1.3.4 and 1.3.5 that $b^{1/n} = x$ only if $x^n = b$ and that $c^{1/n} = y$ only if $y^n = c$. Thus, $bc = x^n y^n = (xy)^n = (b^{1/n} c^{1/n})^n$ follows by substitution and Equation (3). Hence, it follows from Definition 1.3.4 that $b^{1/n} c^{1/n}$ is an nth root of bc. We must now show that $b^{1/n} c^{1/n}$ is the principal nth root of bc. If $bc > 0$, then b and c have the same sign ($+$ or $-$). In either case, the product $b^{1/n} c^{1/n}$ is positive (because $b^{1/n}$ and $c^{1/n}$ will have the same sign). If $bc < 0$, then n is an odd integer. Also, b and c, and consequently $b^{1/n}$ and $c^{1/n}$, have different signs. So the product $b^{1/n} c^{1/n}$ is always negative. Thus, if we apply Definition 1.3.5, it follows that $b^{1/n} c^{1/n} = (bc)^{1/n}$.

DRILL PROBLEMS

S13. Show, by example, that $\sqrt[n]{a} \sqrt[m]{b} \neq \sqrt[nm]{ab}$.

S14. Show that $\sqrt[m]{\sqrt[n]{a}} = \sqrt[mn]{a}$ for integers m and n, $m > 2$ and $n > 2$, provided that the expressions are defined.

Answers

S13. $\sqrt[3]{8} \cdot \sqrt[4]{16} = 2 \cdot 2 = 4$, but $\sqrt[3 \cdot 4]{8 \cdot 16} = \sqrt[12]{128} \neq 4$ since $4^{12} \neq 128$.

S14. The equations $\sqrt[m]{\sqrt[n]{a}} = \sqrt[m]{a^{1/n}} = (a^{1/n})^{1/m}$ follow from Definition 1.3.7. Clearly, $(a^{1/n})^{1/m} = a^{1/nm}$ by Equation (2). So $\sqrt[mn]{a} = a^{1/nm}$ follows again from Definition 1.3.7. Therefore, $\sqrt[m]{\sqrt[n]{a}} = \sqrt[mn]{a}$.

Problem S15. Find the solution set for the equation $(3/\sqrt{x-2}) - 1 = 0$.

Solution: One way to proceed is to add 1 to both sides of the equation. This yields $\dfrac{3}{\sqrt{x-2}} = 1$. Next, square both sides to obtain $\dfrac{3^2}{(\sqrt{x-2})^2} = 1^2$ or $\dfrac{9}{x-2} = 1$.

Multiply both sides by $x - 2$ to get $\dfrac{9}{x-2}(x-2) = 1(x-2)$ or (assuming $x \neq 2$) $9 = x - 2$. Thus, $x = 11$. You should verify that the solution set is $\{11\}$ by direct substitution. (Since we assumed that $x \neq 2$, you should also verify that $x = 2$ is *not* a solution.)

Problem S16. Find the solution set for the equation $x/(\sqrt{2x+3}) = 1$.

Solution: As in Problem S15, there are several valid ways to proceed. One way is to immediately square both sides of the equation. This yields $\dfrac{x^2}{2x+3} = 1$. Next, multiply both sides by $2x + 3$ to obtain $x^2 = 2x + 3$ or $x^2 - 2x - 3 = 0$. (You can assume $2x + 3 \neq 0$. Why?) Factoring yields $x^2 - 2x - 3 = (x + 1)(x - 3) = 0$. Thus, by Theorem 1.2.1, $x + 1 = 0$ or $x - 3 = 0$. Hence, $x = -1$ and $x = 3$ are the apparent solutions. However, direct substitution of $x = -1$ in the original equation yields $\dfrac{-1}{\sqrt{2(-1)+3}} = \dfrac{-1}{\sqrt{1}} = -1 \neq 1$. Hence, $x = -1$ is *not* a solution. You should verify that $x = 3$ is a valid solution and thus the solution set is precisely $\{3\}$.

DRILL PROBLEMS

S17. Find the solution set for the equation $\sqrt{x-1}/\sqrt{x-5} = 2$.
S18. Find the solution set for the equation $\sqrt{x-1} + \sqrt{x+2} = 3$. (Hint: Square both sides of the equation, rearrange terms, and then square both sides of the new equation.)

Answers

S17. $\{19/3\}$
S18. $\{2\}$

III. Simplifying Expressions with Radicals in the Denominator

Problem S19. Show that $\dfrac{1}{\sqrt{a}+\sqrt{b}} = \dfrac{\sqrt{a}-\sqrt{b}}{a-b}$.

Solution: Multiply both numerator and denominator by $\sqrt{a} - \sqrt{b}$ (equivalent to multiplying the fraction by 1) to obtain

$$\frac{1}{\sqrt{a}+\sqrt{b}} = \frac{1(\sqrt{a}-\sqrt{b})}{(\sqrt{a}+\sqrt{b})(\sqrt{a}-\sqrt{b})} = \frac{\sqrt{a}-\sqrt{b}}{a-b}$$

(The process we just went through is commonly called *rationalizing the denominator*.)

Problem S20. Write each of the following without radicals in the denominator and without using negative exponents: (a) $\dfrac{1}{\sqrt{7}+2}$, (b) $\dfrac{\sqrt{x}-\sqrt{y}}{\sqrt{x}+\sqrt{y}}$

Solution: (a) $\dfrac{1}{\sqrt{7}+2} = \dfrac{\sqrt{7}-2}{(\sqrt{7}+2)(\sqrt{7}-2)}$ since $\dfrac{\sqrt{7}-2}{\sqrt{7}-2} = 1$. Notice that $(\sqrt{7}+2)(\sqrt{7}-2) = (\sqrt{7})^2 - 2^2 = 7 - 4 = 3$. So $\dfrac{1}{\sqrt{7}+2} = \dfrac{\sqrt{7}-2}{3}$. (Note that if the denominator is in the form $a + b$ and if we multiply numerator *and* denominator by $a - b$, the denominator becomes $(a + b)(a - b) = a^2 - b^2$. Thus, if a and b are square roots, the denominator becomes free from radicals.)

(b) Multiplying both numerator and denominator by $\sqrt{x} - \sqrt{y}$, we obtain

$$\frac{\sqrt{x} - \sqrt{y}}{\sqrt{x} + \sqrt{y}} = \frac{(\sqrt{x} - \sqrt{y})^2}{(\sqrt{x} + \sqrt{y})(\sqrt{x} - \sqrt{y})} = \frac{x - 2\sqrt{xy} + y}{x - y}$$

DRILL PROBLEMS

S21. Write the following, without using radicals in the denominator and without using negative exponents:

(a) $\dfrac{3\sqrt{2} - 2\sqrt{3}}{4\sqrt{3} - 2\sqrt{6}}$

(b) $\dfrac{\sqrt{2} - 3\sqrt{x}}{2\sqrt{2} - \sqrt{x}}$

Answers

S21. (a) $\dfrac{(3\sqrt{2} - 2\sqrt{3})(4\sqrt{3} + 2\sqrt{6})}{24}$

(b) $\dfrac{(\sqrt{2} - 3\sqrt{x})(2\sqrt{2} + \sqrt{x})}{8 - x}$

SUPPLEMENTARY PROBLEMS

S22. Simplify (a) $\dfrac{(-8)^{1/3}(1/32)^{-2/5}}{-64^{1/2}}$ and (b) $(\sqrt[3]{b} \cdot \sqrt[5]{b^2})^3$.

S23. Prove or disprove $(a^2)^3 = a^{2^3}$.

S24. Find the solution set for the equation $\dfrac{x}{\sqrt{x+2}} - 1$.

S25. Show that $\sqrt{\sqrt{a}} = \sqrt[4]{a}$.

S26. Show that Equation (3) holds for rational exponents.

Answers

S22. (a) 1, (b) $b^{11/5}$

S23. Disprove: Let $a = 2$. Then $(2^2)^3 = 2^2 2^2 2^2 = 4 \cdot 4 \cdot 4 = 4^3 = 64$, but $2^{2^3} = 2^8 = 256$.

S24. $\{2\}$

S25. $\sqrt{\sqrt{a}} = \sqrt{a^{1/2}} = (a^{1/2})^{1/2} = a^{1/4} = \sqrt[4]{a}$.

S26. (Here we assume that Equation (3) holds for exponents that are integers and reciprocals of positive integers. See Problems S11 and S12.) Let $p, q \in Z, q > 0$. Then $(bc)^{p/q} = [(bc)^{1/q}]^p = (b^{1/q}c^{1/q})^p = (b^{1/q})^p(c^{1/q})^p = b^{p/q}c^{p/q}$ follows from Definition 1.3.6, Problem S12, Problem S11, and Definition 1.3.6 again, respectively.

1.4 INEQUALITIES

In Section 1.1 we learned that the statement "a is less than b," written $a < b$, means that the number a lies to the left of the number b on the number line and that $b > a$ (read "b is greater than a") means that the number b lies to the right of the number a on the number scale. The expression $b > a$ is equivalent to the expression $a < b$.

In Section 1.1 we also introduced the notation for an open interval:

$$(a, b) = \{x : a < x < b\}$$

The numbers a and b are the *endpoints* of the open interval (a, b). Sometimes we want to consider intervals that contain one or both of their endpoints. To do so, it is convenient to introduce another order symbol, namely, $\leq$.

Definition 1.4.1
The statement $a \leq b$ (read "a is less than or equal to b") is true if $a < b$ or $a = b$.

For example, $x \leq 6$ is a true statement if x is replaced by any number less than 6, or if x is replaced by 6.

The expression $b \geq a$ (read "b is greater than or equal to a") is equivalent to the expression $a \leq b$. The set $\{x : 0 \leq x\}$ is called the *set of nonnegative real numbers*. Notice that this set is not the set of positive numbers since it includes the number zero.

Q1: List the *integers* that belong to the following sets:
(a) $\{x : -1 \leq x < 2\}$
(b) $\{x : -1 < x \leq 2\}$
(c) $\{x : -1 \leq x \leq 2\}$

An interval that contains both its endpoints a and b is called a *closed interval*, denoted by the symbol $[a, b]$. For example,

$$[1, 4] = \{x : 1 \leq x \leq 4\}$$

In general, the square bracket symbol is used in the interval nota-tion to indicate that an endpoint is to be included in the interval. An interval may contain only one of its endpoints, in which case it is neither open nor closed, for example, $[1, 4) = \{x : 1 \leq x < 4\}$, illustrated in Figure 1.4.1

Figure 1.4.1

Figure 1.4.2 depicts $\{x : x > 1\}$. We shall extend our interval notation by using the symbol ∞ to indicate that an interval has no right endpoint. Thus,

$$(1, \infty) = \{x : x > 1\}$$

Similarly, the symbol $-\infty$ indicates that an interval extends to the left without bound. Thus, $(-\infty, 2) = \{x : x < 2\}$.

Figure 1.4.2

Up to this point we have used a parenthesis with the interval notation to indicate that an endpoint was not included in an interval, and a bracket to indicate that an endpoint was to be included. Since the use of the symbols $-\infty$ and ∞ indicates the absence of endpoints, it makes no sense to discuss the inclusion or exclusion of an endpoint when they are used (∞ is *not* a num-ber!). We adopt the convention that either the parenthesis or the bracket can be used with the symbols ∞ and $-\infty$. Furthermore, we will use either of the words *open* or *closed* in these cases and will avoid using a mixture of parentheses and brackets when possible. Thus, we write $(3, \infty)$ instead of $(3, \infty]$ and $[-\infty, 2]$ instead of $(-\infty, 2]$.

Let us summarize the above discussion in the following definition.

A1: (a) $\{-1, 0, 1\}$ (b) $\{0, 1, 2\}$ (c) $\{-1, 0, 1, 2\}$

Definition 1.4.2

Let $a < b$. Then

$$
\begin{aligned}
(a, b) &= \{x : a < x < b\} & (-\infty, a) &= \{x : x < a\} \\
[a, b) &= \{x : a \le x < b\} & [-\infty, a] &= \{x : x \le a\} \\
(a, b] &= \{x : a < x \le b\} & (a, \infty) &= \{x : x > a\} \\
[a, b] &= \{x : a \le x \le b\} & [a, \infty] &= \{x : x \ge a\}
\end{aligned}
\tag{1}
$$

Q2: Simplify each of the following, and illustrate each on a number scale:

(a) $[-1, 2) \cup (-1, 3]$ (b) $[-1, 2) \cap (-1, 3]$

Q3: Use interval notation to indicate the following sets:
(a) All positive real numbers
(b) All nonnegative real numbers
(c) All real numbers

The following important theorems contain the basic rules for manipulating inequalities. In Appendix A we prove Theorem 1.4.1. The proof of Theorem 1.4.2 is similar and is left for the exercises.

Theorem 1.4.1

Suppose that $a < b$.

(i)   If $b < c$, then $a < c$.
(ii)   For any number c, $a + c < b + c$.
(iii)   If c is a positive number, then $ca < cb$.
(iv)   If c is a negative number, then $ca > cb$.

Theorem 1.4.2

Suppose that $a \le b$.

(i)   If $b \le c$, then $a \le c$.
(ii)   For any number c, $a + c \le b + c$.
(iii)   If $c \ge 0$, then $ca \le cb$.
(iv)   If $c \le 0$, then $ca \ge cb$.

Next we use Theorems 1.4.1 and 1.4.2 to solve inequalities. Solving an inequality means finding the solution set, that is, the set of all real numbers which make the inequality true. For example, to solve the inequality $2x - 4 < 0$, we add 4 to both sides and obtain $2x < 4$. Multiplying both sides of this inequality by $1/2$ (or dividing by 2), we get $x < 2$. Thus,

$$
\{x : 2x - 4 < 0\} = \{x : x < 2\}
\tag{2}
$$

This solution set is shown in Figure 1.4.3.

Figure 1.4.3

Q4: Describe the solution set of $2x - 4 \leq 0$ and illustrate with a picture.

Example 1.4.1

Use interval notation to indicate the solution set of the inequality $5 - 3x \geq 2 - x$.

Solution

Adding x to both sides of this inequality yields $5 - 2x \geq 2$. Subtracting 5 from both sides of this inequality yields $-2x \geq -3$. Since $-2 < 0$, dividing by -2 changes the sense of the inequality [Theorem 1.4.2 (iv)], so that $x \leq 3/2$. Thus, the solution set is

$$\{x : x \leq \tfrac{3}{2}\} = [-\infty, \tfrac{3}{2}]$$

Q5: Using the result of the Example 1.4.1, solve the inequality $5 - 3x < 2 - x$. Express your answer in interval notation.

Example 1.4.2

Use interval notation to indicate the set of all numbers such that $x - 4 > 0$ and $x + 3 > 0$.

Solution

This question asks us to use interval notation to denote the set

$$\{x : x - 4 > 0\} \cap \{x : x + 3 > 0\} \qquad (3)$$

Clearly,

$$\{x : x - 4 > 0\} = \{x : x > 4\} = (4, \infty)$$

and

$$\{x : x + 3 > 0\} = \{x : x > -3\} = (-3, \infty).$$

Thus, Equation (3) becomes

$$(4, \infty) \cap (-3, \infty) = (4, \infty)$$

Example 1.4.3

Solve the inequality $x^2 - x - 12 > 0$.

Solution

Factor the left side of this inequality to obtain:

$$(x - 4)(x + 3) > 0$$

A2: (a) $[-1, 3]$ (b) $(-1, 2)$

A3: (a) $(0, \infty)$ (b) $[0, \infty]$ (c) $(-\infty, \infty)$ or $[-\infty, \infty]$

A4: $\{x : x \le 2\}$. See the figure below.

A5: $(3/2, \infty)$

This inequality states that the product of the numbers $x - 4$ and $x + 3$ is positive. The product of these two factors is positive precisely when they have the same sign. In Example 1.4.2 we found that both factors are positive precisely when $x \in (4, \infty)$. Similarly, you can find that both factors are negative precisely when $x \in (-\infty, -3)$. Thus, the solution set of the inequality is $(-\infty, -3) \cup (4, \infty)$.

Q6: Solve the inequality $x^2 - x - 12 \le 0$.

Another way to solve $(x - 4)(x + 3) > 0$ is to find the points at which the product $(x - 4)(x + 3)$ is zero and then to examine the signs of the factors in each of the intervals created by those points. In this case $(x - 4)(x + 3) = 0$ if, and only if, $x = 4$ or $x = -3$. Now consider the intervals $(-\infty, -3)$, $(-3, 4)$, and $(4, \infty)$. It should be easy for you to convince yourself that for $x \in (-3, 4)$ one factor is positive and the other negative, and for $x \in (-\infty, -3)$ both factors are negative (see Figure 1.4.4). Finally, both factors are positive for $x \in (4, \infty)$. The solution of the inequality then follows from inspection of the figure.

Figure 1.4.4

Example 1.4.4   Solve the inequality $(x - 2)(1 - x) \ge 0$.

multiply by neg. # reverses inequality.

Solution

Following the procedure outlined above, we consider the intervals $(-\infty, 1)$, $(1, 2)$, and $(2, \infty)$. The signs of the factors and of the product for these intervals are shown in Figure 1.4.5. From the figure, we see that the solution set of the inequality is $[1, 2]$. Notice that the endpoints are included because the product is zero at those points.

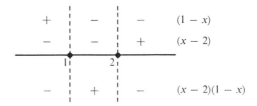

Figure 1.4.5

**Example
1.4.5**

Solve the inequality $(x - 4)/(x + 3) > 0$.

Solution

This inequality states that the quotient of the numbers $x - 4$ and $x + 3$ is positive. Hence, these numbers must have the same sign. In Example 1.4.3 we found that these numbers have the same sign for $x \in (-\infty, -3) \cup (4, \infty)$. Another way to solve the inequality is to find the points at which the quotient is zero or undefined, and then to examine the signs of the numerator and the denominator in each of the intervals created by those points. The quotient is zero precisely when $x = 4$ and undefined precisely when $x = -3$. Again we consider the intervals $(-\infty, -3)$, $(-3, 4)$, and $(4, \infty)$, as in Example 1.4.3. We construct Figure 1.4.6, from which we obtain the solution.

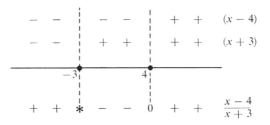

Figure 1.4.6

In addition to solving simple inequalities, we can also use Theorems 1.4.1 and 1.4.2 to prove other basic facts about inequalities, some of which are illustrated in the following examples.

**Example
1.4.6**

Prove that if $c \in (0, 1)$, then $c^2 \in (0, 1)$.

A6: $[-3, 4]$

Solution

We are given that $0 < c < 1$, that is, $c > 0$ and $c < 1$. Using property (iii) of Theorem 1.4.1 with a replaced by c, and b replaced by 1, we see that $c \cdot c < 1 \cdot c$, that is, $c^2 < c$. Since $c < 1$, it follows from property (i) that $c^2 < 1$. Thus, $c^2 > 0$ and $c^2 < 1$, that is, $0 < c^2 < 1$. In other words, $c^2 \in (0, 1)$.

**Example
1.4.7**

Suppose that $1/a < 1/b$. Can you conclude that $b < a$?

Solution

No. For example, $1/(-2) < 1/3$, but $3 > -2$. The following discussion shows under what conditions the statement is true. If $ab > 0$, then we can use Theorem 1.4.1 (iii) to multiply by ab and obtain $ab/a < ab/b$ or $b < a$. On the other hand, if $ab < 0$, Theorem 1.4.1 (iv) applies, and multiplication by ab yields $ab/a > ab/b$ or $b > a$. Thus, $b < a$ only if a and b have like signs.

Q7: Consider Example 1.4.7 if $ab = 0$.

**Example
1.4.8**

If $0 \le a \le b$ and $0 \le c \le d$, show that $ac \le bd$.

Solution

Since $c \ge 0$ and $a \le b$, we use Theorem 1.4.2 (iii) to obtain $ca \le cb$. Similarly, $b \ge 0$ and $c < d$ implies $bc \le bd$. Combining these two results, we have $ac \le bc \le bd$ or $ac \le bd$.

In many applications of mathematics, numbers are used to represent approximate values. For example, suppose that the result of measuring a cylindrical rod are stated as follows: "The length L is 3 feet and the radius R is 2 inches." Since it is impossible to determine whether or not the rod is *exactly* 3 feet long, some indication of the accuracy of the measurements is often stated. For example, one might state that the length L is $3 \pm .01$ feet and the radius R is $2 \pm .01$ inches. Now suppose that we want to calculate the volume V of the rod using the facts that $V = \pi R^2 L$ and that $\pi = 3.14$ to two decimal places. What degree of accuracy can we claim for our volume calculation?

To answer the preceding question, we first express the facts that we are given in terms of inequalities (using feet as units):

$$2.99 \le L \le 3.01$$
$$\frac{1.99}{12} \le R \le \frac{2.01}{12}$$
$$3.135 \le \pi \le 3.145$$

From these inequalities and Example 1.4.8,

$$\frac{(3.135)(1.99)^2(2.99)}{144} \leq \pi R^2 L \leq \frac{(3.145)(2.01)^2(3.01)}{144}$$

Multiplying these products and dividing, we obtain

$$0.257 \leq V \leq 0.266 \qquad \text{(in cubic feet)}$$

PROBLEMS
1.4

1. Picture the following sets on the number line: (a) [1, 3), (b) {1, 3}, (c) [2, 5], (d) (0, 2], (e) [−1, 3) ∪ [2, 3], (f) [0, 2) ∩ [1, 3)

2. Find: (a) {1, 3} ∩ [1, 2), (b) {1, 2} ∪ [0, 3), (c) {0, 1} ∪ (0, 1], (d) {0, 1} ∩ (0, 1]

3. Justify the numbered steps in each of the following: (a) $x < y \overset{1}{\Rightarrow}$ $3 + x < 3 + y \overset{2}{\Rightarrow} 6 + 2x < 6 + 2y$, (b) $5 < x \overset{1}{\Rightarrow} 2 < x \overset{2}{\Rightarrow} -2 > -x$ (The symbol ⇒ means "implies.")

4. Prove that if $x \leq y$, then $-y \leq -x$.

5. Prove that for $a, b, c, d \in R$, $a < b$ and $c < d \Rightarrow a + c < b + d$.

6. The volume of a sphere is given by the formula $V = (4/3)\pi r^3$ where r is the radius. If the radius is 5 inches, accurate to within 1/10 inch, express the volume as an inequality.

7. Find the solution sets for the given inequalities. Express each solution set in interval notation.
(a) $3x + 1 \leq 4x - 2$
(b) $3x^2 + 1 \geq x + 3x^2$
(c) $(x - 1)(x + 1) < 0$
(d) $(x^2 - 5x) \geq 0$
(e) $\dfrac{x + 3}{x^2} \geq 0$
(f) $\dfrac{x - 7}{3x + 6} < 0$

8. Solve the following inequalities:
(a) $(x - 1)(x + 1) < -1$
(b) $(x - 1)(x + 1) \geq -8$
(c) $\dfrac{3x + 5}{2x - 5} \geq 0$
(d) $\dfrac{x^2 + 1}{x} \geq 0$
(e) $\dfrac{3}{x - 1} \leq 1$
(f) $(x - 1)(x - 2)(x - 3) \geq 0$

9. Solve:
(a) $4x - x^2 \leq 4$
(b) $\dfrac{(x + 1)(x - 2)}{x - 4} \leq 0$
(c) $(3x + 2)(2x - 1)(9x - 1) > 0$
(d) $\dfrac{3x + 2}{x^2} > 0$

10. Prove that for $a \in R$, $a \neq 0$, $a^2 > 0$.

11. Prove Theorem 1.4.2.

12. Suppose that $a < b < 0$ and $0 < c < d$. Is $ac < bd$? Is $bd < ac$?

13. Prove that if $a < 0$, then $1/a < 0$.

A7: If $ab = 0$, then either $a = 0$ or $b = 0$, and hence $1/a$ or $1/b$ is undefined.

14. Prove that if $ab > 0$, then either $a > 0$ and $b > 0$ or $a < 0$ and $b < 0$.

15. Prove that $a/b < 0$ if and only if $a > 0$ and $b < 0$ or if $a < 0$ and $b > 0$.

16. Prove that 1 is positive.

17. Prove that if $a > 0$, then $a + (1/a) \geq 2$. (Hint: $(a - 1)^2 \geq 0$.)

18. Prove that if $a, b \geq 0$, then $(a + b)/2 \geq \sqrt{ab}$.

**ANSWERS
1.4**

1. (a) See Figure 1.4.7. (b) See Figure 1.4.8.

Figure 1.4.7

Figure 1.4.8

(c) See Figure 1.4.9. (d) See Figure 1.4.10.

Figure 1.4.9

Figure 1.4.10

(e) See Figure 1.4.11. (f) See Figure 1.4.12.

Figure 1.4.11

Figure 1.4.12

2. (a) $\{1\}$, (b) $[0, 3)$, (c) $[0, 1]$, (d) $\{1\}$

3. (a) Theorem 1.4.1 (ii) and (iii). (b) Since $2 < 5$, Theorem 1.4.1 (i) and (iv).

4. Theorem 1.4.2 (iv) with $c = -1$

5. $a < b \Rightarrow a + c < b + c$ by Theorem 1.4.1 (i) yields $a + c < b + d$.

6. $492.8 \leq V \leq 555.6$

7. (a) $[3, \infty]$, (b) $[-\infty, 1]$, (c) $(-1, 1)$, (d) $[-\infty, 0] \cup [5, \infty]$, (e) $[-3, 0) \cup (0, \infty)$, (f) $(-2, 7)$

8. (a) $\varnothing$, (b) R, (c) $[-\infty, -5/3] \cup (5/2, \infty)$, (d) $(0, \infty)$, (e) $(-\infty, 1) \cup [4, \infty]$, (f) $[1, 2] \cup [3, \infty]$

9. (a) R, (b) $[-\infty, -1] \cup [2, 4)$, (c) $(-2/3, 1/9) \cup (1/2, \infty)$, (d) $(-2/3, 0) \cup (0, \infty)$

Supplement 1.4

I. Properties of the Order Relation

Problem S1.   Picture the following sets on the number line: (a) [1, 4),   (b) [1, 2) ∪ [2, 4),   (c) (1, 3] ∩ [2, 4].

Solution: (a) By Definition 1.4.2, [1, 4) = {x ∈ R : 1 ≤ x < 4}. So [1, 4) is the set of all numbers between 1 and 4, including 1, but *not* including 4. (See Figure 1.4.13.)

Figure 1.4.13

   (b) By Definition 1.4.2 and the definition of the union of two sets, [1, 2) ∪ [2, 4) = {x : 1 ≤ x < 2 or 2 ≤ x < 4}. The union is the set [1, 4), the set of numbers between 1 and 4, including 1, but not including 4. This is true because every number in [1, 4) is in [1, 2) or [2, 4) and, conversely, every number in [1, 2) or [2, 4) is in [1, 4). We can also obtain [1, 2) ∪ [2, 4) = [1, 4) by considering their graphs. Numbers that are "covered" by at least one of the intervals comprise the union. (See Figure 1.4.14.)

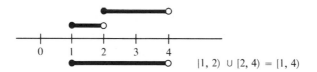

Figure 1.4.14

   (c) By Definition 1.4.2 and the definition of the intersection of two sets, (1, 3] ∩ [2, 4] = {x : 1 < x ≤ 3 and 2 ≤ x ≤ 4} = [2, 3]. To see this, first picture the intervals (1, 3] and [2, 4] on the number line (see Figure 1.4.15). Numbers that are "covered" by *both* intervals are in the intersection. Thus, (1, 3] ∩ [2, 4] = [2, 3].

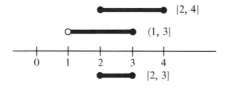

Figure 1.4.15

Problems S2.   List the set of *integers* in the sets (a), (b), and (c) of Problem S1.

Solution: (a) Referring to Figure 1.4.13 of problem S1, we see that the *integers* in the interval $[1, 4)$ are 1, 2, and 3. So $\{1, 2, 3\}$ is the set of integers in the interval $[1, 4)$. Notice that $\{1, 2, 3\} \neq [1, 4)$. $[1, 4)$ contains many other numbers, such as $3/2$, $5/2$, $6/5$, π, $\sqrt{2}$, etc.

(b) Refer to Figure 1.4.14 of problem S1. Since $[1, 2) \cup [2, 4) = [1, 4)$, the set of integers in the set $[1, 2) \cup [2, 4)$ is the same as in (a) above, $\{1, 2, 3\}$. Notice that the answer also follows when we consider the set of integers in $[1, 2)$, $\{1\}$, joined with the set of integers in $[2, 4)$, $\{2, 3\}$. Consequently, $\{1\} \cup \{2, 3\} = \{1, 2, 3\}$.

(c) Referring to Figure 1.4.15 of problem S1, we see that since $(1, 3] \cap [2, 4] = [2, 3]$, the set of integers in $(1, 3] \cap [2, 4]$ is $\{2, 3\}$.

Problem S3.   Justify the steps in each of the following: (a) $3 < 5 \overset{1}{\Rightarrow} 7 < 9 \overset{2}{\Rightarrow} 14 < 18$,   (b) $x < z \overset{1}{\Rightarrow} 2x < x + z \overset{2}{\Rightarrow} -x > -(x + z)/2$.

Solution: (a) Step (1) is derived from Theorem 1.4.1 (ii), since adding 4 to both sides of the inequality yields $3 + 4 < 5 + 4$ or $7 < 9$. Step (2) is from Part (iii) of Theorem 1.4.1, since $7 \cdot 2 < 9 \cdot 2$ or $14 < 18$. (Multiply both sides of the inequality $7 < 9$ by the positive number 2.)

(b) Step (1) is derived from Part (ii) of Theorem 1.4.1 (add x to both sides of $x < z$, obtaining $x + x < x + z$ or $2x < x + z$). Step (2) is part (iv) of Theorem 1.4.1. Multiply both sides of the inequality $2x < x + z$ by the *negative* number $(-1/2)$, obtaining $(-1/2)2x > -(1/2)(x + z)$ or $-x > (-x + z)/2$.

Notice that when we multiply an inequality by a *negative* number, the "direction" of the inequality symbol changes. For example, $2 < 5$, but $-2 \not< -5$ $(-2 > -5$ is the correct result obtained by multiplying both sides of $2 < 5$ by the negative number (-1) *and* changing the "direction" of the symbol $<$).

Problem S4.   Justify each step in the solution of the following problem: Find x if $8 - 2x < 16$.

Solution: If $8 - 2x < 16$, then (1) $-2x < 8$, (2) $x > -4$. Therefore, if $8 - 2x < 16$, then $x > -4$. Step (1) is Part (ii) of Theorem 1.4.1. (Add -8 to both sides, obtaining $(-8) + 8 - 2x < -8 + 16$ or $-2x < 8$.) Step (2) is Part (iv) of Theorem 1.4.1. (Multiply both sides by the *negative* number $-1/2$, obtaining $(-1/2)(-2x) > (-1/2)8$ or $x > -4$. Remember to change the direction of the inequality symbol when multiplying an inequality by a negative number.

Problem S5.   Express, as an inequality, the volume of a box having a rectangular base 3 ft. by 5 ft. and a height of 6 ft. if all the measurements are accurate to within $1/10$ ft.

Solution: Here we use the formula $V = L \cdot W \cdot H$ where $L = 3$, $W = 5$, and $H = 6$. However, since the measurements are accurate only to within $1/10$ foot, $L = 3 \pm 0.1$, $W = 5 \pm 0.1$ and $H = 6 \pm 0.1$. We can express these facts using the inequalities

$$2.9 \leq L \leq 3.1$$
$$4.9 \leq W \leq 5.1$$
$$5.9 \leq H \leq 6.1$$

By Example 1.4.8, it follows that $(2.9)(4.9)(5.9) \leq LWH \leq (3.1)(5.1)(6.1)$, or, multiplying, $83.839 \leq LWH \leq 96.441$. Thus, the volume, expressed as an inequality, is $83.839 \leq V \leq 96.441$.

Problem S6. Prove that if $a < b < 0$ and $c < d < 0$, then $ac > bd$.

Solution: By Part _____ of Theorem 1.4.1, $c < 0$. Multiply $a < b$ by c, obtaining _____ by Part (iv) of Theorem 1.4.1. (Remember to change _____.) Next, observe that b _____. Multiply $c < d$ by b, obtaining _____ by Part _____ of Theorem 1.4.1. Thus, $ac > bc$ and $bc > bd$. So $ac >$ _____ by Part _____ of Theorem 1.4.1.

(i)

$ac > bc$

the "direction" of the inequality sign
< 0
$bc > bd$, (iv)

bd, (i)

DRILL PROBLEMS

S7. Picture $(-1, 2] \cup (0, 3]$ on the number line.
S8. List the set of integers in the set given in Problem S7.
S9. Justify the following implication: $y > 5$ and $3 < x \Rightarrow x + y > 8$.
S10. Prove that if $b < 0$ and $ab \geq 0$, then $a \leq 0$.
S11. If the radius of a circle is 3 ± 0.01 feet, express the area of the circle as an inequality.

Answers

S7. See Figure 1.4.16.

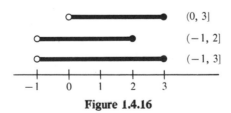

Figure 1.4.16

S8. $\{0, 1, 2, 3\}$
S9. Since $5 < y$ and $3 < x$, then (by Problem 5) $5 + 3 < y + x$ or $8 < x + y$. Thus, $x + y > 8$.
S10. Suppose that $ab = 0$. Since $b \neq 0$ (b is *less* than zero), $a = 0$ by a previous theorem. Next, suppose that $ab > 0$. Since $b < 0$, it follows that $1/b < 0$ (see Problem 13). Therefore, it follows from Theorem 1.4.1 (iv) that

$$(ab)\frac{1}{b} < 0\left(\frac{1}{b}\right) \quad \text{or} \quad a \cdot 1 < 0 \quad \text{or} \quad a < 0$$

Hence, if $ab > 0$, then $a < 0$. So if $ab \geq 0$ and $b < 0$, then either $a < 0$ or $a = 0$ ($a \leq 0$).

S11. $8.9401\pi \leq A \leq 9.0601\pi$, or given that $\pi \cong 3.1416$, $28.086 \leq A \leq 28.464$ (square feet).

II. Solving Linear Inequalities

Problem S12.   Find the solution set for the inequality $3x + 2 < 8$.

Solution: Adding -2 to both sides of the inequality $3x + 2 < 8$, we obtain $3x < 6$. Next we multiply both sides of $3x < 6$ by the positive number $1/3$ to obtain $x < 2$. Hence, the solution set is $\{x : x < 2\}$, or in interval notation, $\{x : x < 2\} = (-\infty, 2)$.

Problem S13.   Solve $2 - 3x \leq 11$.

Solution: Subtracting 2 from each side of the inequality, we get $-3x \leq 9$. Multiply both sides by $-1/3$, a negative number, to obtain $x \geq -3$. The solution set is $\{x \mid x \geq -3\} = [-3, \infty]$.

Problem S14.   Solve $2 + 3x \leq x + 14$.

Solution: Add -2 to both sides to obtain _____. Now add $-x$ to both sides to get _____. Multiply both sides by $1/2$ to get _____. The solution set is $\{x : $ _____ $\}$ or, in interval notation, _____.	$3x \leq x + 12$ $2x \leq 12$ $x \leq 6,$ $x \leq 6$ $[-\infty, 6]$

Problem S15.   Find the solution set for $7x + 1 < 3 + 7x$.

Solution: Add $-7x$ to both sides of the inequality to obtain _____. But this is true _____. So, replacing x in the original inequality $7x + 1 < 3 + 7x$ by any real number gives us a _____ statement. So the solution set is _____.	$1 < 3$ always true $(-\infty, \infty)$

DRILL PROBLEMS

S16. Solve each of the following:
(a) $5x + 1 \geq -9$ (b) $5x + 1 \geq 4x + 3$

(c) $5x + 1 \geq 6x - 3$ (d) $(7x + 1) - x \geq 6x + 3$

(e) $(7x + 1) - x < 6x + 3$

Answer

S16. (a) $[-2, \infty]$,   (b) $[2, \infty]$,   (c) $[-\infty, 4]$,   (d) $\varnothing$,   (e) $(-\infty, \infty) = R$

III. Other Inequalities

Problem S17. Solve $(x + 2)(x - 3) < 0$.

Solution: Following the method of Example 1.4.4, we note that $(x + 2)(x - 3) = 0$ if $x = -2$ or if $x = 3$. The number line is broken into three intervals by these numbers, namely, $(-\infty, -2)$, $(-2, 3)$, and $(3, \infty)$. For $x \in (-\infty, -2)$ both factors of the product $(x + 2)(x - 3)$ are negative and their product, therefore, is positive. For $x \in (3, \infty)$, each factor is positive and their product is positive. For $x \in (-2, 3)$, that is, $-2 < x < 3$, $x + 2$ is positive and $x - 3$ is negative. So for $x \in (-2, 3)$, the product $(x + 2)(x - 3) < 0$. Hence, the solution set is $(-2, 3)$. Figure 1.4.17 summarizes this method of solution.

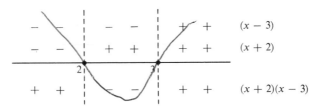

Figure 1.4.17

Alternate Solution: In order for a product of two factors to be negative, one factor must be positive and the other negative. Thus, for $(x + 2)(x - 3)$ to be negative either $x + 2 < 0$ and $x - 3 > 0$ or $x + 2 > 0$ and $x - 3 < 0$. Thus, either $x < -2$ and $x > 3$, or $x > -2$ and $x < 3$. Notice that the first alternative cannot hold because no number x is both less than -2 and, at the same time, greater than 3. The second alternative, $x > -2$ and $x < 3$, holds if $x \in (-2, 3)$. Thus, the solution set is the interval $(-2, 3)$.

Problem S18. Solve $(x + 2)/(x - 3) \leq 0$.

Solution: If the problem were to solve $(x + 2)/(x - 3) < 0$, the analysis of Problem S17 would apply since a fraction is negative when either the numerator is positive and the denominator is negative, or vice versa. Thus, the solution set for $(x + 2)/(x - 3) < 0$ is $(-2, 3)$. We still must consider the case $(x + 2)/(x - 3) = 0$. A fraction is zero precisely when the numerator is zero (and the denominator is not), so we must add the number -2 to our solution set. Thus, the solution set for this problem is $[-2, 3)$.

Problem S19. Solve $(x + 1)(x + 3) < 0$.

Solution: We use the "sign-chart" technique, dividing the number line by the numbers which make the factors zero. These numbers are $x = $ _____ and $x = $ _____. Mark these points on the number line (Figure 1.4.18).

−3,
−1
See Figure 1.4.19.

Figure 1.4.18

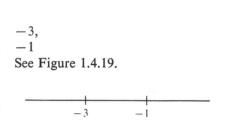

Figure 1.4.19

Now indicate the algebraic signs of each of the factors $(x + 1)$ and $(x + 3)$ (Figure 1.4.20).

See Figure 1.4.21

Figure 1.4.20

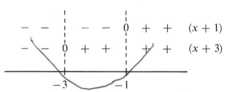

Figure 1.4.21

The product is negative when the factors have opposite signs, namely, the interval _____.

$(-3, -1)$

Problem S20. Solve $(2x - 4)/x^2 \leq 0$.

Solution: First notice that the fraction equals 0 precisely when $x = 2$. We now solve $(2x - 4)/x^2 < 0$. Since $x^2 = 0$ when $x = 0$, the fraction is not defined if $x = 0$, so we consider only nonzero numbers. For $x \neq 0$, $x^2 > 0$. Thus, the fraction can be negative only if $2x - 4 < 0$, that is, $2x < 4$, or $x < 2$. Thus, the solution set is $\{x : x < 2 \text{ but } x \neq 0\}$. In terms of intervals the solution set is $(-\infty, 0) \cup (0, 2)$. Adding the number 2 to this solution set, because the fraction is zero for $x = 2$, we obtain the set $(-\infty, 0) \cup (0, 2]$ as the solution set for this problem.

Problem S21. Solve $(x^2 + 1)/(2x + 6) > 0$.

Solution: The number $x^2 + 1$ is always _____. So the fraction is positive precisely when
_____ > 0. But $2x + 6 > 0$

positive

$2x + 6$

precisely when $x >$ ____.
Hence, the solution set is ____.

-3
$(-3, \infty)$

Problem S22. Solve $(3x + 8)(2 - 7x) \geq 0$.

Solution: We will use the "sign-chart" approach to this problem. The factors $3x + 8$ and $2 - 7x$ are zero if $x =$ ____ or $x =$ ____. Mark these on the number line (Figure 1.4.22).

$-8/3, \quad 2/7$
See Figure 1.4.23

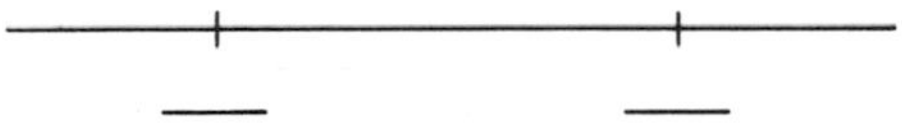

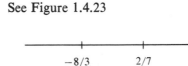

Figure 1.4.22

$-8/3 \qquad 2/7$

Figure 1.4.23

Now indicate in Figure 1.4.24 whether the factors are positive or negative on each of the three intervals.

See Figure 1.4.25

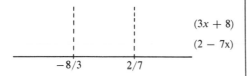

Figure 1.4.24

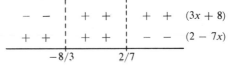

Figure 1.4.25

The product is positive on the interval ____ and negative on the other intervals. Adding the points making the product 0, we have the solution set ____.

$(-8/3, 2/7)$

$[-8/3, 2/7]$

DRILL PROBLEMS

S23. Solve each of the following inequalities:

(a) $(x - 3)(x + 7) > 0$

(b) $(x - 3)(x + 7) \leq 0$

(c) $(3x^2 + 9) \leq 0$

(d) $\dfrac{x - 1}{x^2} \geq 0$

(e) $x^2 - x - 6 \leq 0$ (Factor.)

(f) $x^2 - 5x \leq -6$ (Write in the form $x^2 - 5x + 6 \leq 0$ and factor.)

Answers

S23. (a) $(-\infty, -7) \cup (3, \infty)$, (b) $[-7, 3]$, (c) $\varnothing$, (d) $[1, \infty]$, (e) $[-2, 3]$, (f) $[2, 3]$

SUPPLEMENTARY PROBLEMS

S24. Write $(-1, 2] \cap (0, 3]$ as a single interval, if possible.

S25. List the set of integers in the set given in Problem S24.

S26. Prove, using only the axioms of Appendix A, that $1/a > 0$ if $a > 0$.

S27. Prove that if $0 < x < y$, $xy > 0$ and thus $1/(xy) > 0$.

S28. If $x < y$, is $1/x > 1/y$?

S29. Solve each of the following inequalities:

(a) $2x + 7 \leq x + 5$

(b) $\dfrac{1}{x^2 - 1} \leq 0$

(c) $x^2(x + 1) \geq 0$

(d) $x^2(x + 1) > 0$

(e) $(3x + 2)(x - 1) > 0$

(f) $\dfrac{(2x + 1)(3x - 1)}{(x - 5)} \geq 0$

Answers

S24. $(0, 2]$

S25. $\{1, 2\}$

S26. Use the trichotomy property (see Appendix A) to show that $1/a$ cannot be zero (since $(1/a) \cdot a = 1 \neq 0$) or negative [since, if $1/a < 0$, then $(1/a)a = 1$ would be negative by Theorem 1.4.1 (iv), a contradiction].

S27. Since $x > 0$ and $y > 0$, $xy > 0$ and thus $1/xy > 0$ (see Problem S26). Now multiply $x < y$ by $1/xy$, obtaining $1/y < 1/x$.

S28. No. For example, $-2 < 3$, but $-1/2 < 1/3$.

S29.

(a) $[-\infty, -2]$

(b) $(-1, 1)$

(c) $[-1, \infty]$

(d) $(-1, 0) \cup (0, \infty)$

(e) $(-\infty, -2/3) \cup (1, \infty)$

(f) $[-1/2, 1/3] \cup (5, \infty)$

1.5 ABSOLUTE VALUE

If c is a positive number and the point b is c units to the right of the point a on the number line (see Figure 1.5.1), then $b = a + c$, and the distance between a and b is $c = b - a$ units. On the other hand, if the point a is c units to the right of the point b, then $a = b + c$ and the distance between them is $c = a - b$ (see Figure 1.5.2). Thus, the distance between two points a and b is always either $a - b$ or $b - a$. In fact, the distance is the positive number of the pair, called the *absolute value* and denoted by two vertical bars. Thus, $|a - b|$ is the distance between a and b on a number line. For example,

$$|3 - 1| = 2 \quad \text{and} \quad |1 - 5| = 4$$

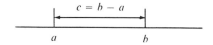

Figure 1.5.1

Figure 1.5.2

Q1: (a) Find the distance between 2 and 5; evaluate $|2 - 5|$.
(b) Find the distance between -2 and 5; evaluate $|-2 - 5|$.

The distance between a number a and the origin 0 is the absolute value of $a - 0 = a$, written in symbols as $|a|$. Notice that $|a|$ is never negative (that is, it is nonnegative), and that $|a|$ is zero if, and only if, a is the number zero. If $a \neq 0$, then $|a|$ is the positive number of the pair a and $-a$. Therefore, if a is positive or zero, then $|a| = a$; but if $-a$ is positive (that is, a is negative), then $|a| = -a$. Let us summarize this discussion with the following definition.

Definition 1.5.1

The absolute value of a number a is defined by the equations

$$|a| = \begin{cases} a & \text{if } a \geq 0 \\ -a & \text{if } a \leq 0 \end{cases} \qquad (1)$$

For example,

$$|3| = 3 \quad \text{and} \quad |-4| = -(-4) = 4$$

If $|x| = 2$, for example, then x is 2 units from the origin and hence is either the number 2 or the number -2. In other words, $\{x : |x| = 2\} = \{2, -2\}$. Thus, the solution set of the equation $|x| = 2$ is $\{2, -2\}$ (Figure 1.5.3).

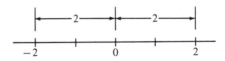

Figure 1.5.3

Q2: Find the solution set of each of the following:

(a) $|y| = 4$ (b) $|z| = 0$ (c) $|u| = -1$

The geometrical interpretation of the equation $|x - 1| = 3$ is that the points x and 1 of the number line are 3 units apart. Thus, the point x is either 3 units to the right of 1 or 3 units to the left of 1. Therefore, either $x = 4$ or $x = -2$ (see Figure 1.5.4). The same result is obtained if Definition 1.5.1 is used as follows:

$$|x - 1| = 3 \Rightarrow x - 1 = 3 \quad \text{or} \quad x - 1 = -3$$
$$\Rightarrow x = 4 \quad \text{or} \quad x = -2$$

Thus, the solution set of the equation $|x - 1| = 3$ is $\{4, -2\}$.

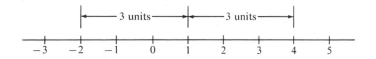

Figure 1.5.4

Q3: Find the solution set of each of the following equations:

(a) $|x - 3| = 4$ (b) $|x - 3| = 0$ (c) $|x - 3| = -1$

The equation $|x + 3| = 2$ can be rewritten as $|x - (-3)| = 2$. The solution set of this equation consists of all numbers that are 2 units from the number -3, namely, $\{-5, -1\}$. Again, the formal definition of absolute value leads to the same results: $|x + 3| = 2$ implies $x + 3 = 2$ or $x + 3 = -2$. Thus, $x = -1$ or -5.

Example 1.5.1

Solve $|2x + 3| = 7$.

Solution

By Definition 1.5.1, $|2x + 3| = 7$ implies $2x + 3 = 7$ or $2x + 3 = -7$. Thus, $x = 2$ or -5.

It is important to remember that the absolute value of a number is equal to the number itself only if that number is nonnegative. Thus, to remove the absolute value bars from an algebraic expression, we must know whether or not the expression represents a negative number.

Example 1.5.2

Solve $|x + 1| = 2x + 3$:

Solution

Using Definition 1.5.1, if $x + 1 \geq 0$ ($x \geq -1$), then $|x + 1| = x + 1$ and the equation to solve becomes $x + 1 = 2x + 3$. Thus, $-x = 2$ or $x = -2$. However, $x = -2$ is not a solution because, in this case, we must have $x \geq -1$.

Next, again using Definition 1.5.1, if $x + 1 < 0$ ($x < -1$), then $|x + 1| = -(x + 1)$ and the equation to solve becomes $-x - 1 = 2x + 3$ or $3x = -4$. Here $x = -4/3$ is a solution since $-4/3 < -1$ as required. Thus, the solution set is $\{-4/3\}$.

Q4: Solve $|x + 1| = \dfrac{1}{2}x + 3$.

A1: (a) 3,3, (b) 7,7

A2: (a) $\{4, -4\}$ (b) $\{0\}$ (c) $\varnothing$

A3: (a) $\{7, -1\}$ (b) $\{3\}$ (c) $\varnothing$

A4: $\left\{4, -\dfrac{8}{3}\right\}$

**Example
1.5.3**

Solve $|x^2 - x - 12| = -(x^2 - x - 12)$.

Solution

From Definition 1.5.1 we know that $|y| = -y$ if, and only if, $y \le 0$. Thus, the solution set is the set of all numbers such that $x^2 - x - 12 \le 0$. In Q6 of Section 1.4 we found that

$$\{x : x^2 - x - 12 \le 0\} = [-3, 4]$$

This closed interval is the solution set of the given equation.

Sometimes we want to specify that an unknown number is within a certain distance of another number. For example, in our discussion of the accuracy of measurements in Section 1.4, we stated that the length L of a certain rod was $3 \pm .01$ feet. In other words, the distance between the point L and the point 3 is no more than .01. We can state this fact by using the concept of absolute value as follows:

$$|L - 3| \le .01$$

This single inequality is equivalent to the two inequalities that we used before, namely,

$$2.99 \le L \le 3.01$$

Now let p be a positive number. The inequality $|z| < p$ means that z is less than p units away from the origin. From Figure 1.5.5, you can see that

$$\{z : |z| < p\} = \{z : -p < z < p\}$$
$$\text{or}\quad |z| < p \quad \text{if, and only if,} \quad -p < z < p \tag{2}$$

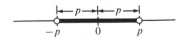

Figure 1.5.5

In the problems following this section you will prove equality (2) using the definition of absolute value. The inequality $|x - a| < p$ means that the distance between the points x and a is less than p units. In other words, x lies between the points $a - p$ and $a + p$ and the number $x - a$ is between $-p$ and p. From Equation (2) with z replaced by $x - a$,

$$\{x : |x - a| < p\} = \{x : -p < x - a < p\}.$$

Thus

$$\{x : |x - a| < p\} = \{x : a - p < x < a + p\}$$
$$= (a - p, a + p) \tag{3}$$

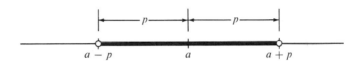

Figure 1.5.6

Example 1.5.4

Solve $|x + 2| \leq 3$.

Solution

$|x + 2| \leq 3$ implies $-3 \leq x + 2 \leq 3$. That is, $-5 \leq x \leq 1$. Thus, the solution set is $[-5, 1]$. Notice that $|x + 2| = |x - (-2)|$, so our inequality states that x is not more than 3 units away from -2. Verify this fact from Figure 1.5.7, which illustrates this example.

Figure 1.5.7

The statement "the point z is greater than 3 units away from the origin" can be written as $|z| > 3$. In other words, $z < -3$ or $z > 3$. If p is any positive number, then

$$\{z : |z| > p\} = \{z : z < -p\} \cup \{z : z > p\}$$
$$= (-\infty, -p) \cup (p, \infty) \tag{4}$$

The inequality $|x - a| > p$ says that the distance between the points x and a is more than p units. In other words, x is more than p units away from a, either to the right or to the left. Using Equation (4) with z replaced by $x - a$, we have (see Figure 1.5.8)

$$\{x : |x - a| > p\} = \{x : x - a < -p\} \cup \{x : x - a > p\}$$
$$= \{x : x < a - p\} \cup \{x : x > a + p\}$$
$$= (-\infty, a - p) \cup (a + p, \infty) \tag{5}$$

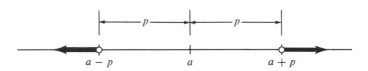

Figure 1.5.8

Example 1.5.5

Solve $|x + 1| \geq 4$.

Solution

$|x + 1| \geq 4$ implies $x + 1 \leq -4$ or $x + 1 \geq 4$. Therefore, $x \leq -5$ or $x \geq 3$ and the solution set is $[-\infty, -5] \cup [3, \infty]$. Because $|x + 1| = |x - (-1)|$, the inequality says that x is at least 4 units away from -1. Verify this fact from Figure 1.5.9, which illustrates this example.

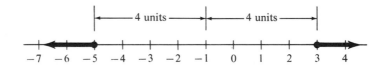

Figure 1.5.9

The length of any interval with endpoints (open, closed, or half-open) is the distance between the endpoints. Since the notation (a, b) for an interval implies that $a < b$, the length of (a, b) is $b - a$. The midpoint of an interval is that point which is equidistant from its endpoints. Thus, the midpoint m of an interval (a, b) is a point such that $b - m = m - a$. Solving for m, we find that

$$m = \frac{a + b}{2} \tag{6}$$

A number x is in the interval (a, b) if, and only if, its distance from the midpoint is less than one-half the length of the interval. The distance from x to the midpoint m is $|x - m|$. Let p denote half the length of the interval. (See Figure 1.5.10.) Then the absolute value symbol can be used to write an interval (a, b) as follows:

If $m = (a + b)/2$ and $p = (b - a)/2$, then
$$(a, b) = \{x : |x - m| < p\} \tag{7}$$

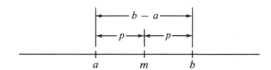

Figure 1.5.10

Q5: Find the midpoint and the length of the interval:

(a) $(-2, 8)$ (b) $\{x : |x + 2| \leq 3\}$

Example
1.5.6

Use the absolute value notation to write the interval $[-1, 5]$.

Solution

The midpoint of the interval $[-1, 5]$ is $m = (-1 + 5)/2 = 2$, and one-half its length is $p = (5 - (-1))/2 = 3$. Therefore,

$$[-1, 5] = \{x : |x - 2| \leq 3\}$$

This solution is illustrated in Figure 1.5.11.

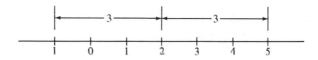

Figure 1.5.11

One useful property involving absolute value is

$$|ab| = |a| \cdot |b| \tag{8}$$

This property follows from the rule for multiplying signed numbers and Definition 1.5.1, because the product ab is either $|a| \cdot |b|$ or $-|a| \cdot |b|$, and the absolute value of both of these numbers is $|a| \cdot |b|$. For example,

$$|-3x| = |-3| \cdot |x| = 3|x|$$

Q6: Prove that $\left|\dfrac{a}{b}\right| = \dfrac{|a|}{|b|}$.

In Section 1.3 we learned that if $b > 0$, then $\sqrt{b}$ is the positive real number whose square is b. In particular, since both $a \cdot a = a^2$, and $(-a) \cdot (-a) = a^2$, the symbol $\sqrt{a^2}$ represents the nonnegative number of the pair a and $-a$. In short,

$$\sqrt{a^2} = |a| \tag{9}$$

A5: (a) 3, 10

(b) $-2, 6$ since $|x + 2| = |x - (-2)| = 3$.

Midpoint   Half-length

A6: The quotient a/b is either $|a|/|b|$ or $-|a|/|b|$, and the absolute value of both of these numbers (and hence of a/b) is $|a|/|b|$.

Thus, $\sqrt{a^2} = a$ if $a \geq 0$, and $\sqrt{a^2} = -a$ if $a \leq 0$. For example, $\sqrt{3^2} = 3$, but $\sqrt{(-3)^2} \neq -3$; instead,

$$\sqrt{(-3)^2} = |-3| = -(-3) = 3$$

We conclude this section by proving a basic inequality called the *triangle inequality*, which appears in various forms in other areas of mathematics.

Theorem 1.5.1

For any two numbers x and y,

$$|x + y| \leq |x| + |y| \qquad \text{(Triangle Inequality)}$$

Q7: Verify the triangle inequality for the following pairs of numbers:

(a) $x = 1, y = 2$ (b) $x = 1, y = -2$ (c) $x = -1, y = -2$

Proof

For any number x, either $x = |x|$, or $x = -|x|$. Since $-|x| \leq 0 \leq |x|$, we can conclude that

$$-|x| \leq x \leq |x| \tag{10}$$

Similarly,

$$-|y| \leq y \leq |y| \tag{11}$$

Using inequalities (10) and (11) and addition, we obtain the inequalities

$$-(|x| + |y|) \leq x + y \leq (|x| + |y|) \tag{12}$$

Using Equality (2) with z replaced by $x + y$ and p replaced by $|x| + |y|$, we see that inequalities (11) and (12) are equivalent to the triangle inequality.

PROBLEMS 1.5

1. Express the following equations and inequalities as statements relating to distance:

(a) $|3 - (-5)| = 8$ (b) $|x| \leq 5$

(c) $|x + 3| > 5$ (d) $|3 - x| \leq 10$

2. Find the solution set for the following equations or inequalities:

(a) $|x + 1| = 5$ (b) $|x - 3| < 5$

(c) $|3x - 2| > 10$ (d) $|8 - x| \le 5$

(e) $x + |x| = 0$ (f) $\dfrac{3}{|x + 4|} > 0$

(g) $\dfrac{|x - 3|}{x + 1} > 0$ (h) $|x^2 - 4| = 4 - x^2$

✗(i) $|x^2 - 5| < 4$ (j) $|x - 2| = \dfrac{1}{2}x + 1$

(k) $|x - 2| = x - 3$

3. (a) Show, by example, that $|a + b| \ne |a| + |b|$. (b) Prove that $|abc| = |a| \cdot |b| \cdot |c|$: (c) Prove that $|a - b| = |b - a|$.

4. Find the midpoint and the length of the indicated intervals:

(a) $\{x : |x + 2| < 6\}$ (b) $\{x : |x - 4| \le 1\}$

5. Use absolute value notation to write the indicated intervals: (a) $(-1, 4)$ and (b) $[0, 10]$.

6. Prove or disprove: $|a| = |b|$ if and only if $a = b$.

7. Prove that for every positive number p,
$$\{z : |z| < p\} = \{z : -p < z < p\}.$$

✗

8. Find the solution set for each of the following inequalities.

(a) $|x + 1| + |x - 1| \le 5$ (b) $|x - 3| + |x + 3| > 8$

(c) $||x| - 3| < 2$ (d) $|x^2 + 2x| \le x|x + 2|$

(e) $|(1/x) - 2| < 3$ (f) $|x - 3| \le |x - 2|$

9. Prove that $|a| \ge |b|$ if and only if $|a| \ge b$ and $|a| \ge -b$.

10. Prove that $|a| - |b| \le |a - b|$. (Hint: Apply Theorem 1.5.1 to $(a - b) + b$.)

11. Prove that $|a - b| \ge -(|a| - |b|)$.

12. Prove that $||a| - |b|| \le |a - b|$. (Hint: Use Problems 9, 10, and 11.)

ANSWERS
1.5

1. (a) The distance between the points 3 and -5 is 8 units. (b) The distance between the origin (0) and the point x is less than or equal to 5 units. (c) The distance between the points x and -3 is greater than 5 units. (d) The distance between the points 3 and x is less than or equal to 10 units.

2. (a) $\{-6, 4\}$, (b) $(-2, 8)$, (c) $(-\infty, -8/3) \cup (4, \infty)$, (d) $[3, 13]$, (e) $[-\infty, 0]$, (f) $\{x : x \ne -4\}$, (g) $(-1, 3) \cup (3, \infty)$, (h) $[-2, 2]$, (i) $(-3, -1) \cup (1, 3)$, (j) $\{6\}$, (k) $\emptyset$

3. (a) For example, $|-8 + 4| \ne |-8| + |4|$, since $|-8 + 4| = |-4| = 4$ and $|-8| + |4| = 8 + 4 = 12$. (b) Apply Equation (8) twice:
$$|(ab)c| = |ab| \cdot |c| = |a| \cdot |b| \cdot |c|$$

(c) Apply Equation (8) again:
$$|b - a| = |-(a - b)| = |-1| \cdot |a - b| = |a - b|$$

A7: (a) $3 \leq 1 + 2$ (b) $1 \leq 1 + 2$ (c) $3 \leq 1 + 2$

4. (a) $m = -2$, $L = 12$; (b) $m = 4$, $L = 2$

5. (a) $\{x : |x - 3/2| < 5/2\}$, (b) $\{x : |x - 5| \leq 5\}$

6. Disprove: $|-1| = |1|$ since $|-1| = 1 = |1|$, but $-1 \neq 1$.

7. Let $|z| < p$. If $z \geq 0$, $|z| = z$, so $0 \leq z < p$. On the other hand, if $z < 0$, then $|z| = -z$, so $0 \leq -z < p$ and thus $-p < z \leq 0$. Combining these statements, we have $-p < z < p$. To prove the "only if" part, let $-p < z < p$. If $z \geq 0$, $|z| = z$ and $|z| < p$. On the other hand, if $z < 0$, $|z| = -z$ or $z = -|z|$, so $-p < -|z|$ or $p > |z|$.

Supplement 1.5

I. Absolute Value and Solution Sets

Problem S1.   Express the following inequality as a statement relating to distance and solve: $|x + 2| > 3$.

Solution: $|x + 2| = |x - (-2)|$. Now $|x - (-2)|$ is the distance between the points x and -2. Hence, $|x + 2| > 3$ is equivalent to stating that x is a number whose distance from -2 is greater than 3 units. On the number line, x could be in either of the regions indicated in Figure 1.5.12.

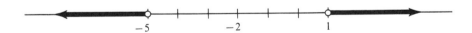

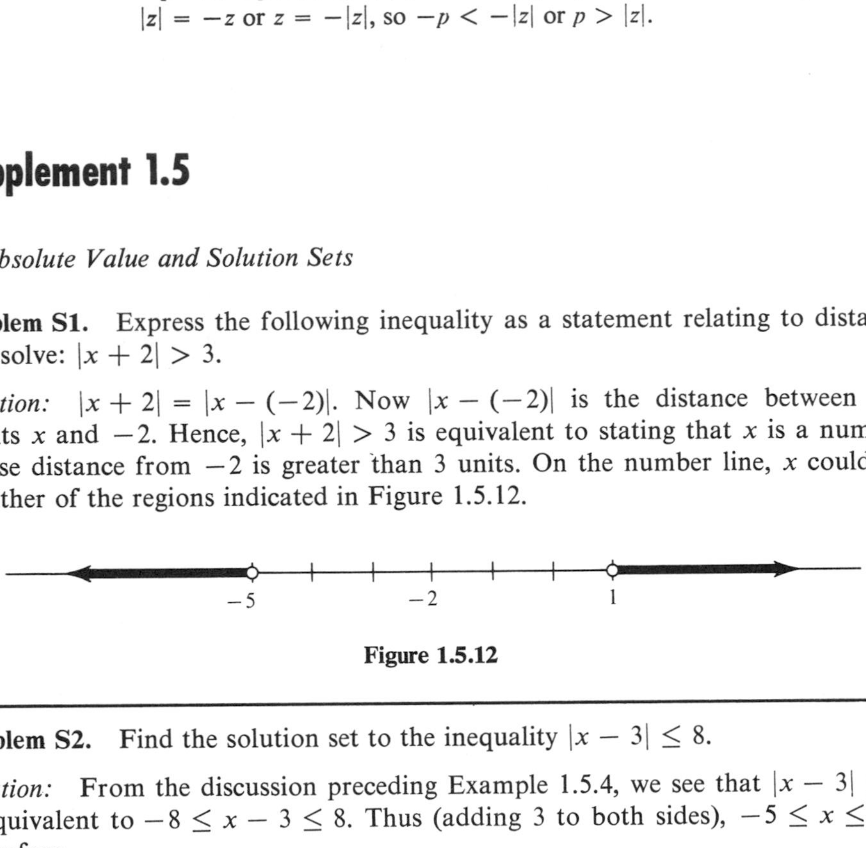

$$-5 \qquad\qquad -2 \qquad\qquad 1$$

Figure 1.5.12

Problem S2.   Find the solution set to the inequality $|x - 3| \leq 8$.

Solution: From the discussion preceding Example 1.5.4, we see that $|x - 3| \leq 8$ is equivalent to $-8 \leq x - 3 \leq 8$. Thus (adding 3 to both sides), $-5 \leq x \leq 11$. Therefore,

$$\{x : |x - 3| \leq 8\} = \{x : -8 \leq x - 3 \leq 8\} = \{x : -5 \leq x \leq 11\} = [-5, 11]$$

Problem S3.   Solve $|2x + 4| \geq 2$.

Solution: If $|2x + 4| \geq 2$, then $2x + 4 \geq 2$ or $2x + 4 \leq -2$. In the first case, $2x \geq -2$ or $x \geq -1$. In the second, $2x \leq -6$ or $x \leq -3$. Therefore,

$$\{x : |2x + 4| \geq 2\} = \{x : x \geq -1\} \cup \{x : x \leq -3\} = [-1, \infty] \cup [-\infty, -3]$$

Problem S4. Solve $\dfrac{(x+2)}{(x-1)^2} > 0$.

Solution: Observe that the denominator, $(x-1)^2$, is always positive if $x \neq 1$. Thus, for the fraction to be positive, the numerator must also be positive. Hence, the problem is to solve $(x+2) > 0$, noting that $x \neq 1$. If $x + 2 > 0$, then $x > -2$. Therefore, the solution set is $\{x : x > -2 \text{ and } x \neq 1\}$ or $(-2, 1) \cup (1, \infty)$.

Problem S5. Express each of the following equations as statements relating to distance: (a) $|5 - x| = 8$ and (b) $|5 - x| = |x - 5|$.

Solution: (a) $\lvert 5 - x \rvert = 8$ is equivalent to the statement that the _____ between _____ and _____ is equal to _____ units. The points 8 units to the right and left of 5 are _____ and _____. Thus, $x \in$ _____.	distance 5, x 8 13 -3, $\{13, -3\}$
(b) $\lvert 5 - x \rvert = \lvert x - 5 \rvert$ says that the distance between 5 and x equals the distance between _____ and _____. Thus, $x \in$ _____.	x, 5 R

Problem S6. Solve $|x - 3| < 10$.

Solution: If $\lvert x - 3 \rvert < 10$, then _____ $< x - 3 <$ _____. By adding 3 to each term, we obtain _____. Therefore, $\{x : \lvert x - 3 \rvert < 10\} =$ $\{x :$ _____$\} =$ _____.	-10, 10 $-7 < x < 13$ $-7 < x < 13$, $\ (-7, 13)$
Geometric Solution: If $\lvert x - 3 \rvert < 10$, then the _____ between x and 3 is _____. The point 10 units to the right of 3 is _____, and 10 units to the left of 3 is _____. Thus, x lies between -7 and 13, and, in interval notation, $\{x : \lvert x - 3 \rvert < 10\} =$ _____.	distance less than 10 13 -7 $(-7, 13)$

Problem S7. Solve $\dfrac{x - 1}{|x - 3|} \geq 0$.

Solution: $\dfrac{x-1}{\lvert x - 3 \rvert} = 0$ if and only if _____ $= 0$ and _____ $\neq 0$. Thus, $\dfrac{x-1}{\lvert x-3 \rvert} = 0$ exactly when _____.	$x - 1 = 0$, $\ \lvert x - 3 \rvert \neq 0$ $x = 1$

Now we must solve the inequality $\dfrac{x-1}{|x-3|} > 0$. Recall that a fraction is positive precisely when both the numerator and the denominator _____. When $x \neq 3$, the denominator ($|x - 3|$) is always _____. Thus, $\dfrac{x-1}{|x-3|} > 0$ is equivalent to _____ and $x \neq 3$. It follows from $x - 1 > 0$ that $x > 1$. Therefore, combining the above results, we have that $\left\{ x : \dfrac{x-1}{|x-3|} \geq 0 \right\}$ is the union of two intervals, _____ $\cup$ _____.

have the same sign

positive

$x - 1 > 0$

$[1, 3),\quad (3, \infty)$

DRILL PROBLEMS

S8. Express the inequality $|x - 5| \leq 9$ as a sentence involving distance.

S9. Solve $|8 - x| = 13$.

S10. Solve $|3x - (1/2)| \geq 4$.

S11. Solve $\dfrac{(x-1)|x+3|}{|x+2|} < 0$.

S12. Solve $x - 1 = |2x - 4|$.

Answers

S8. The distance between x and 5 is less than or equal to 9 units.

S9. $\{-5, 21\}$

S10. $[-\infty, -7/6] \cup [3/2, \infty]$

S11. $(-\infty, -3) \cup (-3, -2) \cup (-2, 1)$

S12. $\{5/3, 3\}$

II. Properties Involving Absolute Value

Problem S13. Find the midpoint and the length of the interval $\{x : |x - 3| < 5\}$.

Solution: $|x - 3| < 5$ is equivalent to $-5 < x - 3 < 5$ or $-2 < x < 8$. Thus, $\{x : |x - 3| < 5\} = (-2, 8)$. The length of $(-2, 8)$ is $|-2 - 8| = 10$ units. The midpoint of $(-2, 8)$ is $m = (-2 + 8)/2 = 3$. These facts are illustrated in Figure 1.5.13. Notice that we can originally read the midpoint 3 and the half-length 5 directly from the inequality (see Q5).

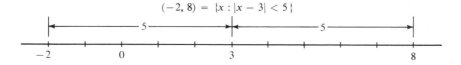

$$(-2, 8) = \{x : |x - 3| < 5\}$$

Figure 1.5.13

Problem S14. Use absolute value notation to write the interval (2, 7).

Solution: The midpoint of the interval (2, 7) is $m = (2 + 7)/2 = 9/2$. The length of the interval (2, 7) is $|2 - 7| = 5$ units. Clearly, the distance between the midpoint $(9/2)$ and either of the endpoints (2 or 7) is $5/2$ (one-half the length of the interval). Thus, $x \in (2, 7)$ if and only if the distance between x and $9/2$ (the midpoint) is less than $5/2$ or, in absolute value notation, $|x - 9/2| < 5/2$. Therefore,

$$(2, 7) = \left\{ x : \left| x - \frac{9}{2} \right| < \frac{5}{2} \right\}$$

Problem S15. Prove that $|a| \geq a$ for every number a.

Solution: If $a \geq 0$, then $|a| = a$ by Definition 1.5.1. Thus, if $a \geq 0$, $|a| \geq a$ as desired. Next, suppose that $a < 0$; then $|a| = -a$ by Definition 1.1.1. But if $a < 0$, then $-a > 0$ and $-a > a$, since a positive number is greater than a negative number. That is, if $a < 0$, then $|a| = -a > a$, so $|a| \geq a$. Therefore, combining the above results, we obtain $|a| \geq a$ for every number a.

Problem S16. Prove that $a^2 = |a|^2$ for every number a.

Solution: If $a \geq 0$, then $|a| = a$. So $|a|^2 = a^2$. If $a < 0$, then $|a| = -a$ and $|a|^2 = (-a)^2 = a^2$. Therefore, when we combine the above, $a^2 = |a|^2$ for any number.

Problem S17. Prove that $a^2 \leq b^2$ if and only if $|a| \leq |b|$.

Solution: Problem S16 shows that $a^2 \leq b^2$ is equivalent to $|a|^2 \leq |b|^2$. Now, add $-|b|^2$ to both sides to obtain $|a|^2 - |b|^2 \leq 0$. Factoring, we obtain $(|a| + |b|) \cdot (|a| - |b|) \leq 0$. So $a^2 \leq b^2$ is equivalent to $(|a| + |b|)(|a| - |b|) \leq 0$. Next, observe that the factor $|a| + |b|$ is always positive or zero. Thus, for the product $(|a| + |b|) \cdot (|a| - |b|)$ to be negative, the factor $|a| - |b|$ must be negative. Therefore, $(|a| + |b|)(|a| - |b|) \leq 0$ is equivalent to $|a| - |b| \leq 0$, or $|a| \leq |b|$. So $a^2 \leq b^2$ if and only if $|a| \leq |b|$.

Problem S18. Find the midpoint and the length of the interval $\{x : |x + 1| \leq 3\}$.

Solution: $|x + 1| \leq 3$ is equivalent to the inequalities _____. In

$$-4 \leq x \leq 2$$

82   *Real Numbers*

| interval notation, ———————. Thus, the length of $[-4, 2]$ is ———————, and its midpoint is ———————. | $x \in [-4, 2]$
 $\|-4 - 2\| = 6$
 $(-4 + 2)/2 = -1$ |

Problem S19.   Use absolute value notation to write $[-3, 5]$ as an inequality.

| *Solution:*   The length of $[-3, 5]$ is ———————. The midpoint of $[-3, 5]$ is ———————. The distance between the midpoint and either of the endpoints is ——. Thus, the distance between a number x in $[-3, 5]$ and the midpoint is ———————. This can be expressed in absolute value notation as ———————. Thus, $[-3, 5] = \{x : ———————\}$. | $\|-3 - 5\| = 8$
 $(-3 + 5)/2 = 1$

 4

 less than or equal to 4

 $\|x - 1\| \leq 4$
 $\|x - 1\| \leq 4$ |

Problem S20.   Solve $\|x + 2\| \leq \|x\| + 3$.

Solution: $\|x + 2\| \leq \|x\| + 2$ by Theorem 1.5.1, the triangle inequality. So $\|x + 2\| \leq \|x\| + 2$. Since $2 < 3$, $\|x\| + 2 < \|x\| + 3$. Therefore, $\|x + 2\| \leq \|x\| + 2 < \|x\| + 3$, or $\|x + 2\| \leq \|x\| + 3$. Thus, the inequality is true for any real number and the solution set is R.

Problem S21.   Solve $\|x - 1\| = \|x\|$.

Solution: If x is a number such that $\|x - 1\| = \|x\|$, then the distance between the numbers x and 1 is exactly the same as the distance between the numbers 0 and x. This is true precisely when $x = 1/2$. So $\{x : \|x - 1\| = \|x\|\} = \{1/2\}$.

Alternate Solution: Apply the fact that $\|a\| = \|b\|$ implies $a = b$ or $a = -b$. Thus, $\|x - 1\| = \|x\|$ is equivalent to $x - 1 = x$ or $x - 1 = -x$. Clearly, $x - 1 = x$ implies $-1 = 0$, which is impossible. So $x - 1 = -x$. Thus, $2x - 1 = 0$ or $x = 1/2$.

DRILL PROBLEMS

S22. Find the midpoint and the length of the interval $\{x : \|x - 1\| < 9\}$.
S23. Use absolute value notation to write the interval $[3, 7]$.
S24. Prove that if $a < b < 0$, then $\|a\| > \|b\|$.
S25. Solve $\|x\| = x$.
S26. For what numbers a and b, if any, is it true that $\|a\| + \|b\| = 0$?

Answers

S22. The length is 18 and the midpoint is 1.
S23. $\{x : \|x - 5\| \leq 2\}$

S24. If $a < b < 0$, then $-a > -b$ (multiply $a < b$ by -1). Now, $a < 0$, so $|a| = -a$. Also, $b < 0$, so $|b| = -b$. Therefore, by substitution, if $a < b < 0$, then $|a| > |b|$.

S25. $[0, \infty]$

S26. Both a and b must be zero.

SUPPLEMENTARY PROBLEMS

S27. Express the inequality $|(1/2)x - 3| < 2$ as a statement relating to distance.

S28. Solve (a) $|3x - 5| \geq 1/2$ and (b) $|3x - 5| \geq -1/2$.

S29. Solve (a) $|(x/2) - 3| < -2$ and (b) $\dfrac{x + 3}{|x - 2|} \geq 0$.

S30. Solve (a) $(x - 4)^2|x - 5| \leq 0$ and (b) $|x^2 - 4x| = 4x - x^2$.

S31. Find the midpoint and the length of the interval $\{x : |2x - 4| < 8\}$.

S32. Prove that $|-x| = |x|$ for every number x.

S33. Solve $|2x + 5| = 3x - 3$.

Answers

S27. The distance between the numbers $x/2$ and 3 is less than 2.

S28. (a) $[-\infty, 3/2] \cup [11/6, \infty]$, (b) R

S29. (a) $\varnothing$, (b) $[-3, 2) \cup (2, \infty)$

S30. (a) $\{4, 5\}$, (b) $[0, 4]$

S31. The length is 8 units and the midpoint is 2.

S32. If $x = 0$, then, clearly, $|x| = |-x|$. If $x > 0$, then $|x| = x$ by definition. Also, if $x > 0$, then $-x < 0$. So $|-x| = -(-x) = x$. Thus, if $x > 0$, $|x| = x = |-x|$. If $x < 0$, then $|x| = -x$ by definition. Also, if $x < 0$, then $-x > 0$. So $|-x| = -x$. Thus, if $x < 0$, then $|x| = -x = |-x|$. Therefore, when the above results are combined, $|-x| = |x|$ for every number x. Of course, this also follows directly from Equation (8) with $a = -1$ and $b = x$.

S33. $\{8\}$

Achievement Test 1

1. Write $(-3, 2) \cap ([-2, 0] \cup (-1, 4])$ as a single interval, if possible.
2. If $E = [-5, 8]$ and $F = \{y : y \in R$ and $|y| \leq x\}$, determine the largest x so that $F \subseteq E$.
3. Find a and p so that $[-4, 10] = \{x : |x - a| \leq p\}$.
4. Find $A \cap B$ if $A = \{x : x \in N$ and $x < 5\}$ and $B = \{2n : n \in Z$ and $n > -1\}$.
5. State the additive identity axiom.

In Problems 6 through 12 simplify the indicated expressions.

6. $\left[1 - \dfrac{1}{1 - (1/4)}\right]^{-2}$

7. $\dfrac{(b^2c^4)^{1/2}}{(b^{1/2}c)^{-2}}$

8. $(1/32)^{-2/5}$

9. $\dfrac{5}{x - 2} - \dfrac{5x - 4}{x^2 - 2x}$

10. $\dfrac{1/(x + h) - 1/x}{h}$

11. $\dfrac{1}{\sqrt{2} + \sqrt{5}}$

12. $\left[\dfrac{(ab^2)^{-1} + (a^2b)^{-1}}{(a^{-1}b)^{-1} - a^{-1}b}\right]^{-1}$

13. True or false: For $c \in Z$, $x < y \Rightarrow -x/c > -y/c$.
14. Order the following real numbers from smallest to largest: $-1, 3^{-1}, 2, \sqrt{5}, -1/2$
15. Find $\left\{x : \dfrac{x + 2}{x(x - 4)}$ is undefined$\right\}$.
16. Write $\dfrac{2(x + h)^2 - 2x^2}{h}$ as an expression that contains no fractions.

84

17. Find the solution set for $2(x - 3) - 3(4 - 2x) = 5 + x$.

18. Find the solution set for $\dfrac{3}{x + 1} - \dfrac{2x + 1}{2x^2 + x - 1} = \dfrac{5}{2x - 1}$.

19. Find the solution set for $\dfrac{x + 5}{4} - \dfrac{2x + 4}{9} = 1$.

20. Find the solution set for $(2y - 1)(y + 3) = 4$.

21. Find the solution set for $\sqrt{x^2 - 11} = 5$.

22. Find the solution sct for $\dfrac{x + 1}{2x - 3} = \dfrac{x - 4}{2x + 5}$.

23. Find the solution set for $3(4 + x) \geq x - 2$.

24. Find $\{x : \sqrt{x^2 - 10x + 25}$ is a real number$\}$.

25. True or false: The *sum* of the solutions to $|2x + 7| = 9$ is -7.

26. Find the solution set for $|x - 6| < 3$.

27. Find the solution set for $|3x + 7| > 2$.

28. Find the solution set for $\dfrac{|x + 3|}{x - 1} \leq 0$.

29. Find the solution set for $2x^2 + 5x + 2 \geq 0$.

30. Find the solution set for $\dfrac{x + 1}{2 - x} \geq 1$.

Answers to Achievement Test 1

1. $[-2, 2)$

✕ 2. 5

3. $a = 3, p = 7$

4. $\{2, 4\}$

✎ 5. If $a \in R$, there exists $0 \in R$ such that $a + 0 = 0 + a = a$.

6. 9

7. $b^2 c^4$

8. 4

9. $\dfrac{4}{x^2 - 2x}$

10. $\dfrac{-1}{x(x + h)}$

✓ 11. $\dfrac{\sqrt{5} - \sqrt{2}}{3}$

12. $ab(a - b)$

13. False

14. $-1, -1/2, 1/3, 2, \sqrt{5}$

15. $\{0, 4\}$

16. $4x + 2h$

17. $\{23/7\}$

18. $\{-9\}$

19. $\{7\}$

20. $\{-7/2, 1\}$

21. $\{-6, 6\}$

22. $\{7/18\}$

23. $\{x \geq -7\}$ or $[-7, \infty]$

24. R

25. True

26. $(3, 9)$

27. $\{x > -5/3, \text{ or } x < -3\}$ or $(-\infty, -3) \cup (-5/3, \infty)$

28. $(-\infty, 1)$

29. $[-\infty, -2] \cup [-1/2, \infty]$

30. $[1/2, 2)$

2 FUNCTIONS

2.0 INTRODUCTION

A ball is thrown into the air. Its height (in feet) above the ground is measured at each second after it is thrown until it hits the ground. The results of the measurements are plotted as points, and a smooth curve is drawn connecting these points, as shown in Figure 2.0.1. This curve depicts the height of the ball at any instant of its flight. For example, at $1\frac{1}{2}$ seconds after it is thrown, it is 60 feet above the ground.

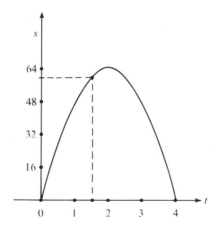

Figure 2.0.1

A balloon contains a fixed amount of helium. If the balloon is squeezed to change its volume while maintaining a constant temperature, the pressure changes. The pressure is recorded and plotted for various volumes. The smooth curve that is drawn in Figure 2.0.2 connecting the plotted points depicts the pressure-volume relationship.

Figures 2.0.1 and 2.0.2 illustrate a special type of relation between two quantities — special in that to each number on the horizontal axis there corresponds exactly one number on the vertical axis. This special relationship is formalized in mathematics as the concept of *function*. This concept and the notation involved are fundamental in any study of mathematics and its applications.

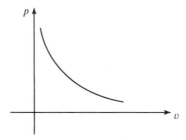

Figure 2.0.2

2.1 BASIC CONCEPTS

Let us return to the curve shown in Figure 2.0.1. We first construct a table showing the measured values of time and height (see Table 2.1.1) and then picture this table as illustrated in Figure 2.1.1. Let $T = \{0, 1, 2, 3, 4\}$, let $S = \{0, 48, 64\}$, and let f be the relationship or *rule of correspondence* between S and T. Here f is termed a function because it has the property that *one and only one* arrow originates from each element in T. The set T is called the domain of f. For the arrow originating at any $x \in T$, the element at which it terminates is called the function value of x, denoted by $f(x)$ and read "f of x." For example, from Figure 2.1.1, $f(0) = 0$, $f(1) = 48$, and $f(4) = 0$.

Table 2.1.1

t	0	1	2	3	4
s	0	48	64	48	0

Q1: Use Figure 2.1.1 to find (a) $f(2)$ and (b) $f(3)$.

The range of a function is the set of function values. In Figure 2.1.1 the range of f is $S = \{0, 48, 64\}$. Notice that $S = \{f(x) : x \in T\}$.

A1: (a) $f(2) = 64$ (b) $f(3) = 48$

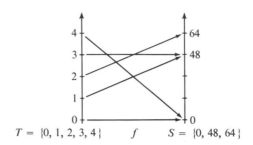

$T = \{0, 1, 2, 3, 4\}$ f $S = \{0, 48, 64\}$

Figure 2.1.1

It is natural to denote this set by the symbol $f(T)$. Notice that in the example, x is a number, $f(x)$ is a number, while T and $f(T)$ are sets of numbers.

Let us summarize our ideas concerning the concept of a function, and at the same time generalize our ideas by allowing arbitrary sets to be domains and ranges. A *function f* is defined if we specify a set D and a rule of correspondence that assigns to each $x \in D$ exactly one element of a second set. The element corresponding to $x \in D$ is denoted by $f(x)$ and is called the *function value* of f at x. The set D is called the *domain* of f, and the set of all function values, namely, $f(D)$, is called the *range* of f. This general concept of a function is illustrated in Figure 2.1.2.

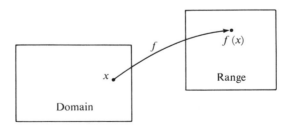

Figure 2.1.2

Notice that f and $f(x)$ are not the same: f is the name of a function, while $f(x)$ is a function value. Notice, also, that the concept of function does not require that the domain and range be sets of numbers. However, most of our work will be with functions whose domains and ranges are subsets of R. Thus, most of the functions we shall encounter will act like "number machines." A number will be the *input* to the machine f, and for each input number x the machine will produce an *output* consisting of a single number $f(x)$.

The rule of correspondence defining a function may be given by a graph, by a table, by a formula, or by a word description. In Table 2.1.1, you can observe that for each $t \in T$, these numbers satisfy the formula $s = f(t) = 64t - 16t^2$. When we define a function by a formula, we normally shorten the phrase "the function f defined by $f(x) = \ldots$" to "the function $f(x) = \ldots$." We usually use letters such as f, g, h, F, etc., to denote functions, and letters such as x, y, z, s, t, etc., to denote inputs to functions. Of course, it is the "form" that is important and not the particular letters used. Thus, all of the following equations represent the same function:

$$f(x) = 2x^2 + 3x \qquad p(q) = 2q^2 + 3q$$
$$g(a + h) = 2(a + h)^2 + 3(a + h)$$

(1)

Before proceeding with our study of functions, let us consider some relationships that are *not* functions. Returning to Table 2.1.1, suppose we were to be asked, "At what time was the ball 48 feet above the ground?" We could not give an unambiguous answer. We could only say that it was either 1 second or 3 seconds after it was thrown. If we turn the arrows around in Figure 2.1.1, we obtain Figure 2.1.3. This figure does *not* depict a function because there is not just one number, but two, that correspond to the number 48, namely, 1 and 3. Likewise, two numbers correspond to the number zero.

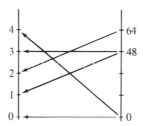

Figure 2.1.3

As a second example of a relationship that is not a function, consider the correspondence defined by the following statement: To each positive real number x let there correspond all numbers whose square is x. This relationship is not a function because to each positive number x there correspond two numbers, namely, $\sqrt{x}$ and $-\sqrt{x}$.

Example 2.1.1

For the function $g(x) = 2x - 1$, find: (a) $g(3 - 1)$, (b) $g(3) - g(1)$, (c) $g(2x - 1)$, (d) x if $g(x) = 7$.

Solution   (a) Since $3 - 1 = 2$, $g(3 - 1) = g(2) = 2 \cdot 2 - 1 = 3$.
(b) Since $g(3) = 2 \cdot 3 - 1 = 5$ and $g(1) = 2 \cdot 1 - 1 = 1$, then
$g(3) - g(1) = 5 - 1 = 4$.
(c) $g(2x - 1) = 2(2x - 1) - 1 = 4x - 3$.
(d) If $g(x) = 2x - 1 = 7$, then $2x = 8$; so $x = 4$.

Q2: For the function $F(t) = t^2 + 1$, find:

(a) $F(3) - 1$ (b) $F(3 + h)$ (c) $F(2x - 1)$

As we have said before, most of the functions that we deal with in elementary mathematics and its applications are functions whose domains and ranges are sets of real numbers. Furthermore, the rule of correspondence is usually given by a formula. Thus, unless otherwise specified, the domain of a function defined by a formula is understood to be the "largest" subset of R for which the formula makes sense. For example, if we say, "Consider the function $f(x) = \sqrt{x} + 1$," we shall understand that the domain is the set of all nonnegative real numbers, $[0, \infty]$.

Q3: What is the range of f if $f(x) = \sqrt{x} + 1$?

**Example
2.1.2**   Find the understood domain for the functions (a) $f(x) = \dfrac{x + 2}{x - 3}$
and (b) $g(x) = \sqrt{x - 1}$.

Solution   (a) Since division by 0 is undefined, x cannot equal 3. Yet for any $x \neq 3$, the algebraic operations involved are all well defined. Thus, the domain is $\{x : x \neq 3\}$.
(b) For $\sqrt{A}$ to be real, we must have $A \geq 0$. Thus, $x - 1 \geq 0$ for g to be defined, so the domain is $\{x : x \geq 1\} = [1, \infty]$.

Q4: Find the domain for:

(a) $f(x) = (5 - x)^{1/2}$ (b) $g(t) = (t - 1)(t - 2)^{-1}$

In Example 2.1.1 and Q2 we dealt with the functions $g(x) = 2x - 1$ and $F(t) = t^2 + 1$. In fact, in Q2(c) you computed $F(t)$ where $t = g(x) = 2x - 1$, obtaining $F(2x - 1) = (2x - 1)^2 + 1 = 4x^2 - 4x + 2$. These equations can be used to define a function h directly, that is, to each x let there correspond $h(x) = 4x^2 - 4x + 2$. Of course, this function is not F, for $F(x) = x^2 + 1 \neq 4x^2 - 4x + 2$. Rather, we obtain the output $4x^2 - 4x + 2$

from the input x by combining F and g in the manner illustrated above and in Figure 2.1.4. To each input x there first corresponds an output of g, the number $g(x) = 2x - 1$. This number then is used as the input to F, and we obtain the output $F(g(x)) = F(2x - 1) = 4x^2 - 4x + 2 = h(x)$. The resulting function h is termed a *composite* function — more precisely, the composition of g by F. Notice that the order in which g and F are joined together is very important. For example, here the function $g(F(x)) = g(x^2 + 1) = 2(x^2 + 1) - 1 = 2x^2 + 1$ is quite different from the function defined by $F(g(x))$.

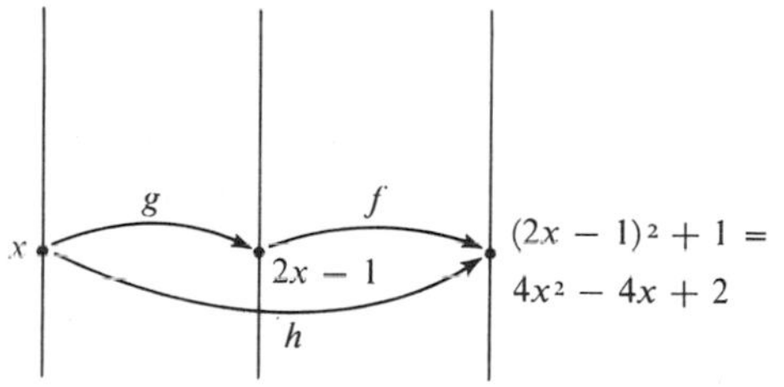

Figure 2.1.4

Example 2.1.3

Let $f(x) = \dfrac{1}{x - 3}$ and $g(x) = x^2 - 1$. Find (a) $h(x) = f(g(x))$ and (b) $H(x) - g(f(x))$.

Solution

(a) $h(x) = f(g(x)) = f(x^2 - 1) = \dfrac{1}{(x^2 - 1) - 3} = \dfrac{1}{x^2 - 4}$.

(b) $H(x) = g(f(x)) = g\left(\dfrac{1}{x - 3}\right) = \left(\dfrac{1}{x - 3}\right)^2 - 1$.

Q5: Using Example 2.1.3, find:

(a) $f(g(2))$ (b) $f(2) \cdot g(2)$ (c) the domain of h (d) the domain of H

PROBLEMS 2.1

1. Consider a colony of amoebas with an initial population of 200. The number of amoebas N after the tth day is given by the following table:

t	0	1	2	3	4	5
N	200	400	800	1600	3200	6400

(a) Construct an arrow representation of this table.
(b) Explain why the correspondence given in the above table defines a function.

A2: (a) 9 (b) $10 + 6h + h^2$ (c) $4x^2 - 4x + 2$

A3: Since $\sqrt{x} \geq 0$, $\sqrt{x} + 1 \geq 1$. For any $y \geq 1$, let $x = (y - 1)^2$. Then $f(x) = y$. Thus, the range is $[1, \infty]$.

A4: (a) $[-\infty, 5]$ (b) $\{x : x \neq 2\}$

A5: (a) Undefined (b) -3 (c) $\{x : x \neq 2, -2\}$ (d) $\{x : x \neq 3\}$

(c) If the letter A denotes this function, compute $A(2)$, $A(3)$, $A(2 + 3)$, $A(2 \cdot 3)$.
(d) Find the domain and range of A.
(e) Find a formula that defines the function A.

2. Consider the following table:

-1	0	1	2
1	0	1	4

(a) To each number in the first row, let there correspond the number below it. Construct the arrow representation of this correspondence.
(b) Does the correspondence in Part (a) define a function?
(c) To each number in the second row, let there correspond the number above it. Construct the arrow representation of this correspondence.
(d) Does the correspondence in Part (c) define a function?
(e) Find a formula that defines the correspondence in Part (a) above.

3. Consider the function $f(x) = 2x - 5$.
(a) Find $f(a)$, $f(2 + h)$, and $f(2x - 5)$.
(b) What is the "understood" domain of f?
(c) What is the range of f?

4. Consider the function $f(x) = \sqrt{3 - x}$ and $g(t) = 4 - t^2$.
(a) Find $g(3)$ and $b(2 + h)$.✗
(b) Find $\dfrac{f(2 + h) - f(2)}{h}$ and $\dfrac{g(a) - g(3)}{a - 3}$

(c) What is the domain of f and the domain of g?
(d) What is the range of f and the range of g?

5. Let the functions f and g be as given in Problem 4.
(a) Find $f(g(3))$ and $g(f(3))$.
(b) Find $h(x) = f(g(x))$ and $t(y) = g(f(y))$.
(c) What is the domain of h and the domain of t?

6. Let H be the correspondence that assigns to each positive number the sum of the number and its reciprocal. Does this correspondence define a function? If so, find $H(2)$, $H(a)$, $H(1/x)$ and the domain of H.

✗ ======

7. A box with no top is to be made from a rectangular piece of aluminum 20 inches long and 18 inches wide by cutting square pieces (side length x inches) from each corner and turning up the sides. Express the volume, V, of the box in terms of x. Does the correspondence associating volume with the size of the removed squares define a function? If so, what is the domain of the function?

8. Suppose that $h(x) = 1/x$. Show that $h(a) = b$ if and only if $h(b) = a$.

9. Let $f(x) = mx + b$, $m \neq 0$. Find $\dfrac{f(a + h) - f(a)}{h}$ and $\dfrac{f(x) - f(a)}{x - a}$.

10. Let $h(x) = \sqrt{x}$, and $G(x) = x^2$. Is $h(G(x)) = x$ for each $x \in R$? Are the functions $t(x) = h(G(x))$ and $T(x) = G(h(x))$ equal?

11. Suppose f is a function (with domain and range subsets of R). Does the correspondence that assigns to each number $f(x)$ in the range of f the number x in the domain define a function?

12. A cylindrical can with a top and bottom is made from 200 square inches of aluminum. Express the volume, V, of the can in terms of the radius, r, of the top. Does the correspondence associating volume with the radius define a function? If so, what is the domain of the function?

ANSWERS 2.1

1. (a) See Figure 2.1.5.

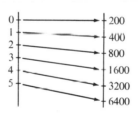

Figure 2.1.5

(b) The correspondence assigns to each day exactly one number representing the number of amoebas in the colony.
(c) $A(2) = 800$, $A(3) = 1600$, $A(2 + 3) = 6400$, $A(2 \cdot 3)$ is undefined.
(d) Domain $= \{0, 1, 2, 3, 4, 5\}$; Range $= \{200, 400, 800, 1600, 3200, 6400\}$.
(e) $N = (200)2^t$.

2. (a) See Figure 2.1.6.   (b) Yes

Figure 2.1.6

(c) See Figure 2.1.7.   (d) No

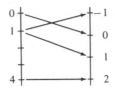

Figure 2.1.7

(e) For example, $f(x) = x^2$ where x is a number in the first row.

3. (a) $f(a) = 2a - 5$, $f(2 + h) = -1 + 2h$, $f(2x - 5) = 4x - 15$
(b) R
(c) R

4. (a) $g(3) = -5$, $f(2 + h) = \sqrt{1 - h}$
(b) $\dfrac{f(2 + h) - f(2)}{h} = \dfrac{\sqrt{1 - h} - 1}{h}$, $\dfrac{g(a) - g(3)}{a - 3} = -(a + 3)$
(c) $[-\infty, 3]$, R
(d) $[0, \infty]$, $[-\infty, 4]$

5. (a) $f(g(3)) = \sqrt{8}$, $g(f(3)) = 4$
(b) $h(x) = \sqrt{x^2 - 1}$, $t(y) = y + 1$
(c) $[-\infty, -1] \cup [1, \infty]$, $[-\infty, 3]$

6. Yes. $H(2) = 5/2$, $H(a) = a + 1/a$, $H(1/x) = x + 1/x$. The domain of H is $(0, \infty)$.

7. $V = 360x - 76x^2 + 4x^3$. Yes. $(0, 9)$.

Supplement 2.1

I. Concepts and Notation

Problem S1.   The results of measuring two quantities are given in the following table:

1	2	3	4
4	3	4	5

(a) To each number in the top row, let there correspond the number below it. Construct an arrow representation of this correspondence. Does this correspondence define a function? (b) To each number in the bottom row, let there correspond the number above it. Construct an arrow representation of this correspondence. Does this correspondence define a function? (c) Denote the function defined in Part (a) by F. Find $F(2) \cdot F(2)$, $F(F(2))$, $F(5)$, the domain of F, and the range of F.

Solution: (a) The correspondence shown in Figure 2.1.8 defines a function since to each number in the set $\{1, 2, 3, 4\}$ there correspond exactly one number from the bottom row.

Figure 2.1.8

(b) The correspondence shown in Figure 2.1.9 does *not* define a function since to the number 4 in the bottom row there correspond two numbers, namely, 3 and 1.

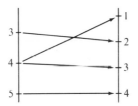

Figure 2.1.9

(c) $F(2) = 3$, so $F(2) \cdot F(2) = 3 \cdot 3 = 9$ and $F(F(2)) = F(3) = 4$. $F(5)$ is not defined by this table. The domain of F is $\{1, 2, 3, 4\}$, and the range of F is $\{3, 4, 5\}$.

Problem S2. Let $f(t) = 2t + 3$. Find $f(2)$, $f(a)$, $f(2 + h)$, $f(t^{-1})$, and $[f(t)]^{-1}$.

Solution: We obtain the output of f by adding 3 to twice the input. For example, $f(2) = 2 \cdot 2 + 3 = 7$ and $f(a) = 2a + 3$. Likewise, $f(2 + h) = 2(2 + h) + 3 = 7 + 2h$. Also, $f(t^{-1}) = 2t^{-1} + 3 = 3 + \dfrac{2}{t}$, and $[f(t)]^{-1} = \dfrac{1}{f(t)} = \dfrac{1}{2t + 3}$.

Problem S3. Let $g(x) = 3x^2 + 4$. Find: $g(3)$, $g(a)$, $\dfrac{g(a + h) - g(a)}{h}$, and $g(\sqrt{x + 1})$.

Solution: Substituting 3 for x in $g(x) = 3x^2 + 4$, we have $g(3) =$ _____. Similarly, $g(a) =$ _____, and $g(a + h) = 3 \cdot ($_____$)^2 + 4 =$

$3 \cdot 3^2 + 4 = 31, \quad 3a^2 + 4$

$a + h$

100   *Functions*

_____. Thus, $\dfrac{g(a+h)-g(a)}{h}$ = | $3a^2 + 6ah + 3h^2 + 4$

$\dfrac{3a^2 + 6ah + 3h^2 + 4 - (3a^2 + 4)}{h} =$

$\dfrac{6ah + 3h^2}{h}$ = _____ . Finally, | $6a + 3h,\qquad h \neq 0$

$g(\sqrt{x+1}) =$ _____ . | $3(x+1) + 4 = 3x + 7$

Problem S4. A rectangular region is enclosed by using an existing wall and 100 feet of new fencing for the other three sides. Let x denote the length of a part of the fence perpendicular to the wall, as shown in Figure 2.1.10. Find a formula for the enclosed area A in terms of x. What is the domain of the function defined by the formula?

Figure 2.1.10

Solution: The total length of fencing is 100 feet, so the length of fencing parallel to the wall is $100 - 2x$ (see Figure 2.1.10). Thus, the enclosed region has area $A = x \cdot (100 - 2x) = 100x - 2x^2$. Since the total length of fencing perpendicular to the wall is $2x$, we must have $0 < 2x < 100$, that is, $0 < x < 50$. Thus, the domain of f is the interval $(0, 50)$.

Problem S5. A man 6 ft. tall walks away from a lamp 10 ft. above the ground. The length L of his shadow increases as his distance D from the lamp increases. Use similar triangles to find f such that $L = f(D)$.

Solution: A sketch illustrating the problem is shown in Figure 2.1.11. Fill in the blanks with the proper letters.

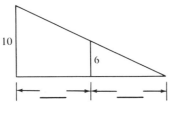

Figure 2.1.11

$D,\quad L$

In Figure 2.1.11 are two similar triangles. The ratio of the length of the base to the height is the same for each triangle. Thus, $L/6 = $ _____ $/10$. Multiplying, we obtain $10L = $ _____ . So $4L = 6D$. Solving, we have $L = $ _____ . If $f(D) = L$, then $f(D) = 3D/2$.

$L + D$
$6L + 6D$
$6D/4 = 3D/2$

DRILL PROBLEMS

S6. Consider the following table:

1	3	2	3
2	1	4	3

(a) To each number in the top row let there correspond the number below it. Does this correspondence define a function? Explain.

(b) To each number in the bottom row let there correspond the number above it. Does this correspondence define a function? Explain.

(c) Let f denote the function defined in (b). Find $f(2)$, $f(f(2))$, $f(2 \cdot 2)$, $f(2) \cdot f(2)$.

(d) Find the domain and range of f.

S7. Let $g(x) = 2/x^2$. Find $g(a)$, $g(a)$ $g(\ a)$, $1/g(x)$, and $g(1/x)$.

S8. A rectangular box with a square base is to have a volume of 20 cubic feet. The material for the bottom costs 4 cents per square foot, while the material for the sides and top costs 3 cents per square foot. Let x be the length of an edge of the base. Find the function f such that the total cost of the box is $f(x)$. What is the domain of f?

Answers

S6. (a) not a function, (b) a function, (c) 1, 3, 2, 1, (d) $D = \{1, 2, 3, 4\}$, the range is $\{1, 2, 3\}$.

S7. $2/a^2$, 0, $x^2/2$ if $x \neq 0$, $2x^2$ if $x \neq 0$

S8. $f(x) = 7x^2 + (240/x)$, $\{x \in R : x > 0\}$

II. Understood Domain

Problem S9. Find the domain of f if $f(x) = 1/(x^2 + 1)$.

Solution: For any real numbers x, the numbers x^2 and $x^2 + 1$ are defined. Since $x^2 \geq 0$ and $x^2 + 1 \geq 1 > 0$, the quotient $1/(x^2 + 1)$ is defined for any real number x. Thus, the domain of f is R, the set of real numbers.

Problem S10.   Find the domain of f if $f(x) = \sqrt{(x - 2)(1 - x)}$.

Solution: The product $(x - 2)(1 - x)$ is defined for any real number x. However, this product can be a negative number, in which case $\sqrt{(x - 2)(1 - x)}$ is undefined. For example, if $x = 0, (x - 2)(1 - x) = -2$ so $f(0)$ is undefined. Thus, 0 is not in the domain of f. The domain of f is the set $\{x : (x - 2)(1 - x) \geq 0\}$, in which case $f(x) = \sqrt{(x - 2)(1 - x)}$ is defined. If you have forgotten how to solve the inequality $(x - 2)(1 - x) \geq 0$, reread Example 1.4.4 where the solution set was $[1, 2]$.

Problem S11.   Find the domain of f if $f(x) = 3/(x^2 - 2x)$.

Solution: The expression $x^2 - 2x$ (is/is not) defined for each real number x. A quotient is defined except when its _____ is zero.	is
	denominator
The expression $x^2 - 2x = x(x - 2)$ equals zero precisely when $x \in$ _____. Thus, the domain of f is $\{x : $_____$\}$, or, in interval notation, _____.	$\{0, 2\}$ $x \neq 0, 2$ $(-\infty, 0) \cup (0, 2) \cup (2, \infty)$

DRILL PROBLEMS

S12 Find the domain of each of the following functions:

(a) $f(x) = \sqrt{\dfrac{x - 4}{x - 3}}$

(b) $f(x) = (x - 3)^{-2}$

(c) $f(x) = \dfrac{x - 1}{x^2 + 4x + 5}$

Answer

S12. (a) $(-\infty, 3) \cup [4, \infty]$,   (b) $\{x : x \neq 3\}$,   (c) R

III. Composite Functions

Problem S13.   Let $f(x) = 3 + 2x$ and $g(x) = \sqrt{x}$. (a) Find $f(3)$ and $g(f(3))$. (b) Let $h(x) = g(f(x))$. Find $h(3)$ and write a formula for h. (c) Let $F(x) = f(g(x))$. Write a formula for F. (d) What is the domain of h? Of F?

Solution: (a) $f(3) = 3 + 2 \cdot 3 = 9$. Thus, $g(f(3)) = g(9) = \sqrt{9} = 3$.
(b) $h(3) = g(f(3)) = 3$ from (a). Furthermore, $g(f(x)) = \sqrt{f(x)} = \sqrt{3 + 2x}$. That is, $h(x) = \sqrt{3 + 2x}$.
(c) $f(g(x)) = 3 + 2 \cdot g(x) = 3 + 2\sqrt{x} = F(x)$. Notice that $f(g(x)) \neq g(f(x))$ in this case.

(d) The domain of h is the set $\{x : 3 + 2x \geq 0\} = \{x : x \geq -3/2\}$, or, in interval notation, $[-3/2, \infty]$. The domain of F is the set $\{x : x \geq 0\} = [0, \infty]$. Notice that the domains of h and F are not equal.

Problem S14. The area A of a circle of radius r can be computed using the formula $A = \pi r^2$. This formula defines a function f such that $A = f(r)$. The radius can be computed from the circumference C according to the formula $r = C/2\pi$, which defines a function g such that $r = g(C)$. Discuss the composite function $h(C) = f(g(C))$.

Solution: We shall first find a formula for $h(C)$. Since $g(C) =$ _____ and $f(r) =$ _____, $f(g(C)) =$ $f(\_\_\_) = \pi(\_\_\_)^2 =$ _____. Now, $h(C) = f(g(C)) = f(r)$, so the number $h(C)$ is the _____ of a circle of circumference C. That is, we can compute the area A of a circle of circumference C from the formula _____.	$C/2\pi, \quad \pi r^2$ $C/2\pi, \quad C/2\pi, \quad C^2/4\pi$ area $A = C^2/4\pi$

DRILL PROBLEMS

S15. Let $f(x) = \sqrt{x}$ and $g(x) = x^2$. (a) Does $f(g(x)) = g(f(x))$? (b) Do these composite functions have the same domain?

S16. Let $a(t) = (t - 1)^{-2}$ and $b(u) = u^{1/2} + 1$. Find formulas for f and g if $f(x) = a(b(x))$ and $g(y) = b(a(y))$. What is the domain of f? Of g?

Answers

S15. (a) no, (b) no; $f(g(x)) = |x|$ with domain R, but $g(f(x)) = x$ with domain $[0, \infty]$.

S16. $f(x) = x^{-1}$, $x > 0$. $g(y) = |y - 1|^{-1} + 1$ for $y \neq 1$ [not $(y - 1)^{-1} + 1$].

SUPPLEMENTARY PROBLEMS

S17. Let $f(x) = 2x^2 + 1$ and $g(x) = \sqrt{x} - 1$. Simplify $\dfrac{f(z) - f(x)}{z - x}$. Find $2f(x) + 1$, $f(2x^2 + 1)$, $f(g(x))$, and $g(f(x))$. What is the domain of g?

S18. A large, spherical snowball is melting at a rate such that its radius decreases $1/2$ inch per minute. If the radius of the snowball is 20 inches at noon, find a function f such that at t minutes after noon its radius r is given by the equation $r = f(t)$. The volume V of a sphere of radius r is given by $V = 4\pi r^3/3$. Find a function g such that the volume V of the snowball t minutes after noon is $g(t)$. What is the domain of f?

S19. A rectangle is inscribed in a triangle with base b and height h as shown in Figure 2.1.12. Using similar triangles, find the function f such that $A = f(x)$ where A is the area of the rectangle.

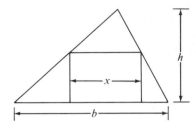

Figure 2.1.12

Answers

S17. $2(z + x)$, $4x^2 + 3$, $8x^4 + 8x^2 + 3$, $2x - 4\sqrt{x} + 3$, $\sqrt{2x^2 + 1} - 1$, $[0, \infty]$

S18. $f(t) = 20 - t/2$, $g(t) = \dfrac{4\pi(20 - t/2)^3}{3}$, $[0, 40]$

S19. $f(x) = \dfrac{hx(b - x)}{b}$

2.2 THE COORDINATE PLANE, R², AND GRAPHS

We have already referred to curves and graphs in the introduction to this chapter. Let us discuss briefly a coordinate system for points in the plane, called the *Cartesian coordinate system*.

We begin by drawing two number lines perpendicular to each other such that they intersect at the zero points of both number lines. These lines are called *coordinate axes*, and their point of intersection is called the *origin*. Typically, one number scale is horizontal with the positive direction to the right and is called the *X-axis*. The vertical axis, with positive direction up, is called the *Y-axis*. Given any point P in the plane, we construct lines through P parallel to the axes. The numbers where these lines intersect the X- and Y-axes are labeled x and y, respectively, and are called the *coordinates* of P (see Figure 2.2.1). We write these coordinates

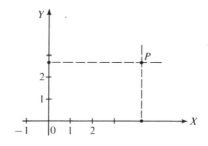

Figure 2.2.1

105

as the *ordered pair* (*x*, *y*). The first number is always the *X*-coordinate and the second the *Y*-coordinate, thus, the word ordered. It is a geometric fact that each point determines exactly one ordered pair of real numbers in this fashion.

Our notation for the coordinates of a point (*a*, *b*) is the same as that for an interval $\{x : a < x < b\}$. However, if you carefully note the context surrounding the notation, you will not become confused. As with the number scale, we often ignore the distinction between a point and its coordinates and say, "the point (*x*, *y*)," rather than "the point whose coordinates are (*x*, *y*)."

Q1: Which point in Figure 2.2.2 has the following coordinates:

(a) (2, 1)   (b) (−1, 1)

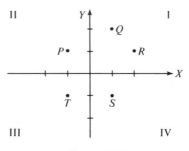

Figure 2.2.2

In Figure 2.2.2 the Roman numerals I, II, III, and IV are used to indicate the four regions called *quadrants* formed by the coordinate axes.

Q2: In what quadrant does the point lie whose coordinates are as follows:

(a) (−2, 1)   (b) (1, −2)

Instead of *R* we sometimes use R^1 to denote the set of all real numbers. We use R^2 to denote the set of all ordered pairs of real numbers.

**Definition
2.2.1**

$R^2 = \{(x, y) : x \in R, y \in R\}$.

Given any element of R^2, say, (*a*, *b*), then there is a unique line parallel to the *Y*-axis that contains the point *a* of the *X*-axis, and a unique line parallel to the *X*-axis containing the point *b* of the

Y-axis; furthermore, these lines intersect in a unique point. Thus, there corresponds a unique point of a coordinate plane to each element of R^2. Conversely, as we saw before, there corresponds a unique element of R^2 to each point of a coordinate plane. Thus, just as the points of a number line are in one-to-one correspondence with R^1, so, too, are the points of a coordinate plane in one-to-one correspondence with R^2.

We have previously observed that the distance between two points x_1 and x_2 in R^1 is the nonnegative number $|x_2 - x_1|$. We can use this fact and the Pythagorean theorem to compute the distance between two points in R^2. In Figure 2.2.3 we have indicated the points P_1 and P_2 with coordinates (x_1, y_1) and (x_2, y_2), respectively. We have also indicated a third point Q with coordinates (x_1, y_2). Clearly, distance a between Q and P_2 is $|x_2 - x_1|$, and distance b between Q and P_1 is $|y_2 - y_1|$. Furthermore, $P_1 Q P_2$ is a right triangle. Thus, the Pythagorean theorem implies that the distance d between P_1 and P_2 satisfies the equation

$$d^2 = a^2 + b^2 = |x_2 - x_1|^2 + |y_2 - y_1|^2$$

Since $|z|^2 = z^2$ (proved in Supplement 1.5), we can write the *distance formula* as follows: The distance d between points (x_1, y_1) and (x_2, y_2) is given by

$$d = \sqrt{(x_2 - x_1)^2 + (y_2 - y_1)^2} \qquad\qquad (1)$$

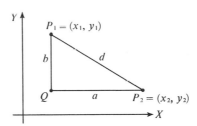

Figure 2.2.3

Q3: Find the distance between the points $(1, 2)$ and $(-3, 4)$.

Example 2.2.1

Write an equation which has as its graph the circle with center $(1, 0)$ and radius 2.

Solution

A point (x, y) lies on the circle if the distance between (x, y) and $(1, 0)$ is 2, and conversely (see Figure 2.2.4). We use the distance formula; thus, all such points must satisfy the equation $(x - 1)^2 + y^2 = 4$. [Here we have squared both sides of Equation (1).]

A1: (a) R (b) P

A2: (a) II   (b) IV

A3: $d = \sqrt{(-3-1)^2 + (4-2)^2} = \sqrt{16+4} = \sqrt{20} = 2\sqrt{5}$

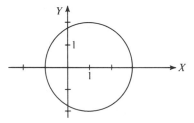

Figure 2.2.4

Q4: Find an equation for the circle with radius 2 and center $(0, -1)$.

A Cartesian coordinate system helps us picture functions. Recall that a function f is defined when there is a rule that assigns to each x in its domain exactly one number $f(x)$ in the range. We can use the ordered pair $(x, f(x))$ to describe the input-output of f; the set of all such ordered pairs then defines f.

Q5: The function g is defined by the following table. Describe g by writing a set of ordered pairs.

x	0	1	2	3	4
$g(x)$	0	48	64	48	0

For the function $h(x) = 2 - x$, it is obviously impossible to list all such pairs separately, but we can use set notation to write

$$h = \{(x, h(x)) : h(x) = 2 - x, x \in R\}$$
$$= \{(x, 2 - x) : x \in R\}$$

In any case, the *graph of a function* f means the set of all points of a coordinate plane with coordinates $(x, f(x))$ where x is any number in the domain of f. The Y-coordinate of a point of the graph of f is the function value of that X-coordinate, so we often write $y = f(x)$.

Q6: Describe the graph of *g* in Q5.

Example 2.2.2

Sketch the graph of $y = f(x) = \dfrac{1}{x}$.

Solution

Some points of the graph of *f* are (1, 1), (2, 1/2), (1/2, 2). (3, 1/3), (1/3, 3), etc. Also, $(-1, -1)$, $(-2, -1/2)$, $(-1/2, -2)$, etc., lie on the graph. If we plot many of these points and connect them with a smooth curve, we get the graph shown in Figure 2.2.5. Notice that the domain of *g* is $\{x : x \neq 0\}$. It is also easy to see that the range of *g* is $\{y : y \neq 0\}$. You should remember this graph, as we shall use it again.

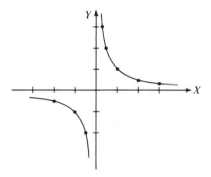

Figure 2.2.5

Example 2.2.3

Sketch the graph of $y < \dfrac{1}{x}$.

Solution

In Example 2.2.2 we sketched the graph of $y = f(x) = 1/x$. If we consider any point (a, b) "below" the graph of $y = f(x)$, then $b < f(a)$, and (a, b) satisfies $y < f(x)$. (See Figure 2.2.6.) Therefore,

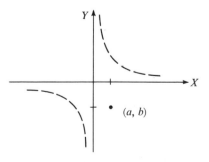

Figure 2.2.6

A4: $x^2 + (y + 1)^2 = 4$

A5: {(0, 0), (1, 48), (2, 64), (3, 48), (4, 0)}

A6: The five points (0, 0), (1, 48), (2, 64), (3, 48), (4, 0)

the graph of $y < f(x)$ consists of all points below the "boundary" $y = f(x)$ as shown in Figure 2.2.7. We have used a dashed line to indicate that points of the boundary are not included in the graph.

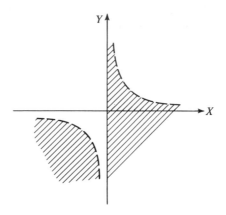

Figure 2.2.7

Example 2.2.4

The graph of a function f is shown in Figure 2.2.8. Use this graph to estimate (a) $f(0)$, (b) $f(f(0))$, (c) $f(3) - f(2)$, (d) $\{x : f(x) = 1\}$, and (e) $\{x : f(x) < 1\}$.

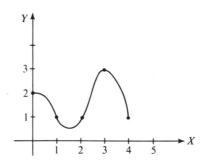

Figure 2.2.8

Solution

(a) $f(0) = 2$.
(b) Since $f(0) = 2$, $f(f(0)) = f(2) = 1$.
(c) $f(3)$ is 3 and $f(2)$ is 1, so $f(3) - f(2) = 3 - 1 = 2$.

(d) The solutions to $f(x) = 1$ are 1, 2, and 4 because $f(1) = f(2) = f(4) = 1$. Thus, $\{x : f(x) = 1\} = \{1, 2, 4\}$.

(e) If x is between 1 and 2, then $f(x) < 1$, and conversely. Thus, $\{x : f(x) < 1\} = (1, 2)$.

Not all graphs of subsets of R^2 are graphs of functions, as shown in Example 2.2.5 and Q7.

Example 2.2.5

Graph $\{(x, y) : xy > 0\}$.

Solution

For the product xy to be positive, x and y must have like signs, both positive or both negative. Now, any point in Quadrant I has coordinates that are both positive, and conversely. Likewise, points in Quadrant III have coordinates that are both negative. Thus, the graph is Quadrant I ∪ Quadrant III, as indicated by the shaded area in Figure 2.2.9.

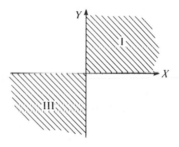

Figure 2.2.9

Q7: Is the subset of R^2 graphed in Example 2.2.5 a function? Explain.

A function can be used to describe a physical situation in which the value of one quantity can be determined from the value of another. For example, the height of the ball as depicted in Figure 2.0.1 can be found if the elapsed time is specified. To each time there is "paired" exactly one height. Similarly, Figure 2.0.2 depicts a physical situation in which to each volume there is paired exactly one pressure. The idea that to each number in the domain of a function there is paired exactly one number in the range suggests the following definition of a function in terms of ordered pairs.

Definition 2.2.2

A *function* is a set of ordered pairs such that no two pairs have the same first member and different second members. The set of all first members is called the *domain* of the function. The set of

A7: No, to each $x \neq 0$ there correspond *many* y values, not just one.

all second members is called the *range* of the function. If the function is denoted by f, then $f = \{(x, f(x)\}$.

As we have said before, most of the time we deal with functions whose domains and ranges are subsets of R. According to Definition 2.2.2, such a function f is a special type of subset of R^2, namely, one such that if (x_1, y_1) and (x_2, y_2) are in f and if $x_1 = x_2$, then $y_1 = y_2$.

Example 2.2.6

Graph each of the following subsets of R^2 and determine which of them are functions: (a) $\{(x, y) : y = |x|\}$, (b) $\{(x, y) : |y| = x\}$, and (c) $\{(x, y) : x^2 + y^2 = 1\}$.

Solution

Part (a) defines a function $y = f(x)$ since to each x there corresponds one y. From the definition of absolute value,

$$y = \begin{cases} x & \text{if} \quad x \geq 0 \\ -x & \text{if} \quad x \leq 0 \end{cases}$$

Some points on the graph are $(0, 0)$, $(1, 1)$, $(-1, 1)$, $(2, 2)$, $(-2, 2)$, . . . , and the graph is the two straight-line segments pictured in Figure 2.2.10. Notice that the domain is R and the range is the set of all nonnegative real numbers.

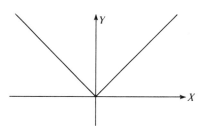

Figure 2.2.10

(b) To each positive real number x there correspond not one, but two, numbers. For example, if $x = 1$, $|y| = 1$, so y could be either 1 or -1. Hence, (b) is not a function. Notice that

$$x = \begin{cases} y & \text{if} \quad y \geq 0 \\ -y & \text{if} \quad y \leq 0 \end{cases}$$

The graph is shown in Figure 2.2.11.

For Part (c) we use the distance formula to recognize the graph in Figure 2.2.12 as that of a circle of radius 1 with center $(0, 0)$.

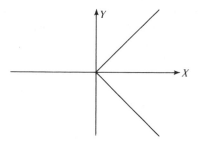

Figure 2.2.11

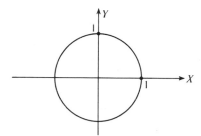

Figure 2.2.12

(See Example 2.2.1.) Clearly, this does not define a function, since to each $x \in (-1, 1)$ correspond two numbers y, namely, $y = \sqrt{1 - x^2}$ and $y = -\sqrt{1 - x^2}$.

Q8: Which of the following define(s) a function $y = f(x)$:

   (a) $\{(x, y) : x^2 = y^2\}$ (b) $\{(x, y) : x^3 = y^3\}$

Example 2.2.6 illustrates how we can tell from a graph whether or not a subset of R^2 is a function. If every *vertical* line intersects the graph in at most one point, then it is the graph of a function; otherwise it is not. On the other hand, a *horizontal* line may intersect the graph of a function in more than one point. In Figure 2.2.8 a horizontal line containing the point 1 of the Y-axis intersects the graph of the function f in 3 points. Thus, the equation $f(x) = 1$ has 3 solutions. There is not a one-to-one correspondence between numbers in the domain and range in that case. However, each horizontal line may intersect the graph of a function in at most one point. Such a function is depicted in Figure 2.0.2. In this case a given number p in the range corresponds to exactly *one* number in the domain — there is a one-to-one correspondence between numbers in the domain and range. Such a function is called a one-to-one function according to the following definition.

A8:Part (a) does not; (b) does.

**Definition
2.2.3**

A function f is called a *one-to-one* function if, for each number y in the range of f, the equation $f(x) = y$ has exactly one solution for x. In other words, suppose that (x_1, y_1) and (x_2, y_2) belong to $f = \{(x, y) : y = f(x)\}$. The function f is a one-to-one function if $x_1 \neq x_2$ implies $y_1 \neq y_2$.

If D is the domain of a function f, the set $f(D)$ is the range. That is, $f(D)$ is the set $\{f(x) : x \in D\}$. We use this same notation when A is any subset of D. Thus,

$$f(A) = \{f(x) : x \in A\} \tag{2}$$

**Example
2.2.7**

Let f be the function whose graph is shown in Figure 2.2.13. Sketch $f([1, 3])$.

Solution

In this example we are using the notation of Equation (2) with $A = [1, 3]$. Thus, $f([1, 3]) = \{f(x) : x \in [1, 3]\}$. The set $f([1,3])$ is depicted in Figure 2.2.13. From the figure we see that $f([1, 3]) = [2, 3]$.

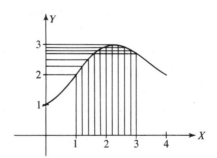

Figure 2.2.13

The set $f(A)$ may contain just one number, although A contains more than one number. For the function depicted in Figure 2.2.8, if $B = \{1, 2, 4\}$, then $f(B) = \{1\}$. Notice that this situation cannot occur if f is a one-to-one function. We will have more to say about one-to-one functions in a later section.

**PROBLEMS
2.2**

1. In what quadrants are the points $(-1, 2)$, $(1, 2)$, $(-2, -3)$, and $(1, -2)$?

2. Find the distance between $(-1, 2)$ and (a) $(5, 2)$, (b) $(5, 3)$, and (c) $(2, -2)$.

3. Consider the function g given by the following table:

x	0	1	4	9
$g(x)$	0	1	2	3

(a) Describe g as a set of ordered pairs. Draw the graph of g.
(b) Is g a one-to-one function?
(c) Find a formula for $g(x)$.

4. Consider the subset h of R^2 whose graph is shown in Figure 2.2.14 where $h = \{(x, y) : y = h(x)\}$.

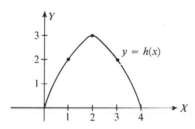

Figure 2.2.14

(a) Is h a function? If so, find its domain and range.
(b) Use the graph of estimate $h(0)$, $h(2 + 1)$, and $h(h(2))$.
(c) Is h a one-to-one function? (Prove your answer.)

5. Graph each of the following subsets of R^2 and determine which of them are functions:
(a) $\{(x, y) : x = 3\}$ (b) $\{(x, y) : y = 1\}$
(c) $\{(x, y) : y = 2x\}$ (d) $\{(x, y) : x = y^2\}$
(e) $\{(x, y) : y + 1 \geq 0\}$ (f) $\{(x, y) : |x| = |y|\}$
(g) $\{(x, y) : x^2 + y^2 = 4\}$ (h) $\{(x, y) : x^2 + y^2 < 4\}$
(i) $\{(x, y) : (x - 1)^2 + (y + 2)^2 = 1\}$

6. Find an equation for each of the following circles:
(a) center at $(0, 0)$, radius 3
(b) center at $(1, -3)$, radius 2
(c) center at (h, k), radius a

7. Consider the function $f(x) = 4 - x^2$.
(a) Plot the points $(2, f(2))$, $(0, f(0))$, $(-1, f(-1))$, $(1, f(1))$, and $(-3, f(-3))$.
(b) Draw a smooth curve that contains the points in Part (a) to obtain a sketch of the graph of f.
(c) What, if any, is the maximum value of f?
(d) Is f a one-to-one function? (Prove your answer.)

8. Let h be the function given in Problem 4, and let f be the function in Problem 7.

(a) Use the graph of h to find $h([0, 4])$, $h((1, 3])$, and $h([1, 4])$.

(b) Find $f((-2, 2))$, $f([0, 3])$, and $f([2, \infty])$.

9. Let $A = \{(x, y) : x^2 + y^2 = 4 \text{ and } y \geq 0\}$, and let $B = \{(x, y) : y = \sqrt{1 - x^2}, x \geq 0\} \cup \{(x, y) : y = -\sqrt{1 - x^2}, x < 0\}$.

(a) Is A a function? If so, find its domain and range, and determine whether or not it is a one-to-one function.

(b) Is B a function? If so, find its domain and range, and determine whether or not it is a one-to-one function.

10. How are the graphs of the functions f, g, and h related if $f(x) = x^2$, $g(x) = x^2 + 5$, and $h(x) = x^2 - 3$.

11. Graph each of the following subsets of R^2 and determine which of them are functions:

(a) $\{(x, y) : y + |y| = x + |x|\}$ (b) $\{(x, y) : y = |x - 3|\}$

(c) $\{(x, y) : |x| + |y| = 1\}$ (d) $\{(x, y) : |y| = |x^2|\}$

12. Consider the function $f = \{(x, y) : y = f(x)\}$. If f is one-to-one, then is the set $g = \{(y, x) : y = f(x)\}$ a function? Why?

13. Consider the function $f(x) = 2x$. Let $A = [-4, 0) \cup (2, 3)$. What is $f(A)$?

14. Consider the function $f(x) = 3 - 5x$. Prove that f is one-to-one.

15. Let f be a function with domain D. Explain why the statement, "If $x_1, x_2 \in D$ such that $f(x_1) = f(x_2)$, then $x_1 = x_2$," is equivalent to the assertion that f is a one-to-one function.

16. Consider a function f with domain D. Suppose that A and B are subsets of D. Is $f(A \cup B) = f(A) \cup f(B)$? Is $f(A \cap B) = f(A) \cap f(B)$?

17. Let $f(x) = x(5 - x)$. Graph f and find $f[0, 5]$.

ANSWERS 2.2

1. II, I, III, IV, respectively.

2. (a) 6, (b) $\sqrt{37}$, (c) 5

3. (a) $g = \{(0, 0), (1, 1), (4, 2), (9, 3)\}$. See Figure 2.2.15.

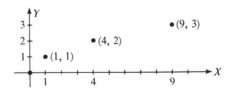

Figure 2.2.15

(b) Yes, (c) $g(x) - \sqrt{x}$

4. (a) Yes, $[0, 4]$, $[0, 3]$

(b) $h(0) = 0$, $h(2 + 1) = 2$, $h(h(2)) = h(3) = 2$

(c) No; for example, $h(1) = 2 = h(3)$ but $1 \neq 3$.

5. (a) See Figure 2.2.16.

(b) See Figure 2.2.17.

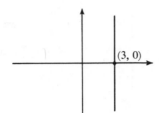

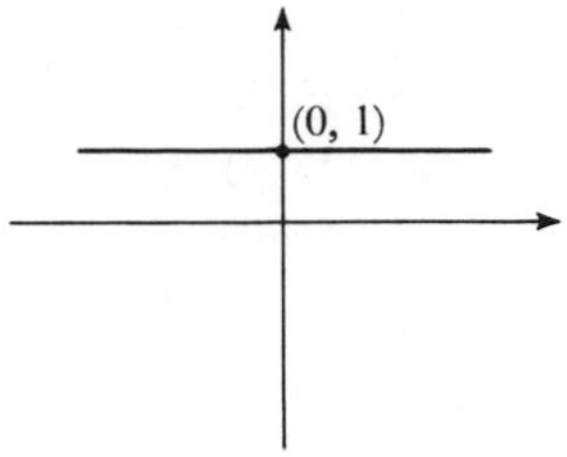

$(3, 0)$

Figure 2.2.16

$(0, 1)$

Figure 2.2.17

(c) See Figure 2.2.18.

(d) See Figure 2.2.19.

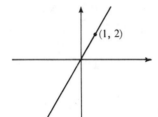

$(1, 2)$

Figure 2.2.18

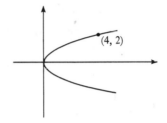

$(4, 2)$

Figure 2.2.19

(e) See Figure 2.2.20.

(f) See Figure 2.2.21.

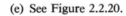

$(0, -1)$

Figure 2.2.20

Figure 2.2.21

(g) See Figure 2.2.22.

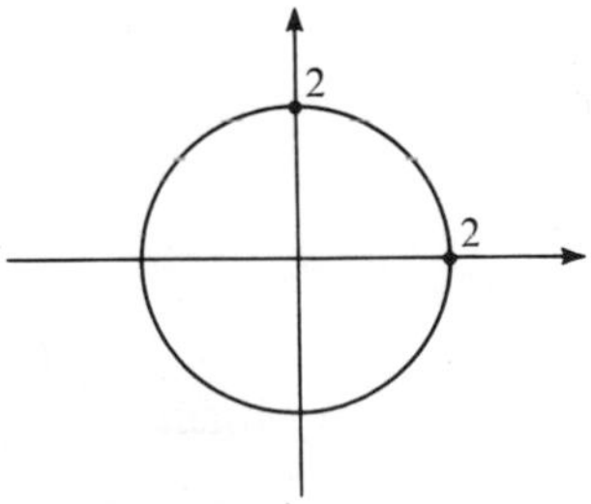

(h) See Figure 2.2.23.

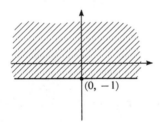

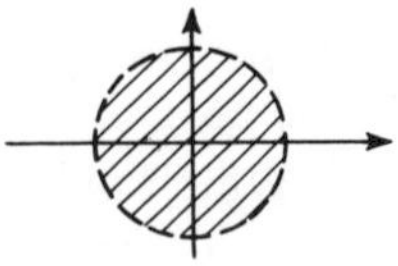

2

2

Figure 2.2.22

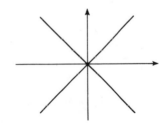

Figure 2.2.23

(i) See Figure 2.2.24.

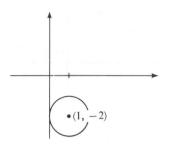

Figure 2.2.24

6. (a) $x^2 + y^2 = 9$, (b) $(x - 1)^2 + (y + 3)^2 = 4$, (c) $(x - h)^2 + (y - k)^2 = a^2$

7. For (a) and (b), see Figure 2.2.25.

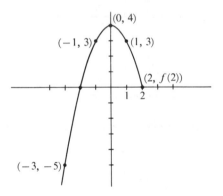

Figure 2.2.25

(c) $f(0) = 4$, (d) no; for example, $f(-2) = 0 = f(2)$ and $-2 \neq 2$.

8. (a) [0, 3], [2, 3], and [0, 3]

(b) (0, 4], [-5, 4], and [-∞, 0]

Supplement 2.2

I. The Coordinate Plane

Problem S1. Which point in Figure 2.2.26 has coordinates $(-1, 2)$?

Solution: *P* is the correct answer. *Q* is the point $(-2, 1)$, *R* is the point $(-2, -1)$, and *S* is $(-1, -2)$.

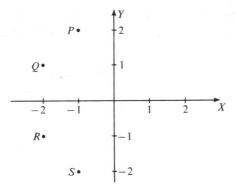

Figure 2.2.26

Problem S2. The point (u, v) is shown in Figure 2.2.27. Identify by letter each of the following points: (a) $(-u, v)$, (b) $(-u, -v)$, (c) (v, u), and (d) $(v, -u)$.

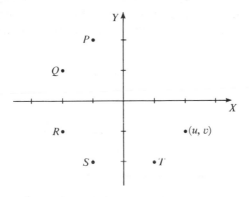

Figure 2.2.27

Solution: (a) R. Point R has the same Y-coordinate as (u, v), namely v. Also, R is the same distance to the Y-axis as (u, v) and is in the opposite direction; thus, its X-coordinate is $-u$. R is the point $(-u, v)$.

(b) Q. Similarly, since R has coordinates $(-u, v)$, Q has coordinates $(-u, -v)$.

(c) P. Notice that $u > 0$ and $v < 0$, so (v, u) lies in Quadrant II. Also P is as far from the X-axis as (u, v) is from the Y-axis, namely, u.

(d) S. From (c), S has the same X-coordinate as P, and its Y-coordinate is the negative of that of P. The coordinates of T appear to be $(-v, -u)$.

Problem S3. Let $A = \{(x, y) : x \geq 2\}$ and $B = \{(x, y) : |y - 2| < 1\}$. Sketch the graph of each of the following subsets of R^2: (a) A, (b) B, (c) $A \cup B$, and (d) $A \cap B$.

Solution: (a) A can be described as the set of all points with X-coordinate greater than or equal to 2, and with *any* Y-coordinate. [For example, $(2, 0)$, $(2, 27)$, $(2, -10)$, $(4, 0)$, and $(27, 0)$ all belong to A, while $(0, 2)$, $(1.9, 2)$, and $(-27, 0)$ do not.] Thus, the shaded region in Figure 2.2.28 indicates A. We use a solid line as the boundary to indicate that these points with X-coordinates that equal 2 are included in A.

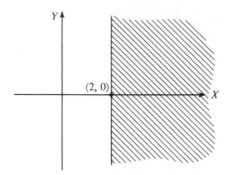

Figure 2.2.28

(b) Similarly, since $|y - 2| < 1$ is equivalent to $-1 < y - 2 < 1$ or $1 < y < 3$, B contains points with any X-coordinate, if the Y-coordinates are between 1 and 3. We use dotted lines to indicate that the points on the boundary are not included (see Figure 2.2.29).

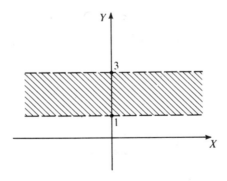

Figure 2.2.29

(c) $A \cup B$ contains all points in the plane which were shaded in either A or B above. (d) $A \cap B$ contains all points shaded in *both* A and B above.

DRILL PROBLEMS

S4. The points $(0, 1)$ and $(2, 1)$ are two vertices of a square. Find three sets of other possible vertices.

S5. If (p, q) is in Quadrant II, find the quadrant containing (a) (q, p), (b) $(-q, p)$, and (c) $(-p, q)$.

S6. Let $C = \{(x, y) : |x| < 2\}$, $D = \{(x, y) : y + 1 \le 3\}$, and $E = \{(x, y) : xy \ge 0\}$. Sketch each of the following: (a) $C \cup D$, (b) $C \cap D$, (c) E, and (d) $(C \cup D) \cap E$.

Answers

S4. $\{(0, -1), (2, -1)\}$, $\{(0, 3), (2, 3)\}$, and $\{(1, 0), (1, 2)\}$
S5. (a) IV, (b) III, (c) I
S6. (a) See Figure 2.2.30.   (b) See Figure 2.2.31.

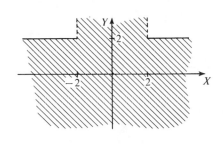

Figure 2.2.30

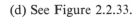

Figure 2.2.31

(c) See Figure 2.2.32.   (d) See Figure 2.2.33.

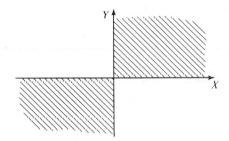

Figure 2.2.32

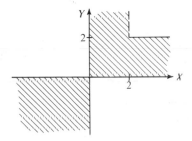

Figure 2.2.33

II. Graphs of Functions

Problem S7.   The complete graph of a function f is in Figure 2.2.34. Use this graph to estimate (a) $f(1)$, (b) $f(f(1))$, (c) x if $f(x) = 1$, (d) the domain of f, and (e) the range of f.

Solution: (a) It appears that $f(1) = 2$ or, equivalently, that $(1, 2)$ is a point of the graph. We "begin" at $x = 1$ and "end" at $f(1) = 2$, as illustrated in Figure 2.2.35.
   (b) Similarly, since $f(1) = 2$, $f(f(1)) = f(2) = 4$ (estimated). (c) The only x such that $f(x) = 1$ appears to be $x = 0$. (d) The domain of f is $[0, 4]$. (It appears that $x = 0$ and $x = 4$ are included.) (e) The range of f is $[1, 4]$.

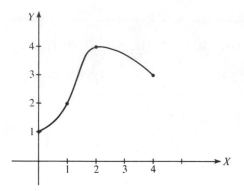

Figure 2.2.34

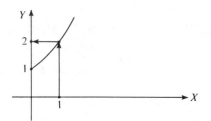

Figure 2.2.35

Problem S8. Which of the subsets of R^2 shown in Figure 2.2.36 are graphs of functions $\{(x, y) : y = f(x)\}$?

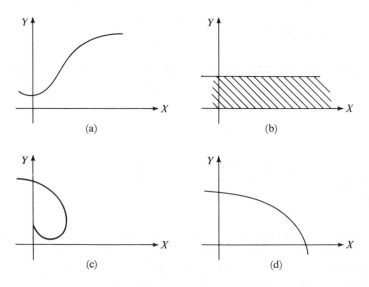

Figure 2.2.36

Solution: (a) and (d) are graphs of functions, since each vertical line intersects the graph at most once. (See Example 2.2.6 and Figure 2.2.37.) For example, (c) is not the graph of a function since it portrays a number x as having two numbers which correspond to it, not just one.

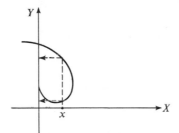

Figure 2.2.37

Problem S9. Plot several points of the graph of the function $f = \{(x, 2 - x^2)\}$. Draw a smooth curve connecting these points.

Solution: Let us arbitrarily choose as X-coordinates the numbers $-2, -1, 0, 1,$ and 2. The corresponding points are then $(-2, -2), (-1, 1), (0, 2), (1, 1),$ and $(2, -2)$. [For example, $f(2) = 2 - 2^2 = -2$, so $(2, -2)$ is a point of the graph.] In Figure 2.2.38 is shown the graph of f, with domain $(-\infty, \infty)$. The range of f is $[-\infty, 2]$, although we shall not prove that here.

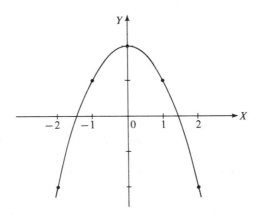

Figure 2.2.38

Problem S10. If (u, v) is a point on the graph of $f(x) = 2x^2$, which of the following points are on the graph of f: (a) $(u, -v)$, (b) $(-u, v)$, (c) $(1/u, 1/v)$, and (d) $(1/(2u), 1/v)$.

Solution: (u, v) an element of f means that $v = f(u) = $ _____. We now $\quad 2u^2$

check every other pair to see if it
expresses an equivalent relationship
involving u and v.

(a) Since $-v$ [=, $\neq$] $2u^2$, $(u, -v)$ [is, is not] an element of f. (b) Since v [=, $\neq$] $2(-u)^2$, $(-u, v)$ [is, is not] a member of f. (c) Since $1/v$ [=, $\neq$] $2\cdot(1/u^2)$, $(1/u, 1/v)$ [$\in$, $\notin$]f. (d) Since $1/v$ [=, $\neq$] $2[1/(2u)^2]$, $(1/(2u), 1/v)$ [$\in$, $\notin$]f.	$\neq$ is not =, is $\neq$, $\notin$ = $\in$

DRILL PROBLEMS

S11. Graph each of the following subsets of R^2, and determine which are graphs of functions:

(a) $\{(x, y) : 2x - y = 1\}$ (b) $\{(x, y) : x^4 = y^4\}$

(c) $\{(x, y) : x < y\}$

Answers

S11. Only (a) is a function. See Figure 2.2.39.

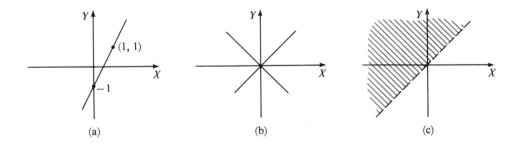

(a)   (b)   (c)

Figure 2.2.39

III. Sets of Function Values

Problem S12. Let $f(x) = x^2$, $A = \{1, 2\}$ and $B = \{-1, 1\}$. Find $f(A)$ and $f(B)$. [Recall Equation (2): $f(A) = \{f(x) : x \in A\}$.]

Solution: Since $f(1) = 1^2 = 1$ and $f(2) = 2^2 = 4$, then $f(A) = \{1, 4\}$. Since $f(-1) = (-1)^2 = 1 = f(1)$, then $f(B) = \{1\}$. Notice that $f(B)$ is a set; thus, $f(B) \neq 1$.

Problem S13.   Use Figure 2.2.34 to find $f([1, 2])$ and $f([1, 4])$.

Solution: Each number $x \in [1, 2]$ "maps" onto a number $f(x) \in [2, 4]$; in fact, the entire set $[1, 2]$ maps into $[2, 4]$ as shown in Figure 2.2.40. Thus, $f([1, 2]) = [2, 4]$.

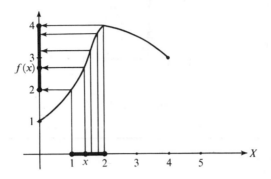

Figure 2.2.40

Similarly, $f([1, 4]) = [2, 4]$. To $x \in [1, 4]$ there corresponds an $f(x) \in [2, 4]$, and to each number $y \in [2, 4]$, there is at least one x such that $f(x) = y$. (See Figure 2.2.41.)

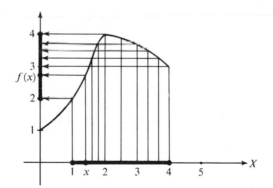

Figure 2.2.41

Problem S14.   Let $f(x) = x^2$, $A = [2, 3)$, $B = [-1, 0]$, and $C = [-1, 2]$. Find $f(A), f(B)$, and $f(C)$.

Solution: If $0 < a < b$, then $a^2 < b^2$. Thus, $f(2) = 4$ will be the smallest number in $f(A)$. Furthermore, for any $y \subset (4, 9)$, let $x = \sqrt{y}$. Then $x \in (2,3)$ and $f(x) = y$. Thus, f takes on all values between 4 and 9. Hence, $f(A) = [4, 9)$. Similarly, $f(B) = [0, 1]$ (not $[1, 0]!$), since for $x \in B$ the smallest value of f is $f(0) = 0$, the largest is $f(-1) = 1$, and f takes on all values between 0 and 1.

Using similar methods, you can find that $f(C) = [0, 4]$. The situation is illustrated in Figure 2.2.42.

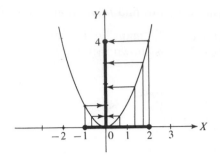

Figure 2.2.42

DRILL PROBLEMS

S15. Let $f(x) = 2 - x^2$ (see Problem S9), $A = [0, 1)$, $B = \{-1, 1\}$, and $C = [-2, 2]$. Find $f(A), f(B),$ and $f(C)$.

Answers

S15. $f(A) = (1, 2], f(B) = \{1\}, f(C) = [-2, 2]$

IV. One-to-One Functions
Review Definition 2.2.3 and the paragraph preceding it.

Problem S16. Which of the graphs in Figure 2.2.43 is the graph of a one-to-one function?

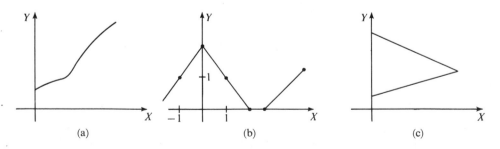

(a)   (b)   (c)

Figure 2.2.43

Solution: (a) represents a one-to-one function. To each y in the range there corresponds exactly one number x in the domain such that $f(x) = y$. (b) does *not* portray a one-to-one function. To each y in the range there correspond *two* numbers in the domain such that $f(x) = y$. For example, for $y = 1$ it appears that both $f(1) = 1$ and $f(-1) = 1$. (c) does not even represent a function, since to each x in the domain there correspond *two* numbers.

Problem S17. Which of the following subsets of R^2 are one-to-one functions $y = f(x)$: $A = \{(x, y) : 2x + y = 1\}$, $B = \{(x, y) : y = |x|\}$, $C = \{(x, y) : y = x^3\}$, and $D = \{(x, y) : y^2 = x\}$.

Solution: A is a one-to-one function $f(x) = 1 - 2x$. If $x_1 \neq x_2$, then $-2x_1 \neq -2x_2$. Thus, $1 - 2x_1 \neq 1 - 2x_2$ or $f(x_1) \neq f(x_2)$. That is, two different inputs always yield two different outputs. Thus, f is one-to-one by Definition 2.2.3. B is *not* a one-to-one function. For example, both $x = 1$ and $x = -1$ yield $y = 1$. C is a one-to-one function. There is exactly one solution x to the equation $f(x) = y$, namely, $x = y^{1/3}$. D is not even a function. To each x there correspond two numbers. For example, to $x = 4$ correspond 2 and -2.

DRILL PROBLEMS

S18. Which of the subsets or graphs shown in Figures 2.2.44, 2.2.45, and 2.2.46 and listed in (d), (e), and (f) represent functions? Which are one-to-one?

(a)   (b)

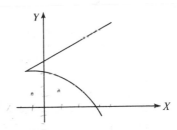

Figure 2.2.44   Figure 2.2.45

(c)   (d) $\{(x, y) : 2y - x = 3\}$

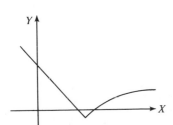

Figure 2.2.46

(e) $\{(x, y) : y^3 = x^2\}$ (f) $\{(x, y) : y^2 = x^3\}$

Answers

S18. (a) and (d) are one-to-one; (c) and (e) are also functions.

V. Distance in R^2 and Circles

Problem S19. Find the distance between the points $(2, -1)$ and $(3, 0)$.

Solution: Using the distance formula [Equation (1)] with $(x_1, y_1) = (2, -1)$ and $(x_2, y_2) = (3, 0)$, we have $d = \sqrt{(3-2)^2 + (0-(-1))^2} = \sqrt{1^2 + 1^2} = \sqrt{2}$. Of course, if we interchange the names of the points we get the same result:

$$d = \sqrt{(2-3)^2 + (-1-0)^2} = \sqrt{(-1)^2 + (-1)^2} = \sqrt{2}$$

Problem S20. Find all points on the Y-axis which are 5 units from the point $(3, 2)$.

Solution: Let $P_1 = (3, 2)$, and let P_2 lie on the Y-axis. Thus P_2 has coordinates $(\underline{\qquad}, y)$. Using the distance formula, we have $5 = \underline{\qquad}$. To solve this equation, we square both sides, obtaining the equation $\underline{\qquad}$. Simplifying and factoring yields $0 = y^2 - 4y - 12 = (y-6)(y+2)$, so y equals either $\underline{\qquad}$ or $\underline{\qquad}$. Thus, the possible solutions are the points $\underline{\qquad}$ and $\underline{\qquad}$. You should check to see that each point is indeed a solution.

> 0
> $\sqrt{(0-3)^2 + (y-2)^2}$
>
> $25 = 9 + y^2 - 4y + 4$
>
> $6, \quad -2$
> $(0, 6), \quad (0, -2)$

Problem S21. Sketch the graph of the circle given by $(x-4)^2 + (y+3)^2 = 9$. Indicate the center and the radius.

Solution: Using Equation (1), we find that the center of the circle is the point $(4, -3)$ and the radius is 3. Using this information, we can easily sketch the circle, as in Figure 2.2.47.

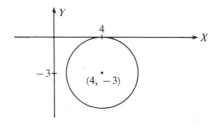

Figure 2.2.47

Problem S22. Write the equation of the circle with center $(-3, 0)$ and radius 5.

Solution: The equation of a circle with center (h, k) and radius r is _____ $= r^2$. In this case, $h =$ ____, $k =$ ____ and $r =$ ____. Therefore, the desired equation is _____.

$(x - h)^2 + (y - k)^2$
$-3, \quad 0, \quad 5$

$(x + 3)^2 + y^2 = 25$

DRILL PROBLEMS

S23. Find the distance between $(-1, 3)$ and $(2, 6)$.
S24. Find the distance between $(a + b, b)$ and $(a - b, 2b)$.
S25. Sketch the graph of the circle given by $(x + 3)^2 + (y - 2)^2 = 3$. Indicate the center and the radius.

Answers

S23. $\sqrt{18} = 3\sqrt{2}$
S24. $|b|\sqrt{5}$
S25. Center, $(-3, 2)$; radius, $\sqrt{3}$. See Figure 2.2.48.

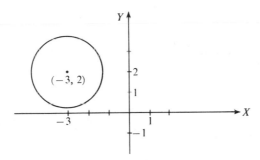

Figure 2.2.48

SUPPLEMENTARY PROBLEMS

S26. Find the distance between (a, b) and (b, a).
S27. Find the domain of $(x + 1)(x^2 - 4)^{-1}$.
S28. Find $f([-1, 3])$ if $f(x) = 2 - x^2$.
S29. Sketch the graph of each of the following:
(a) $\{(x, y) : |y| \leq 2\}$ (b) $(1 - x)^2 = 1 - y^2$
(c) $(1 - x)^2 > 1 - y^2$
S30. Which of the following are one-to-one functions?
(a) $f(x) = |x + 2|$ (b) $g(x) = |x| + 2$
(c) $h(x) = |x| + 2x$ (d) $F(x) = x^2 + 2$
(e) $G(x) = x^3 + 2$

Answers

S26. $\sqrt{2}|a - b|$
S27. $\{x \neq 2, -2\}$
S28. $[-7, 2]$
S29. (a) See Figure 2.2.49.   (b) See Figure 2.2.50.

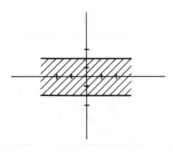

Figure 2.2.49

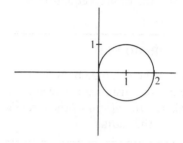

Figure 2.2.50

(c) See Figure 2.2.51.

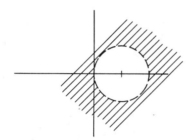

Figure 2.2.51

S30. (c) and (e)

2.3 POLYNOMIAL FUNCTIONS

Many of the functions that we have defined thus far are *polynomial* functions. Polynomial functions are relatively easy to deal with mathematically and have many useful properties. Furthermore, they arise naturally in many physical situations such as in the first example presented in Section 2.0. A polynomial function is defined formally as follows

Definition 2.3.1

Let $a_0, a_1, \ldots, a_n$ be $n + 1$ numbers with $a_n \neq 0$. Then the function

$$P(x) = a_n x^n + a_{n-1} x^{n-1} + \cdots + a_1 x + a_0 \qquad (1)$$

is called a *polynomial function of degree n*. The numbers $a_0, \ldots, a_n$ are called *coefficients of P*.

For example, $f(x) = 4 - 5x^3 + 2x^2$ is a polynomial function of degree three. However, $g(x) = (4 - 5x^3 + 2x^2)/(2x^2 + 1)$ is not a polynomial function; rather, it is an example of a *rational* function, that is, a quotient (or ratio) of two polynomial functions. Polynomial functions of degree one are also called *linear functions;* polynomials of degrees two and three are called *quadratics* and *cubics*, respectively. (Of course, the polynomial functions of degree zero are simply constant functions.) In Figure 2.3.1 are shown the graphs of two polynomial functions, $f(x) = x^2$ and $g(x) = x^3$.

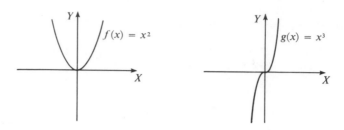

Figure 2.3.1

Let us now consider linear functions in more detail. A polynomial function of degree one, or a linear function, is often written in the form

$$f(x) = mx + b \qquad (2)$$

In the problems at the end of this section, you will show that the graph of any linear function is a straight line. Accepting this fact, we notice that this line cannot be vertical and still be the graph of a function. (Why not?) Thus, it must intersect the Y-axis at the point $(0, f(0))$. Since $f(0) = m \cdot 0 + b = b$, the point b on the Y-axis is called the Y-intercept. (See Figure 2.3.2.)

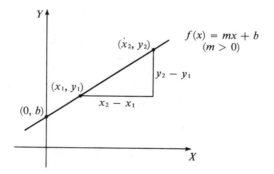

Figure 2.3.2

Q1: Plot the Y-intercept and a second point of the graph of $f(x) = 2 - 3x$, and sketch the graph of f.

We can observe the significance of the number m, called the *slope*, by considering two distinct points (x_1, y_1) and (x_2, y_2) of the line. Now, $y_2 = f(x_2) = mx_2 + b$ and $y_1 = f(x_1) = mx_1 + b$. Subtracting these two equations yields $y_2 - y_1 = mx_2 + b - (mx_1 + b) = m(x_2 - x_1)$. Since $x_2 \neq x_1$,

$$m = \frac{y_2 - y_1}{x_2 - x_1} \qquad (3)$$

Thus, the slope m represents the ratio of the *change in output* of the linear function to the *change in input*. Furthermore, this ratio of changes is constant for a given linear function. That is, the ratio does not depend on the particular points used to compute it. The linear function graphed in Figure 2.3.2 has a positive slope ($m > 0$), since an increase in x values generates a corresponding increase in y values. The graph in Figure 2.3.3, however, has a negative slope, since an increase in x generates a decrease in y. In fact, if the increase in x equals 1, then Equation (3) implies that the change in y equals m as shown in Figure 2.3.3 — an increase if $m > 0$ and a decrease if $m < 0$.

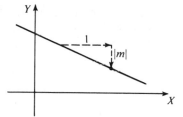

Figure 2.3.3

Q2: Write the equation of the linear function with slope 2 and Y-intercept 1. Sketch its graph using the slope and Y-intercept.

If $m = 0$, the graph of the function $f(x) = mx + b = b$ is a line parallel to the X-axis through $(0, b)$. If a line is parallel to the Y-axis, it does not represent the graph of a function and the quotient in Equation (3) is undefined. Thus, the slope of a line parallel to the Y-axis is undefined (Figure 2.3.4).

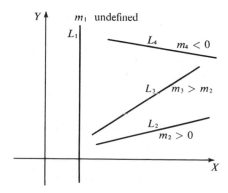

Figure 2.3.4

A1: $f(0) = 2$ and, for example, $f(1) = -1$ determines the straight-line graph. See the figure below.

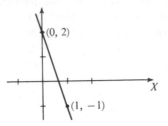

A2: Using the Y-intercept, we find that the point is $(0, 1)$. Increasing x by 1 unit changes y by $m = 2$ units, yielding a second point $(1, 3)$. Also, $f(x) = 2x + 1$. See the figure below.

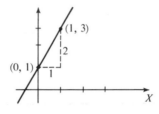

If $b = 0$, the graph of $y = f(x) = mx + b = mx$ passes through the origin. In this case the output is directly proportional to the input, and we often say that y is *proportional* to x or y *varies directly* with x.

There are many different ways that are commonly used to write an equation for a straight line. If we know the slope and the Y-intercept, we can use the *slope-intercept* form and we write $y = mx + b$.

Knowing two points on a nonvertical line enables us to compute the slope of the line using Equation (3). Then any other point (x, y) on the line and the known point (x_1, y_1) must satisfy Equation (3). Thus, $m = (y - y_1)/(x - x_1)$. Multiplying, we have

$$y - y_1 = m(x - x_1) \qquad\qquad (4)$$

the *point-slope* form for the equation of a line.

Q3: Find the equation of the line which passes through $(1, -2)$ and $(3, 4)$.

Every line not parallel to the Y-axis is the graph of some linear function $y = f(x) = mx + b$. However, it is useful to be able to write a "general" equation for any straight line, whether that equation can be used to define a function or not. In addition, the next theorem enables us to conclude immediately that certain graphs are lines so that we can sketch them quickly.

Theorem 2.3.1

Let A, B, and C be numbers with not both A and B equal to 0. Then any line in the Cartesian plane is the graph of the equation

$$Ax + By + C = 0 \tag{5}$$

Conversely, the graph of this equation is a straight line.

Proof

A line parallel to the Y-axis is the graph of an equation $x = c$ for some number c, so let $A = -1$, $B = 0$, and $C = c$ in Equation (5). A line not parallel to the Y-axis is the graph of an equation $y = mx + b$ where m is computed from Equation (3) using any two points of the given line and $b = y_1 - mx_1$. In this case we let $A = m$, $B = -1$, and $C = b$ in Equation (5). Conversely, if $B \neq 0$, Equation (5) can be written as $y = -(A/B)x - C/B = mx + b$ with $m = -A/B$ and $b = -C/B$. We know that the graph of $y = mx + b$ is a line. If $B = 0$ and $A \neq 0$, then the graph of Equation (5) is the graph of $\{(x, y) : x = -C/A\}$, which is a line parallel to the Y-axis.

Q4: Sketch the graph of each of the following:

    (a) $2x - 3y + 6 = 0$ (b) $2x - 3y + 6 \leq 0$

It is easy to see from a picture that parallel lines must have the same slope, and conversely. Theorem 2.3.2 states this fact formally.

Theorem 2.3.2

Two lines $y = m_1 x + b_1$ and $y = m_2 x + b_2$ are parallel if, and only if, they have equal slopes, that is, if $m_1 = m_2$.

Proof

Two lines $y = m_1 x + b_1$ and $y = m_2 x + b_2$ are *not* parallel precisely when there is a point that belongs to both lines, that is, when there is a pair of numbers (x, y) that satisfies both equations. Equating the two expressions for y, we see that the two lines are *not* parallel precisely when $m_1 x + b_1 = m_2 x + b_2$, that is, when

$$(m_1 - m_2)x = b_2 - b_1$$

has exactly one solution. If $m_1 \neq m_2$, then we can solve the equation, so the lines are not parallel. If $m_1 = m_2$, then the equation has no solution unless $b_1 = b_2$; but if $b_1 = b_2$, the equations are the same and thus describe the same line.

A3: $y + 2 = 3(x - 1)$ or $y = 3x - 5$

A4: (a) See the figure below.   (b) See the figure below.

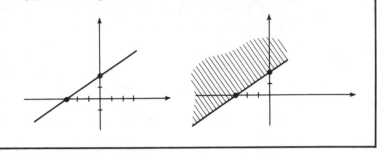

The relationship between two perpendicular lines is given by the following theorem.

Theorem 2.3.3

The two lines $y = m_1 x + b_1$ and $y = m_2 x + b_2$ (with $m_2 \neq 0$) are perpendicular if, and only if, $m_1 m_2 = -1$.

Proof

Suppose that the lines intersect in the point $P = (c, d)$. The vertical line one unit to the right of P intersects the two lines in points $R = (c + 1, \; d + m_1)$ and $S = (c + 1, \; d + m_2)$. (See Figure 2.3.5.) Using the distance formula and the Pythagorean theorem, we obtain $\overline{PR}^2 + \overline{PS}^2 = \overline{RS}^2$, that is,

$$(1 + m_1^2) + (1 + m_2^2) = (m_1 - m_2)^2$$

Expanding and simplifying yields $2 = -2m_1 m_2$, from which the result follows.

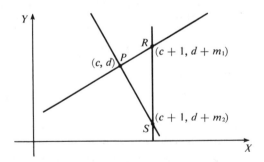

Figure 2.3.5

Q5: Find an equation for the line which passes through (1, 2) and (a) which is parallel to $3x - y - 2 = 0$, and (b) which is perpendicular to $3x - y - 2 = 0$.

If two lines are not parallel or coincident, they intersect in precisely one point. We can find the coordinates of this point algebraically by solving the equations of the two lines simultaneously.

Example 2.3.1

Solve the system

$$3x - y = 1$$
$$x + 2y = 5$$

Solution

The solution set for the above system is the set of pairs (x, y) which satisfy each equation. One way to proceed is as follows: Rewrite the first equation as $y = 3x - 1$. Substituting this expression for y into the second equation and simplifying yields $x + 2(3x - 1) = 5$ or $7x = 7$ or $x = 1$. Using either equation, when $x = 1$, then $y = 2$. This result is illustrated in Figure 2.3.6.

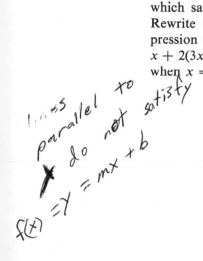

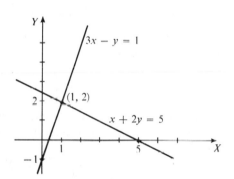

Figure 2.3.6

If f is a linear function and $m \neq 0$, let $(x_1, 0)$ denote the point where the graph of f intersects the X-axis. Thus, $f(x_1) = 0 = mx_1 + b$. Solving this last equation for x_1 gives $x_1 = -b/m$. Thus, $mx + b = m[x + (b/m)] = m(x - x_1)$. Hence, for any linear function f with $m \neq 0$,

$$f(x) = m(x - x_1) \quad \text{where} \quad f(x_1) = 0 \qquad (6)$$

In general, the X-coordinates of the points of intersection of the X-axis and the graph of any function f are called the *zeros* of f. This concept is stated more concisely in the following definition.

Definition 2.3.2

$\{x : f(x) = 0\}$ is the set of *zeros of* f.

A5: (a) $3x - y - 1 = 0$, (b) $x + 3y - 7 = 0$

Q6: State the degree and find the real number zeros of the following:

(a) $h(x) = 2 - 3x$ (b) $f(x) = x^2 - 4x + 3$

(c) $g(x) = 16 - x^4$

The zeros of a quadratic function are found by solving the quadratic equation $ax^2 + bx + c = 0$. To solve this equation we first write

$$x^2 + \frac{b}{a}x = -\frac{c}{a}$$

Next, we complete the square (see Supplement 2.3):

$$x^2 + \frac{b}{a}x + \frac{b^2}{4a^2} = \frac{b^2}{4a^2} - \frac{c}{a} \quad \text{or} \quad \left(x + \frac{b}{2a}\right)^2 = \frac{b^2 - 4ac}{4a^2}$$

It follows from this equation that $x + b/2a$ is equal to either $\sqrt{b^2 - 4ac}/2a$ or $-\sqrt{b^2 - 4ac}/2a$. We write this fact as follows:

$$x + \frac{b}{2a} = \frac{\pm\sqrt{b^2 - 4ac}}{2a}$$

From these equations we obtain the quadratic formula for the zeros of the quadratic function $f(x) = ax^2 + bx + c$:

$$x = \frac{-b \pm \sqrt{b^2 - 4ac}}{2a} \tag{7}$$

Notice that this formula does not yield a real number if $b^2 - 4ac < 0$, so there are no real zeros of a quadratic function in this case. If $b^2 - 4ac = 0$, then the quadratic function has one real zero, and if $b^2 - 4ac > 0$, it has two.

Example 2.3.2

Solve $2x^2 - 3x - 4 = 0$.

Solution

Applying the quadratic formula mentioned above, with $a = 2$, $b = -3$, and $c = -4$, we have solutions

$$x = \frac{-(-3) \pm \sqrt{(-3)^2 - 4 \cdot 2 \cdot (-4)}}{2 \cdot 2}$$

$$= \frac{3 \pm \sqrt{9 + 32}}{4}$$

$$= \frac{3 \pm \sqrt{41}}{4}$$

Thus, since $b^2 - 4ac = 41 > 0$, the solution set is

$$\left\{ \frac{3 + \sqrt{41}}{4}, \frac{3 - \sqrt{41}}{4} \right\}$$

Suppose that our quadratic function f has two real zeros x_1 and x_2 given by the quadratic formula. Adding yields

$$x_1 + x_2 = -\frac{b}{a} \tag{8}$$

and multiplying yields

$$x_1 x_2 = \frac{c}{a} \tag{9}$$

Q7: Verify Equations (8) and (9).

Thus,

$$ax^2 + bx + c = a\left(x^2 + \frac{b}{a}x + \frac{c}{a} \right)$$
$$= a[x^2 - (x_1 + x_2)x + x_1 x_2]$$
$$= a(x - x_1)(x - x_2)$$

Thus, if $f(x) = ax^2 + bx + c$ has real zeros x_1 and x_2, then

$$f(x) = a(x - x_1)(x - x_2) \tag{10}$$

Let us compare the results of Equations (6) and (10). Equation (6) states that if x_1 is a zero of a first-degree polynomial function f, then $f(x)$ can be written as the product $x - x_1$ times m (a zeroth-degree polynomial). Conversely, if $f(x) = m(x - x_1) = mx - mx_1$, then f is a first-degree polynomial when $m \neq 0$, and x_1 is a zero of f. Equation (10) states that if x_1 is a zero of a second-degree polynomial function f, then $f(x)$ can be written as the product $x - x_1$ times $a(x - x_2)$ (a first-degree polynomial). Conversely, if $f(x) = (x - x_1) \cdot g(x)$ where $g(x) = a(x - x_2)$, then f is a quadratic function (since $f(x) = (x - x_1)a(x - x_2) = ax^2 - a(x_1 + x_2)x + ax_1 x_2$), and x_1 is a zero of f. We state the fact that a similar statement is true for any polynomial of degree n as Theorem 2.3.4, but we shall not prove it. Its proof can be found in a more advanced text.

Theorem 2.3.4

Let f be a polynomial of degree $n > 0$. Then there is a polynomial function g such that $f(x) = (x - x_1)g(x)$ if, and only if, x_1 is a zero of f. Furthermore, g is of degree $n - 1$.

Example 2.3.3

Verify $f(2) = 0$ if $f(x) = x^3 - 2x^2 + 3x - 6$, and find the zeros of f to write f in factored form.

A6: (a) 1, $\{2/3\}$; (b) 2, $\{1, 3\}$; (c) 4, $\{2, -2\}$

A7:

$$x_1 + x_2 = \frac{-b + \sqrt{b^2 - 4ac}}{2a} + \frac{-b - \sqrt{b^2 - 4ac}}{2a} = -\frac{2b}{2a} = -\frac{b}{a}$$

$$x_1 x_2 = \frac{1}{4a^2}[(b^2 - (b^2 - 4ac)] = \frac{4ac}{4a^2} = \frac{c}{a}$$

Solution

$f(2) = 8 - 8 + 6 - 6 = 0$, so 2 is a zero of f. Theorem 2.3.4 says that $x - 2$ is a factor of $f(x)$. By the process of long division we obtain $x^3 - 2x^2 + 3x - 6 = (x - 2)(x^2 + 3)$. Since $x^2 + 3 > 0$, f has no zeros other than 2, and $x^2 + 3$ cannot be factored.

Q8: Verify that $f(2) = 0$ for $f(x) = x^3 - 3x^2 + 4$. Find the zeros of f, and write f in factored form.

Theorem 2.3.4 tells us that the problems of factoring a polynomial and finding the zeros of a polynomial function are equivalent. Neither problem is easy if $n > 4$, and research is still being conducted to discover practical methods for computing the zeros of a polynomial.

PROBLEMS 2.3

1. Write slope-intercept equations of the following lines:
(a) $3x - y = 2$ (b) $3x + 6y = 12$
(c) $x - 3 = 0$ (d) $2y + 4 = 0$

2. Sketch each line in the preceding problem.

3. Find an equation $Ax + By + C = 0$ of the line if it
(a) contains the points $(-1, 3)$ and $(2, -1)$.
(b) contains the points $(3, -2)$ and $(-1, -2)$.
(c) has slope -2 and contains the point $(0, 3)$.
(d) contains the point $(-2, -1)$ and is parallel to the line $6y - 3x = 4$.
(e) contains the point $(-1, 2)$ and is perpendicular to the line $2x - 3y - 4 = 0$.

4. Determine whether or not the following sets of points are collinear:
(a) $\{(4, -3), (-5/2, 2), (0, 0)\}$
(b) $\{(3, 2), (2, 1), (-1, -2)\}$

5. A function f is said to be *increasing* if for every x_1, x_2 in its domain, $x_1 > x_2$ implies $f(x_1) > f(x_2)$, Similarly, a function f is *decreasing* if $x_1 > x_2$ implies $f(x_1) < f(x_2)$. Show that a linear function $f(x) = mx + b$ is (a) increasing if $m > 0$ and (b) decreasing if $m < 0$.

6. (a) Is P proportional to V if $P = 3$ when $V = 1$, and if $P = 4$ when $V = 3$? Explain.

(b) Can g be a linear function if $g(1) = 3$, $g(3) = 4$, and $g(-1) = 2$? Explain.

7. Let f and g be linear functions. Which of the following functions are linear:

(a) $h(x) = f(x) + g(x)$ (b) $F(x) = f(x) \cdot g(x)$

(c) $G(x) = f(g(x))$

8. For the linear function $f(x) = mx + b$, compute $f(k + 1) - f(k)$ and $f(x + h) - f(x)$. Does the resulting expression involve x? Interpret geometrically.

9. Find the set of zeros for each of the following functions:

(a) $f(x) = x^2 - 6x$ (b) $f(x) = x^2 - 6x + 8$

10. Use the quadratic formula to solve the following:

(a) $4x^2 + 2x - 1 = 0$ (b) $2x^2 + 2x + 5 = 0$

(c) $4x^4 + 3x^2 - 1 = 0$

11. Let f be a quadratic function with zeros 2 and -3. Write a formula for f.

12. Find the solution set for each of the following equations, and write each equation as a product of factors.

(a) $x^3 + x^2 - 4x - 4 = 0$ (b) $x^5 + x^4 - 8x^3 - 12x^2 = 0$

(c) $x^5 - 4x^3 + x^2 - 4 = 0$

13. Sketch the graph of A, B, and $A \cap B$ if $A = \{(x, y) : 2x + 3y < 8\}$ and $B = \{(x, y) : y \geq |2x|\}$. Label all corners.

14. The Fahrenheit and Centigrade temperature scales are related by a linear function $F = g(C)$. The freezing point of water is $F = 32$ and $C = 0$, and the boiling point is $F = 212$ and $C = 100$. Find the formula for $g(C)$. Find a formula for the function h if $C = h(F)$. Is h a linear function?

$\times$ ════════════════════════

15. Let f and g be linear functions. What can you conclude if $f(|x|) = |f(x)|$ for every number x? If $f(g(a)) = g(f(a))$ for one number a?

16. Let $f(x) = ax^2 + bx + c$ with $a > 0$. Under what conditions are the two zeros of f both positive? Both negative? Of opposite sign?

17. Solve $x^2 + 2x - 1 < 0$.

18. Find all possibilities for the fourth vertex of a parallelogram with three vertices $(-1, -1)$, $(1, 4)$, and $(2, 1)$.

19. Let f and g be linear functions. Are the graphs of the composite functions $h(x) = f(g(x))$ and $t(x) = g(f(x))$ parallel lines?

20. Show that the graph of a linear function is a straight line.

ANSWERS   **1.** (a) $y = 3x - 2$, (b) $y = -(1/2)x + 2$, (c) not defined, (d) $y = $
2.3 $0 \cdot x - 2$

A8: $\{2, 2, -1\}$, $f(x) = (x - 2)^2(x + 1)$

2. See Figure 2.3.7.

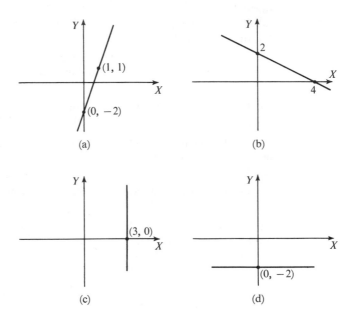

(a)

(b)

(c)

(d)

Figure 2.3.7

3. (a) $4x + 3y - 5 = 0$, (b) $y + 2 = 0$, (c) $2x + y - 3 = 0$, (d) $3x - 6y = 0$, (e) $3x + 2y - 1 = 0$

4. (a) not collinear, (b) collinear

5. (a) $m > 0$ and $x_1 > x_2$, then $mx_1 > mx_2$ and $mx_1 + b > mx_2 + b$. (b) If $m < 0$ and $x_1 > x_2$, then $mx_1 < mx_2$ and $mx_1 + b < mx_2 + b$.

6. (a) No. If $P = f(V) = mV$, $f(1)/f(3) = (m \cdot 1)/(m \cdot 3) = 1/3 \neq 3/4$. (b) Yes, $g(x) = (x + 5)/2$.

7. h and G are linear.

8. m, mh

9. (a) $\{0, 6\}$, (b) $\{2, 4\}$

10. (a) $\{(-1 + \sqrt{5})/4, (-1 - \sqrt{5})/4\}$, (b) $\varnothing$, (c) $\{1/2, -1/2\}$

11. For example, $f(x) = (x - 2)(x + 3) = x^2 + x - 6$.

12. (a) $(x + 1)(x + 2)(x - 2)$, $\{-1, -2, 2\}$
(b) $x^2(x + 2)^2(x - 3)$, $\{0, 0, -2, -2, 3\}$
(c) $(x + 1)(x + 2)(x - 2)(x^2 - x + 1)$, $\{-1, -2, 2\}$

13.
A. See Figure 2.3.8.   *B*. See Figure 2.3.9.

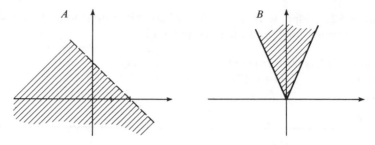

Figure 2.3.8   Figure 2.3.9

$A \cap B$. See Figure 2.3.10.

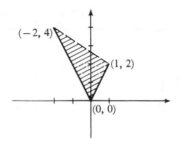

Figure 2.3.10

14. $F = g(C) = (9C/5) + 32$, $C = h(F) - (5/9)(F - 32)$, yes.

Supplement 2.3

I. Direct Variation

Problem S1. Suppose that y varies directly as x and that $y = 6$ when $x = 3$. Find the constant of proportionality, and sketch the graph of $y = f(x)$.

Solution: By definition, $y = mx$ for some constant m. Since $y = 6$ when $x = 3$, $6 = m \cdot 3$. Hence, $m = 2$. The graph of f is sketched in Figure 2.3.11.

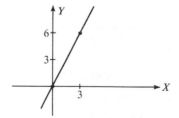

Figure 2.3.11

Problem S2.   Suppose that $y = f(x)$ where y is proportional to x and $f(-10) = 5$. Find a formula for f, and then compute $f(3)$.

Solution: f has the form $f(x) = $ ___.	mx
Since $f(-10) = 5$, $m \cdot (-10) = $ ___.	5
Thus, $m = $ ___. So $f(x) = $ ___	$-1/2$, $-(1/2)x$
and $f(3) = $ ___.	$-3/2$

DRILL PROBLEMS

S3. Suppose that y varies directly as x and $y = f(x)$. Find the proportionality constant for each of the following: (a) $f(2) = 4$, (b) $f(-1) = -1$, (c) $f(9) = 1$, and (d) $f(100) = 1/100$.

S4. If $y = f(x)$ and $f(0) = 2$, does y vary directly as x? Explain.

Answers

S3. (a) $m = 2$, (b) $m = 1$, (c) $m = 1/9$, (d) $m = 1/10,000$
S4. No. If y varies directly as x, then $f(0) = m \cdot 0 = 0$.

II. Linear Functions

Problem S5. Find the zero and the Y-intercept of the linear function $f(x) = 3x + 6$. Sketch the graph of f.

Solution: We obtain the given function by setting $m = 3$ and $b = 6$ in the equation form $f(x) = mx + b$. Thus, the Y-intercept is $b = 6$. To find the zero of f, we set $3x + 6 = 0$ and solve for x, obtaining $x = -6/3 = -2$. Thus, the zero is -2. The graph is easy to obtain because we know it is a straight line. We have two points of the line, namely, $(0, 6)$ and $(-2, 0)$. Plot these and draw the line (see Figure 2.3.12).

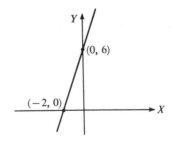

Figure 2.3.12

Problem S6. Let f be a linear function such that $f(0) = 3$ and $f(2) = 7$. Write its equation.

Solution: We know that $f(x) = mx + b$ for certain numbers m and b. Since $f(0) = 3$, $m \cdot 0 + b = 3$, or $b = 3$. So, $f(x) = mx + 3$. Also, $f(2) = 7$, so $m \cdot 2 + 3 = 7$, or $2m = 4$, that is, $m = 2$. Thus, $f(x) = 2x + 3$.

Problem S7. Let f be a linear function such that $f(-1) = 2$ and $f(2) = 8$. Write an equation defining f, and sketch its graph.

Solution: The graph is the straight line through the two points $(-1, f(-1))$ and $(2, f(2))$, or $(-1, 2)$ and $(2, 8)$. Plot these and draw the line (see Figure 2.3.13). To find the equation, start with $f(x) = mx + b$. Now, $f(-1) = m(-1) + b = 2$, and $f(2) = m(2) + b = 8$. We can solve these two equations for m and b as follows: Solve the first equation for b, namely, $b = 2 + m$. Then replace b with $2 + m$ to obtain $2m = 8 - (2 + m)$, or $3m = 6$. Thus, $m = 2$, and since $b = 2 + m$, $b = 4$. Thus, $f(x) = 2x + 4$.

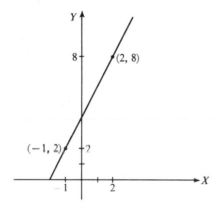

Figure 2.3.13

Problem S8. Write the equation of the linear function f with Y-intercept -3 when $f(2) = 3$.

Solution: $f(x) = $ _____. Here we know that $b = $ _____ because b is the _____. So $f(x) = mx - 3$. Since $f(2) = 3$, then _____ $= 3$. If we solve this equation for m, we get $m = $ _____. Hence, $f(x) = $ _____.	$mx + b$ -3 Y-intercept $m \cdot 2 - 3$ 3 $3x - 3$

Problem S9. The graph of a linear function f has Y-intercept 12, and -3 is a zero of f. Find an equation defining f.

Solution: From $f(x) = mx + b$, $b = $ _____, so $f(x) = mx + 12$. But	12

since $f(-3) =$ _____, $m(-3) + 12 =$ 0
0. Thus, $m =$ _____. So the equation   4
is $f(x) =$ _____. $4x + 12$

Problem S10. A linear function f is such that $f(3) = -1$ and $f(1) = 3$. Find an equation defining f, and sketch its graph.

Solution: We know two points
_____ and _____ of the line graph $(3, -1),$ $(1, 3)$
of f. Plot these and draw the line.   See Figure 2.3.14.

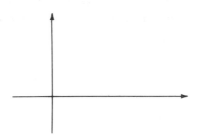

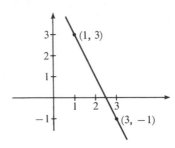

Figure 2.3.14

   We know that there are numbers
m and b such that $f(x) =$ _____. $mx + b$
Because $f(3) = -1,$ _____ $=$ $3m + b$
-1. Because $f(1) = 3,$ _____ $=$ $m + b$
3. Solving the second of these
equations for b, we get $b =$ _____. $3 - m$
Using this value of b in the first
equation yields the equation
_____. Solving for m $3m + 3 - m = -1$
yields $m =$ _____. Substituting this -2
value for m in the second equation,
we obtain _____ $= 3$ or $b =$ _____. $-2 + b,$ 5
So the function is given by $f(x) =$
_____. $-2x + 5$

DRILL PROBLEMS

S11. For each of the following linear functions determine the Y-intercept of its line graph and find a zero (if it exists):
(a) $f(x) = 2x - 4$ (b) $f(x) = -(1/2)x + 4$
(c) $f(x) = 6$ (d) $f(x) = 3(x - 1)$
(e) $f(x) = -x$

S12. Sketch the graph of each of the functions in Problem S11.

S13. Find an equation that defines a linear function satisfying the following conditions:

(a) f is a direct variation function and $f(2) = 8$

(b) The number 2 is a zero of f and its graph has Y-intercept -1.

(c) The graph of f has Y-intercept 3 and $f(1) = 0$.

(d) The number -2 is a zero of f and $f(3) = 1$.

(e) $f(2) = 5$ and $f(-2) = 5$.

(f) $f(2) = 6$ and $f(10) = 30$.

(g) $f(1) = 1, f(2) = 2$, and $f(3) = 4$.

S14. A linear function v satisfies the condition that for some number k, $v(k) = -v(-k)$. Show that the Y-intercept of v is zero.

Answers

S11. (Y-intercepts are listed first.) (a) -4, $x = 2$, (b) 4, $x = 8$, (c) 6, no zero, (d) -3, $x = 1$, (e) 0, $x = 0$

S12. (a) See Figure 2.3.15.   (b) See Figure 2.3.16.

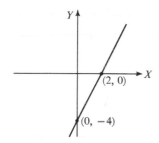

Figure 2.3.15

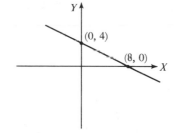

Figure 2.3.16

(c) See Figure 2.3.17.   (d) See Figure 2.3.18.

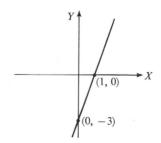

Figure 2.3.17

(e) See Figure 2.3.19.

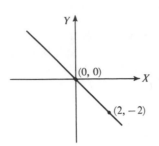

Figure 2.3.19

S13.
(a) $f(x) = 4x$ (b) $f(x) = (1/2)x - 1$
(c) $f(x) = -3x + 3$ (d) $f(x) = (1/5)x + 2/5$
(e) $f(x) = 5$ (f) $f(x) = 3x$
(g) No such linear function exists. $f(1) = 1$ and $f(2) = 2$ determine the linear function $f(x) = x$. But $f(3) = 3$, *not* 4.
S14. If $-v(-k) = -[m(-k) + b] = mk - b$ equals $v(k) = mk + b$, then $b = 0$.

III. Lines

Problem S15. Write the slope-intercept form of the equation of the line $2y + x = 4$, and sketch the line.

Solution: We can change the given equation to slope-intercept form by solving $2y + x = 4$ for y. We obtain $2y = -x + 4$ or $y = -(1/2)x + 2$. The coefficient of x, $-1/2$, is the slope of the line. The constant term, 2, is the Y-intercept, and thus $(0, 2)$ is one point of the line. A slope of $-1/2$ means that we may obtain a second point of the graph by moving 1 unit to the right and one-half unit *down* from the point $(0, 2)$. (See Figure 2.3.20.) Equivalently, we can move 2 units to the

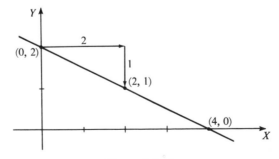

Figure 2.3.20

right and 1 unit down. Thus, (2, 1) is another point on the graph of $2y + x = 4$. Since two points determine a line, we can easily obtain the graph of $2y + x = 4$ as shown in Figure 2.3.20. Notice that the X-intercept (when $y = 0$) is easily determined. Since $2y + x = 4$, if $y = 0$, then $x = 4$; so the point (4, 0) is on the graph. This additional point serves as a check of our previous findings.

Problem S16. Find an equation $Ax + By + C = 0$ of the line that contains the points (2, −1) and (0, 4).

Solution: We first apply Equation (3) to obtain the slope of the desired line, $m = \dfrac{4 - (-1)}{0 - 2} = -\dfrac{5}{2}$. Next, we use the point-slope equation of a line [Equation (4)] and the point (2, −1) to obtain $y - (-1) = -\dfrac{5}{2}(x - 2)$. However, we desire an equation of the form $Ax + By + C = 0$. Thus, we write $2(y + 1) = -5x + 10$, or $2y + 2 + 5x - 10 = 0$, and finally $5x + 2y - 8 = 0$.

You should verify that if we use (0, 4) rather than (2, −1) in the point-slope form above, we would obtain the same result.

Problem S17. Write the slope-intercept equation of the line $y - 4 = 0$. Sketch the line.

Solution: We first solve $y - 4 = 0$ for y, obtaining _____, or equivalently, $y = 0 \cdot x + 4$. Thus, the slope is _____. The Y-intercept is _____, and thus one point of the graph is _____. A zero slope means that the line is _____. Sketch the line.

$y = 4$

0, 4

(0, 4)

horizontal   See Figure 2.3.21.

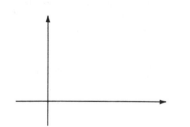

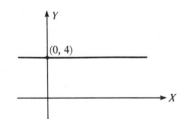

Figure 2.3.21

Problem S18. Find an equation $Ax + By + C = 0$ of the line that contains the points (2, −1) and (2, 4).

Solution: We first apply Equation (3) to find the slope of the line containing the given points. Since

$\dfrac{4 - (-1)}{2 - 2}$ is not a real number, the	
slope is _____. Lines with	undefined
undefined slope are _____ to	parallel
the *Y*-axis. Thus, $x =$ _____ is	2
the equation of the line containing	
points $(2, -1)$ and $(2, 4)$. Written in	
the desired form, the equation is	
_____.	$x - 2 = 0$

Problem S19. Find an equation $Ax + By + C = 0$ of the line that contains the point $(-3, 1)$ and is (a) parallel and (b) perpendicular to the line $y - 3x = 9$.

Solution: (a) We first find the slope of the given line, since any line parallel to it must have the same slope. Solving for y, we obtain $y = 3x + 9$, so the slope is 3. We use Equation (4) of a line with $m = 3$ and the given point $(-3, 1)$. Thus, $y - 1 = 3(x - (-3)) = 3x + 9$. Therefore, an equation is $3x - y + 10 = 0$.

(b) Since the slope of the given line is 3, any line perpendicular to it must have slope $-1/3$, the negative reciprocal of 3. Thus, using Equation (4) with $m = -1/3$ and point $(-3, 1)$, we obtain $y - 1 = -(1/3)(x + 3)$ or $3y - 3 = -x - 3$ or $x + 3y = 0$.

Problem S20. Determine whether or not the points $(-1, 2)$, $(1, 3)$, and $(-3, 0)$ are collinear.

Solution: Three different points P, Q, and R are collinear (contained by the same line) if the slope of the line through PQ is equal to the slope of the line through QR. Applying Equation (3) to find the slope of the line containing $(-1, 2)$ and $(1, 3)$, we obtain $m = (3 - 2)/[1 - (-1)] = 1/2$. Similarly, we find that the slope of the line through the points $(1, 3)$ and $(-3, 0)$ is $m = (0 - 3)/(-3 - 1) = 3/4$. Thus, the given points are *not* collinear.

Problem S21. Show that the set of points equidistant from the origin O and the point $P = (2, 1)$ is a line perpendicular to the line OP.

Solution: The slope of the line OP	
is _____, so a line is perpendicular	$1/2$
to it if it has slope _____. Let	-2
$Q = (x, y)$. The distance $\overline{OQ}$ equals	
$\overline{PQ}$ if, and only if, $\overline{OQ}^2 = \overline{PQ}^2$.	
Using the distance formula, we have	
$\overline{OQ}^2 =$ _____ and $\overline{PQ}^2 =$	$x^2 + y^2$
_____. Equating these	$(x - 2)^2 + (y - 1)^2$
expressions and simplifying yields the	
equation _____. The graph	$4x + 2y - 5 = 0$

of this equation is a line with
slope _____, so this line is _____ to -2,   perpendicular
the line *OP*.

DRILL PROBLEMS

S22. Write slope-intercept equations of the following lines:
(a) $y - 2x - 6 = 0$ (b) $2x + 3y = 4$
S23. Sketch each line in the preceding problem and find its slope.
S24. Find an equation $Ax + By + C = 0$ of the line if it
(a) contains the points $(2, 1)$ and $(-3, -4)$.
(b) contains the point $(-1, 0)$ and is parallel to the line $6y - 3x = 12$.
(c) contains the point $(2, -1)$ and is perpendicular to the line $y - 2x = 4$.

Answers

S22. (a) $y = 2x + 6$, (b) $y = -(2/3)x + (4/3)$
S23. (a) See Figure 2.3.22.   (b) See Figure 2.3.23.

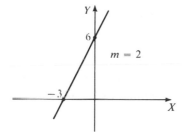

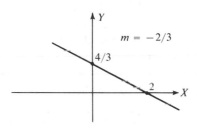

Figure 2.3.22   **Figure 2.3.23**

S24. (a) $x - y - 1 = 0$, (b) $x - 2y + 1 = 0$, (c) $x + 2y = 0$

IV. Quadratic Functions

Recall that the zeros of a quadratic function $f(x) = ax^2 + bx + c$ are given by the
quadratic formula, $x = \dfrac{-b \pm \sqrt{b^2 - 4ac}}{2a}$, if $b^2 - 4ac \geq 0$.

Problem S25.   Find the zeros of the quadratic function $f(x) = 2x^2 + 10x + 12$.

Solution: Since f is a quadratic function, we can use the quadratic formula. In
this case, a, the coefficient of x^2, is 2; b, the coefficient of x, is 10; and c, the constant
term, is 12. Computing $b^2 - 4ac$, we obtain $10^2 - 4 \cdot 2 \cdot 12 = 100 - 96 = 4$. Substituting into the quadratic formula yields

$$x = \frac{-10 \pm \sqrt{4}}{2(2)} = \frac{-10 \pm 2}{4}$$

Thus, $x = -3$ and $x = -2$ are zeros of f. You should verify these answers by evaluating $f(-2)$ and $f(-3)$.

Problem S26.   Solve the quadratic equation $x^2 + x + 10 = 0$.

Solution: The problem is to determine the zeros of the quadratic function given by $f(x) = x^2 + x + 10$. If we attempt to find the zeros by applying the quadratic formula with $a = 1$, $b = 1$, and $c = 10$, we find that the zeros are

$$x = \frac{-1 \pm \sqrt{-39}}{2}$$

But since $\sqrt{-39}$ is not a real number, there are no (real) zeros, and the solution set of our equation is the empty set, $\varnothing$.

Problem S27.   Solve $x^2 - x + 1 = 0$.

Solution: We shall find the zeros using the quadratic formula. In this case we will let $a = $ ____, $b = $ ____, and $c = $ ____, so $b^2 - 4ac = $ ____. Since $b^2 - 4ac$ is less than zero, there are ____ zeros of f. Thus, the solution set for our equation is ____.

1,	-1
1,	-3
no	
$\varnothing$	

Problem S28.   Find the zeros of the quadratic function $f(x) = x^2 - x - 3$.

Solution: We will use the quadratic formula with $a = $ ____, $b = $ ____, and $c = $ ____. Then $b^2 - 4ac = $ ____, so the zeros are given by $x = $ ____. The set of zeros is ____.

1,	-1
-3	
13	
$(1 \pm \sqrt{13})/2$	
$\{(1 + \sqrt{13})/2, (1 - \sqrt{13})/2\}$	

Problem S29.   Solve $x + \dfrac{1}{x} = 1$.

Solution: If $x \neq 0$, then the given equation is equivalent to the equation ____ or ____ $= 0$. But this is Problem S27, so the solution set is $\varnothing$.

$x^2 + 1 = x, \quad x^2 - x + 1$

DRILL PROBLEMS

S30. Find the zeros of the following functions without using the quadratic formula:
(a) $f(x) = (x - 2)(x - 3)(x + 4)$ (b) $f(x) = 2x^2 + 3x$
(c) $f(x) = (x + 1)^2$ (d) $f(x) = x^2 + 10$
(e) $f(x) = x^2 + 6x + 9$ (f) $f(x) = 3x^2 - 27$

S31. Find the zeros of each of the following quadratic functions. Use the quadratic formula.
(a) $f(x) = 2x^2 + x - 1$
(b) $f(x) = -x^2 - 2x + 2$
(c) $f(x) = 3x^2 + x - 8$
(d) $f(x) = x^2 + 10x + 9$

S32. Each of the following equations may be considered as a quadratic equation in some variable. Make an appropriate substitution and change the equation to a quadratic equation.
(a) $1/t^2 + 1/t + 1 = 0$ (Let $x = 1/t$.)
(b) $u + 3\sqrt{u} + 2 = 0$
(c) $(\sin t)^2 + 3 \sin t + 2 = 0$
(d) $1/u + 3/\sqrt{u} - 4 = 0$
(e) $a^4 + 1 = 0$
(f) $w^{2/3} + 5w^{1/3} = 0$

Answers

S30. (a) $\{2, 3, -4\}$, (b) $\{-3/2, 0\}$, (c) $\{-1\}$, (d) $\varnothing$, (e) $\{-3\}$, (f) $\{-3, 3\}$
S31. (a) $\{-1, 1/2\}$, (b) $\{-1 - \sqrt{3}, -1 + \sqrt{3}\}$, (c) $\{(-1 \pm \sqrt{97})/6\}$, (d) $\{-9, -1\}$
S32. (a) Let $x = 1/t$. $x^2 + x + 1 = 0$
(b) Let $x = \sqrt{u}$. $x^2 + 3x + 2 = 0$
(c) Let $x = \sin t$. $x^2 + 3x + 2 = 0$
(d) Let $x = 1/\sqrt{u}$. $x^2 + 3x - 4 = 0$
(e) Let $x = a^2$. $x^2 + 1 = 0$
(f) Let $x = w^{1/3}$. $x^2 + 5x = 0$

V. Completing the Square

When we work with quadratic functions or equations, it is sometimes desirable to change expressions of the form $x^2 + bx$ to expressions of the form $(x + a)^2 - a^2$. If we want these expressions to be equal, we must have $x^2 + bx = (x + a)^2 - a^2 = x^2 + 2ax + a^2 - a^2 = x^2 + 2ax$. It is easy to see that $b = 2a$ or that $a = b/2$. This process of writing $x^2 + bx$ as $x^2 + bx + (b/2)^2 - (b/2)^2 = (x + b/2)^2 - (b/2)^2$ is called *completing the square*.

Problem S33. Solve the quadratic equation $x^2 + 4x - 2 = 0$ by completing the square.

Solution: For convenience we rewrite the equation as $x^2 + 4x = 2$. We have an expression of the form $x^2 + bx$ on the left. We add $(b/2)^2$ to this expression to complete the square. In this particular case, $b = 4$, so $b/2 = 2$ and $(b/2)^2 = 4$. We add 4 to both sides of $x^2 + 4x = 2$ to obtain $x^2 + 4x + 4 = 6$. In other words, $(x + 2)^2 = 6$. (Completing the square was done precisely to obtain a perfect square on the left.) Thus, $x + 2 = \pm\sqrt{6}$. So either $x = \sqrt{6} - 2$ or $-\sqrt{6} - 2$.

Problem S34. Graph $x^2 - 6x + y^2 - 2y = 6$ by completing *two* squares and writing the equation in the form $(x - h)^2 + (y - k)^2 = a^2$.

Solution: There are two expressions to treat, namely, $x^2 - 6x$ and $y^2 - 2y$. Since $6/2 = 3$, we add $3^2 = 9$ to the first expression. Since $2/2 = 1$, we add $1^2 = 1$ to the second. We add both numbers not only to the left side to complete the squares but also to the right side of the equation. We get $x^2 - 6x + 9 + y^2 - 2y + 1 = 6 + 9 + 1 = 16$ or $(x - 3)^2 + (y - 1)^2 = 4^2$.

The graph of this equation is now easily obtained. It is a circle centered at $(3, 1)$ and with radius 4 (see Section 2.2 and Figure 2.3.24).

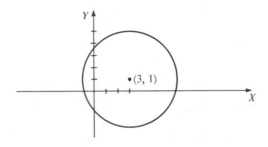

Figure 2.3.24

Problem S35.   Use the technique of completing the square to solve $3x^2 + 4x - 1 = 0$.

Solution: Rewrite the equation as $3[x^2 + (4/3)x] = 1$. The expression that we consider is $x^2 + (4/3)x$. We add the square of $1/2\cdot$____ or

$4/3$

_____ to $x^2 + (4/3)x$. Thus, we

$(2/3)^2$

will change the left side of our equation to $3[x^2 + (4/3)x +$

_____]. We must add

$4/9$ or $(2/3)^2$

_____ to the right side of

$3\cdot(4/9)$ or $4/3$ (because we distribute

the equation. So the equation

the 3 across the sum)

becomes $3[x^2 + (4/3)x + 4/9] = 1 + 4/3 = 7/3$ or $3(\_\_\_\_)^2 = 7/3$.

$x + 2/3$

Dividing by 3, we get $(x + 2/3)^2 = 7/9$. Thus, $x + 2/3 = \pm\_\_\_$, so

$\sqrt{7}/3$

$x = -2/3 \pm \sqrt{7}/3$.

DRILL PROBLEMS

S36. Solve the quadratic equations below by completing the square. Check your answers either by factoring or by using the quadratic formula.

(a) $x^2 - 2x = 0$ (b) $x^2 - 3x + 2 = 0$

(c) $x^2 + 3x + 10 = 0$ (d) $x^2/2 + 2x - 4 = 0$

(e) $2x^2 + 6x - 3 = 0$ (f) $2x^2 + 6x + 5 = 0$

Answers

S36. (a) 0, 2, (b) 1, 2, (c) no solutions in R, (d) $-2 \pm 2\sqrt{3}$, (e) $(-3 \pm \sqrt{15})/2$, (f) no solutions in R

VI. Systems of Linear Equations

Problem S37. Find the solution set for the system $2x + y - 3 = 0$ and $y - x + 3 = 0$.

Solution: We solve $2x + y - 3 = 0$ for y, obtaining $y = -2x + 3$. Now we substitute this expression for y into the other equation and simplify, obtaining $0 = y - x + 3 = (-2x + 3) - x + 3 = -3x + 6 = 0$. Thus, $x = 2$. Since we earlier obtained $y = -2x + 3$, $y = -2 \cdot 2 + 3 = -1$. Thus, the solution set is $\{(2, -1)\}$.

Of course, we could have begun by solving the first original equation for x rather than for y, or we could have used several other variations.

Problem S38. Solve the system $4x - 2y + 3 = 0$ and $y = 2x + 1$.

Solution: This time we begin by solving the second equation for y, $y = 2x + 1$. Substituting this expression for y into the other equation yields $4x - 2(2x + 1) + 3 = 0$ or $4x - 4x - 2 + 3 = 0$. Now, of course, $1 \neq 0$, so the equation $4x - 4x - 2 + 3 = 0$ has no solution x; that is, there is no real number x such that $4x - 4x - 2 + 3 = 0$, since $4x - 4x - 2 + 3 = 1$. Thus, this system has no solution. Geometrically, we see that these two lines have the same slope and thus are parallel. Illustrate this problem with a sketch.

Problem S39. Find the solution set for the system $2x + 3y - 4 = 0$ and $3x - 2y = 3$.

Solution: Let us solve $2x + 3y - 4 = 0$ for x, obtaining $2x = \underline{\quad}$, and $x = \underline{\quad}$. Now substitute this expression for x into the equation $\underline{\quad}$, obtaining $3[2 - (3y/2)] - 2y = 3$. Simplifying yields $6 - (9y/2) - 2y = 3$ or $6 - 3 = 3 = (9/2)y + 2y = \underline{\quad}y$. Thus, $y = (2/13) \cdot 3 = \underline{\quad}$. Now, since $x = 2 - (3/2)y$, we have $x = 2 - (3/2)(6/13) = 2 - (9/13) = \underline{\quad}$. Thus, the solution set is $\underline{\quad}$.

$4 - 3y$
$2 - (3y/2)$
$3x - 2y = 3$
$(13/2)$
$6/13$
$17/13$
$\{(17/13, 6/13)\}$

DRILL PROBLEMS

S40. Find the solution set for each of the following systems of linear equations.
(a) $5x - 3y = -11$ (b) $2x - y = 3$
   $y - 3x = 5$ $2y + 6 = 4x$
(c) $x -$ $1 = 2y$
   $4y - 2x = 3$

Answers

S40. (a) $\{(-1, 2)\}$, (b) $\{(x, 2x - 3) : x \in R\}$, (c) $\emptyset$

VII. Zeros, Roots, and Factors

Problem S41.   Find a zero of $f(x) = x^3 + 4x - 5$ by the "guess and check" method. (Hint: Try $x \in \{0, \pm 1\}$.)

Solution: Clearly, $f(1) = 1^3 + 4 \cdot 1 - 5 = 0$, so 1 is a zero of f. However, $f(0) = -5, f(-1) = -10$, so 0 and -1 are not zeros of f.

Problem S42.   Find all of the zeros of $f(x) = x^3 + 4x - 5$, and write the equation as a product of factors.

Solution: Using the result of Problem S41 and Theorem 2.3.4, since 1 is a zero of f, we know that $x - 1$ is a factor of $x^3 + 4x - 5$. By "long division," the quadratic factor is found to be $x^2 + x + 5$ (you should verify this). Thus, $f(x) = (x - 1) \cdot (x^2 + x + 5)$. Other zeros must be obtained from the equation $x^2 + x + 5 = 0$. However, this equation has no real solutions (you should also verify this) and, therefore, 1 is the only real zero of f.

Problem S43.   Find all real zeros of $f(x) = x^4 - x^3 - 2x^2$, and write the equation as a product of factors.

Solution: First factor x^2 from the expression $f(x) = x^2(x^2 - x - 2)$. Clearly, 0 is a zero since $f(0) = 0$. Other zeros must be obtained from the second factor. Hence, we let $x^2 - x - 2 = 0$. Factoring, we obtain $(x - 2)(x + 1) = 0$, so $x = 2$ or $x = -1$ are the other zeros. Thus, $f(x) = x^2(x - 2)(x + 1)$ in factored form.

DRILL PROBLEMS

S44.   Find all real zeros of $f(x) = x^3 + 1$, and write the equation as a product of factors.

S45.   Find all real zeros of $f(x) = x^4 - 3x^2 - 4$, and write the equation as a product of factors.

Answers

S44. $\{-1\}, f(x) = (x + 1)(x^2 - x + 1)$
S45. $\{-2, 2\}, f(x) = (x - 2)(x + 2)(x^2 + 1)$

SUPPLEMENTARY PROBLEMS

S46. Sketch the graph of each of the following:
(a) $4x = 2y + 6$ (b) $3x + 5y > 15$
S47. Find the equation of the line that
(a) contains the point $(2, 0)$ and is parallel to the line $x - 3y = 6$.
(b) contains the point $(-1, 3)$ and is perpendicular to the line $y = 2$.
(c) contains the points $(-1, 2)$ and $(3, -2)$.
S48. Find the solution set for each of the following:
(a) $2x^2 - 2x - 5 = 0$ (b) $x^2 + x + 5 = 0$
S49. Solve the system $2x - 3y = 13$ and $2y + x + 4 = 0$.
S50. Find all real zeros of $f(x) = x^3 - 2x^2 - x + 2$.

Answers

S46. (a) See Figure 2.3.25.   (b) See Figure 2.3.26.

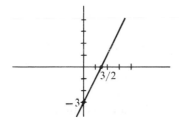

                       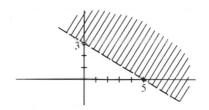

           Figure 2.3.25   **Figure 2.3.26**

S47. (a) $x - 3y - 2 = 0$, (b) $x = -1$, (c) $x + y = 1$
S48. (a) $\{1/2 - (1/2)\sqrt{11}, 1/2 + (1/2)\sqrt{11}\}$, (b) $\varnothing$
S49. $x = 2, y = -3$
S50. $\{-1, 1, 2\}$

2.4 TRANSLATIONS

In the preceding section we learned that if m and b are given numbers, the graphs of $y = mx$ and $y = mx + b$ are parallel lines. In particular, we can regard the graph of $y = mx + b$ as being the result of shifting the graph of $y = mx$ up or down. This shifting is an example of what is called a translation. For example, if the graph of $y = 2x$ is translated 3 units upward in the Y-direction (that is, in the direction parallel to the Y-axis), the result is the graph of $y = 2x + 3$ (Figure 2.4.1). In this section we shall formalize the concept of translations and use the results to help us sketch certain graphs quickly.

A translation is a particular kind of function. It takes a point as input and produces a point as output. Thus, a translation is a function with domain R^2 and range R^2. Such a function is called

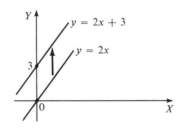

Figure 2.4.1

a *mapping* of R^2 onto R^2. If the point Q of the range corresponds to the point P of the domain, then P *is mapped onto* Q, or Q is the image of P. For example, if every point of R^2 is translated 3 units in the Y-direction, the point $(1, 2)$ is mapped onto the point $(1, 5)$; and each point of the line $y = 2x$ is mapped onto a point on the line $y = 2x + 3$. This particular mapping is called a *Y-translation by* 3, according to the following definition.

Definition 2.4.1

Let k be a real number. The translation YT/k (read "Y-translation by k") is a mapping in which each point (x, y) of a Cartesian plane is mapped onto the corresponding point $(x, y + k)$. We represent the above statement in symbols as follows:

$$(x, y) \xrightarrow{YT/k} (x, y + k)$$

Q1: Complete the following and illustrate geometrically:

(a) $(2, 1) \xrightarrow{YT/3}$ (——, ——) (b) $(2, 1) \xrightarrow{YT/-3}$ (——, ——)

Notice that a translation YT/k shifts a point or a graph vertically. If $k > 0$, the graph is moved k units upward; if $k < 0$, the graph is moved $|k|$ units downward. Figure 2.4.2 shows the result of a translation YT/k with $k > 0$ applied to the graph of a function $y = f(x)$. Clearly, the Y-coordinate of a point on the top graph is k units greater than the corresponding point below it on the lower graph. Thus, an equation satisfied by the top graph is $y = f(x) + k$. We summarize the above discussion symbolically as follows:

$$y = f(x) \xrightarrow{YT/k} y = f(x) + k \qquad (1)$$

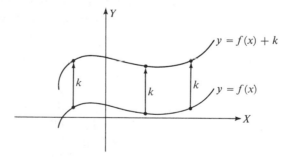

Figure 2.4.2

Q2: (a) Sketch the graph of $y = x^2$. (b) Apply the translation $YT/2$ to the above graph and sketch this result. (c) Write an equation for the new graph.

A1: (a) $(2, 1) \xrightarrow{YT/3} (2, 4)$ (b) $(2, 1) \xrightarrow{YT/-3} (2, -2)$

A2:
(a) See the following figure.   (b) See the following figure.

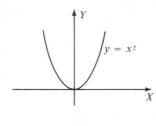

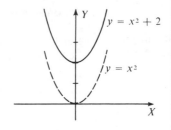

(c) $y = x^2 + 2$

Equation (1) should seem quite straightforward. Nevertheless, it is instructive to rephrase this result so that we can apply it to more complicated equations. Rather than write $y = f(x) \xrightarrow{YT/k} y = f(x) + k$, let us write $y = f(x) \xrightarrow{YT/k} y - k = f(x)$ and say that we have replaced y by $y - k$ everywhere in the original equation to obtain the new equation.

Example 2.4.1

Sketch the result of the translation $YT/2$ applied to the graph of $x^2 + y^2 = 1$, and write an equation satisfied by the new graph.

Solution

The graph of $x^2 + y^2 = 1$ is a circle of radius 1 centered at the origin. The translation $YT/2$ moves the circle 2 units vertically as shown in Figure 2.4.3. An equation for the top circle can be found using the distance formula and the fact that all points on this circle are 1 unit from the point $(0, 2)$. Thus, $(x - 0)^2 + (y - 2)^2 = 1^2$ or $x^2 + (y - 2)^2 = 1$ is an equation for the top circle.

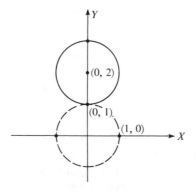

Figure 2.4.3

Notice in the example above that the effect of the translation $YT/2$ is to replace y by $y - 2$ everywhere in the original equation $x^2 + y^2 = 1$, yielding the equation $x^2 + (y - 2)^2 = 1$ for the translated circle. This result is true in general as we now illustrate.

By definition, a translation YT/k maps (x, y) onto $(x, y + k)$ or (x, y') where $y' = y + k$. Now, (x, y) satisfies the original equation (let us call it E), and (x, y') satisfies some new (as yet unknown) equation E' (see Figure 2.4.4). Since (x, y) satisfies E and $y' = y + k$, the point $(x, y' - k)$ satisfies E. That is, if we replace y with $y' - k$ everywhere in the original equation E, then the pair of numbers (x, y') will satisfy the new equation E'. In Example 2.4.1 $(x, y) = (1, 0)$ satisfies the equation $E : x^2 + y^2 = 1$, and $(x, y') = (1, 2)$ satisfies $E' : x^2 + (y - 2)^2 = 1$. We generally drop the symbol $'$ and restate the next to the last sentence above as follows: If we replace y with $y - k$ everywhere in the original equation, we obtain the equation of the translated curve.

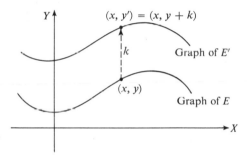

Figure 2.4.4

Q3: Sketch the graph and write an equation for the image of $x^2 + y^2 = 4$ under (a) $YT/1$ and (b) $YT/-2$.

Example 2.4.2

Sketch the graph of $x^2 + y^2 + 4y = 0$.

Solution

We complete the square involving the y terms as follows: $x^2 + y^2 + 4y + 4 = 4$ or $x^2 + (y + 2)^2 = 4$. Notice that this equation can be regarded as the result of replacing y by $y + 2 = y - (-2)$ everywhere in the equation $x^2 + y^2 = 4$. Thus, the desired graph is a $YT/-2$ of the circle $x^2 + y^2 = 4$, as shown in Figure 2.4.5.

We define horizontal shifts in a similar manner.

Definition 2.4.2

The translation XT/h is a mapping in which each point (x, y) of a Cartesian plane is mapped onto the corresponding point $(x + h, y)$. We write

$$(x, y) \xrightarrow{XT/h} (x + h, y)$$

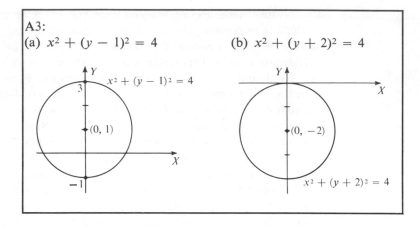

A3:
(a) $x^2 + (y - 1)^2 = 4$ (b) $x^2 + (y + 2)^2 = 4$

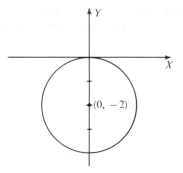

Figure 2.4.5

Q4: Complete the following and illustrate geometrically:

(a) $(2, 1) \xrightarrow{XT/3}$ (____, ____)   (b) $(2, 1) \xrightarrow{XT/-1}$ (____, ____)

A translation XT/h shifts graphs horizontally. If $h > 0$, the graph is translated $|h|$ units to the right; if $h < 0$, the graph is translated $|h|$ units to the left. Figure 2.4.6 shows the result of a translation XT/h (for $h > 0$) applied to the graph of a function $y = f(x)$. Let us find the relationship between the equation $y = f(x)$ of the original graph and the equation of the image (new) graph [let us call it $y = g(x)$]. As shown in Figure 2.4.6 the g function value of the number a must be the same as the f function value of the number $a - h$ (which is h units to the left of a). Thus, for any x in the domain, $g(x) = f(x - h)$. That is, the image of the graph of $y = f(x)$ under a translation XT/h is the graph of $y = f(x - h)$. For example, Figure 2.4.7 shows the graph of $y = f(x) = \sqrt{x}$ as well as the image of that graph under a translation $XT/2$. The equation of the translated curve is $y = f(x - 2) = \sqrt{x - 2}$, obtained by re-

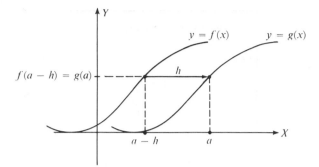

Figure 2.4.6

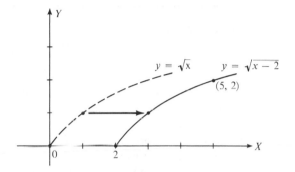

Figure 2.4.7

placing x with $x - 2$ everywhere in the original equation $y = \sqrt{x}$. Notice that $x \geq 0$ for the original graph and $x \geq 2$ for its image.

The discussion above suggests the relationship between the equation of a graph and the equation of the image of that graph under a translation XT/h. If we apply a translation XT/h to the graph of an equation E, we obtain a graph which satisfies a new equation obtained by *replacing x with $x - h$* everywhere in the original equation E. This result is analogous to the result obtained earlier for a translation YT/k. We leave the proof of this fact for the problems.

Q5: (a) Sketch the graph of $y = x^2$. (b) Apply the translation $XT/-2$ to the above graph and sketch the result. (c) Write an equation for this new graph.

Let us review the effects of translations on points and on certain equations. For *points*,

$$(x, y) \xrightarrow{XT/h} (x + h, y)$$
$$(x, y) \xrightarrow{YT/k} (x, y + k)$$

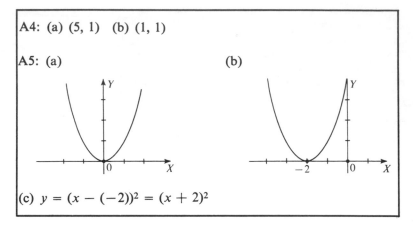

A4: (a) (5, 1)   (b) (1, 1)

A5: (a)   (b)

(c) $y = (x - (-2))^2 = (x + 2)^2$

For *equations*,

$$y = f(x) \xrightarrow{XT/h} y = f(x - h)$$
$$y = f(x) \xrightarrow{YT/k} y - k = f(x)$$

Notice that the *coordinates* of points transform one way (x becomes $x + h$, etc.); yet the *equations* satisfied by these points transform in different ("opposite") ways (replace x by $x - h$, etc.). This result is very important, and, if not observed carefully, can lead to considerable confusion.

As we mentioned at the beginning of this section, the main reason for studying mappings is to help us sketch certain graphs quickly. The previous examples illustrated that we often proceed by manipulating a given equation into a certain form and then ascertaining what mappings are necessary to relate the desired graph to a familiar graph. The following examples illustrate some of the "reverse thinking" useful in these situations.

Example 2.4.3

Sketch the graph of $y = 1/(x - 3)$.

Solution

Notice that $y = 1/(x - 3)$ is obtained from $y = 1/x$ by replacing x with $x - 3$. Thus, we simply apply the translation $XT/3$ to the graph of $y = 1/x$ (sketched in Figure 2.2.5), as shown in Figure 2.4.8.

Example 2.4.4

Sketch the graph of $y = (1/x) - 3$.

Solution

We simply move the graph of $y = 1/x$ down 3 units to obtain the graph of $y = (1/x) - 3$, as shown in Figure 2.4.9. Formally, we have used the translation $YT/-3$; the replacement of y with $y + 3$ yields $y + 3 = 1/x$ or $y = (1/x) - 3$. (In this simple case you probably do not need this formality, as your intuition should guide you correctly.)

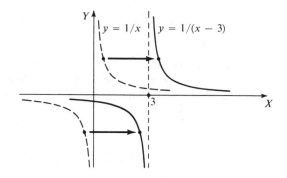

Figure 2.4.8

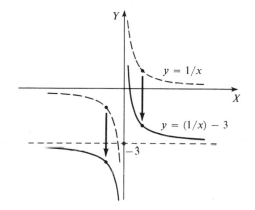

Figure 2.4.9

**Example
2.4.5**

Sketch the graph of $y = x^2 + 2x + 1$.

Solution

Since $y = x^2 + 2x + 1 = (x + 1)^2 = (x - (-1))^2$, the translation $XT/-1$ applied to the graph of $y = x^2$ will produce the graph of $y = (x - (-1))^2 = x^2 + 2x + 1$ as shown in Figure 2.4.10.

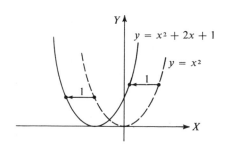

Figure 2.4.10

Example   Sketch the graph of $y = x^2 + 2x - 3$.
2.4.6

Solution   Completing the square yields $y = x^2 + 2x + 1 - 1 - 3 = (x + 1)^2 - 4$ or $y + 4 = (x + 1)^2$. We can again start with the graph of $y = x^2$ and translate it as follows:

$$y = x^2 \xrightarrow{XT/-1} y = (x + 1)^2 \xrightarrow{YT/-4} y + 4 = (x + 1)^2$$
$$\text{or } y = x^2 + 2x - 3$$

(See Figure 2.4.11.) Notice that $y = (x + 3)(x - 1)$, so $y = 0$ when $x \in \{1, 3\}$. Also, $x = 0$ implies that $y = -3$.

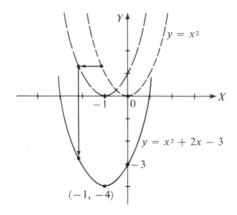

Figure 2.4.11

Example   Sketch the graph of $x^2 + y^2 - 8x + 2y + 8 = 0$.
2.4.7

Solution   Completing the square of both x and y yields $0 = (x^2 - 8x) + (y^2 + 2y) + 8 = (x^2 - 8x + 16) - 16 + (y^2 + 2y + 1) - 1 + 8 = 0$ or $(x - 4)^2 + (y + 1)^2 = 9$. This is a circle with radius 3 and center $(4, -1)$, since $x^2 + y^2 = 9 \xrightarrow{XT/4} (x - 4)^2 + y^2 = 9 \xrightarrow{YT/-1} (x - 4)^2 + (y + 1)^2 = 9$ (see Figure 2.4.12).

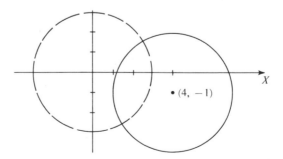

Figure 2.4.12

PROBLEMS 2.4

1. Complete each of the following:

(a) $(4, -2) \xrightarrow{XT/3}$?

(b) $(4, -2) \xrightarrow{YT/-2}$?

(c) $(4, -2) \xrightarrow{?} (3, -2)$

(d) $y = x^3 \xrightarrow{?} y = (x - 1)^3$

(e) $y = x^3 \xrightarrow{YT/2}$?

(f) $y = x^3 \xrightarrow{?} y = x^3 - 1$

2. Sketch each of the following sets of graphs:

(a) $y = x^2$, $y = (x - 2)^2$, $y = (x - 2)^2 - 3$

(b) $y = \sqrt{x}$, $y = \sqrt{x} + 1$, $y = \sqrt{x + 1}$

(c) $y = x^3$, $y = (x - 2)^3$, $y = x^3 - 2$

(d) $y = 1/x$, $y = 1/(x + 2)$, $y = (1/x) + 2$

3. Sketch the graph of each of the following:

(a) $y = 1/(x - 2)^2$

(b) $y = x^2 - 6x + 3$

(c) $x^2 + 2y + y^2 - 3$

(d) $x + 2y + y^2 = 3$

4. Sketch the graph of each of the following:

(a) $y = x^2 - 3$

(b) $y \le x^2 - 3$

(c) $xy > 0$

(d) $(x - 2)(y + 1) > 0$

(e) $1 \le x^2 + y^2 < 4$

(f) $x^2 + y^2 < 2x$

---

5. Show that the equation $y = x^2 + bx + c$ (where b and c are constants) can always be put in the form $y = (x - h)^2 + k$, for some constants h and k. Use this fact to show that the graph of $y = x^2 + bx + c$ is the image of translations of the parabola $y = x^2$.

6. Show that the equation $Ax^2 + Ay^2 + Cx + Dy + E = 0$ (A, C, D, and E are constants, $A \ne 0$) can always be put in the form $(x - h)^2 + (y - k)^2 - p$ for some constants h, k, and p). Use this fact to show that the graph of $Ax^2 + Ay^2 + Cx + Dy + E = 0$ is a circle (if $p > 0$), a "degenerate" circle (if $p = 0$), or empty (if $p < 0$).

7. Show that the translation XT/h applied to the graph of an equation E yields a graph which satisfies the equation obtained by replacing x with $x - h$ everywhere in E.

ANSWERS 2.4

1. (a) $(7, -2)$, (b) $(4, -4)$, (c) $XT/-1$, (d) $XT/1$, (e) $y = x^3 + 2$, (f) $YT/-1$

2. (a) See Figure 2.4.13.

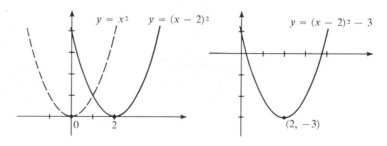

Figure 2.4.13

(b) See Figure 2.4.14.

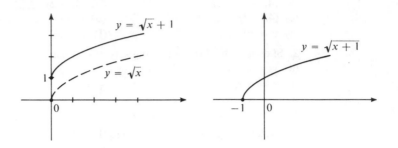

Figure 2.4.14

(c) See Figure 2.4.15.

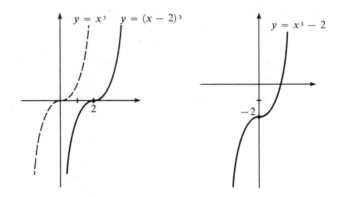

Figure 2.4.15

(d) See Figure 2.4.16.

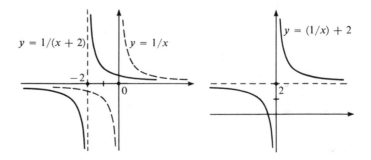

Figure 2.4.16

3. (a) See Figure 2.4.17.   (b) See Figure 2.4.18.

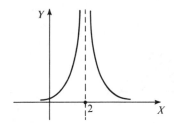

Figure 2.4.17

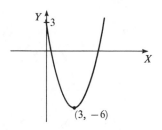

Figure 2.4.18

(c) See Figure 2.4.19.   (d) See Figure 2.4.20.

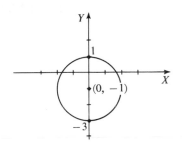

Figure 2.4.19

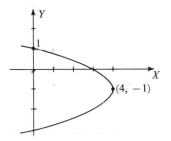

Figure 2.4.20

4. (a) See Figure 2.4.21.   (b) See Figure 2.4.22.

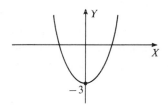

Figure 2.4.21

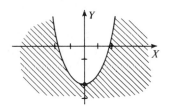

Figure 2.4.22

(c) See Figure 2.4.23.   (d) See Figure 2.4.24.

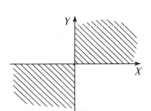

Figure 2.4.23

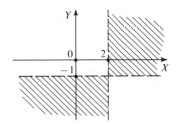

Figure 2.4.24

(e) See Figure 2.4.25.   (f) See Figure 2.4.26.

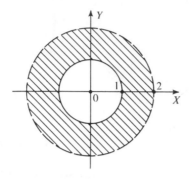

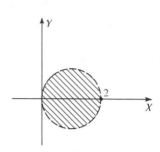

Figure 2.4.25   Figure 2.4.26

Supplement 2.4

I. The Effects of X- and Y-Translations

Problem S1.   Sketch the images of the point $(-1, 1)$ and the curve $y = x^2$ under (a) the translation $YT/3$ and (b) the translation $XT/3$.

Solution: (a) The translation $YT/3$ increases each Y-coordinate by 3, so $(-1, 1) \xrightarrow{YT/3} (-1, 4)$. In addition, the translation $YT/3$ shifts the graph of $y = x^2$ up 3 units. Observe that $y = x^2 \xrightarrow{YT/3} y = x^2 + 3$ (see Figure 2.4.27).

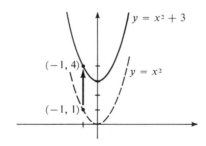

Figure 2.4.27

(b) The translation $XT/3$ increases each X-coordinate by 3, so $(-1, 1) \xrightarrow{XT/3} (2, 1)$. Also, the translation $XT/3$ shifts the graph of $y = x^2$ to the right 3 units. Observe that $y = x^2 \xrightarrow{XT/3} y = (x - 3)^2$ (see Figure 2.4.28).

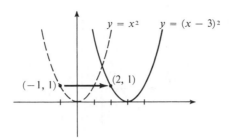

Figure 2.4.28

Problem S2. Sketch the images of the point (4, 2) and the curve $y = \sqrt{x}$ under (a) the translation $YT/-2$ and (b) the translation $XT/-2$.

Solution: (a) $(4, 2) \xrightarrow{YT/-2} (4, 0)$. The translation $YT/-2$ shifts the graph of $y = \sqrt{x}$ *down* 2 units. Observe that $y = \sqrt{x} \xrightarrow{YT/-2} y = \sqrt{x} - 2$ (see Figure 2.4.29).

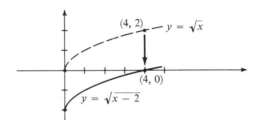

Figure 2.4.29

(b) $(4, 2) \xrightarrow{XT/-2} (2, 2)$. The translation $XT/-2$ shifts the graph of $y = \sqrt{x}$ to the *left* 2 units. Observe that $y = \sqrt{x} \xrightarrow{XT/-2} y = \sqrt{x + 2}$ (see Figure 2.4.30).

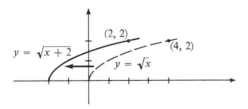

Figure 2.4.30

Problem S3. Sketch the image of the curve $y = x^2$ under the translation $YT/2$ followed by the translation $XT/-3$.

Solution: The translation $YT/2$ shifts the graph of $y = x^2$ *up* 2 units (the dashed curve in Figure 2.4.31). The translation $XT/-3$ is now applied to the graph, shifting it 3 units to the *left* (the solid curve). Observe that

$$y = x^2 \xrightarrow{YT/2} y = 2 + x^2 \xrightarrow{XT/-3} y = 2 + (x + 3)^2$$

What happens in this case if the order of the mappings is reversed?

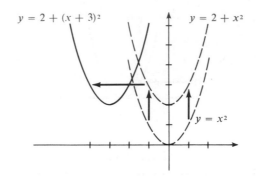

Figure 2.4.31

DRILL PROBLEMS

S4. Sketch the images of the point $(-1, -1)$ and the curve $y = 1/x$ under the translation $XT/2$.

S5. Sketch the images of the point $(1, 1)$ and the curve $y = x^3$ under the translation $YT/-3$.

Answers

S4. See Figure 2.4.32.

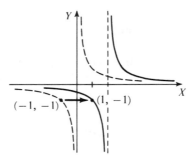

Figure 2.4.32

S5. See Figure 2.4.33.

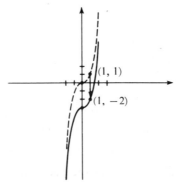

Figure 2.4.33

II. Graphing Using Translations

Problem S6.   Sketch the graph of $y = x^2 + 4$.

Solution: Observe that $y = x^2 \xrightarrow{YT/4} y = x^2 + 4$ (y is replaced by $y - 4$ in the equation $y = x^2$, so $y - 4 = x^2$ or $y = 4 + x^2$). Thus, we obtain the graph of $y = x^2 + 4$ by shifting the graph of $y = x^2$ up 4 units (see Figure 2.4.34). Notice, for example, that $(0, 0) \xrightarrow{YT/4} (0, 4)$.

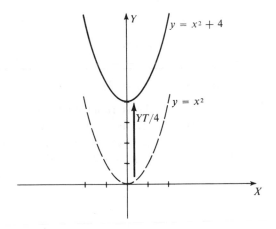

Figure 2.4.34

Problem S7.   Sketch the graph of $y = \sqrt{x + 4}$.

Solution: Observe that $y = \sqrt{x} \xrightarrow{XT/-4} y = \sqrt{x + 4}$. (Here x is replaced by $x - (-4) = x + 4$ in the equation $y = \sqrt{x}$.) Thus, the graph of $y = \sqrt{x + 4}$ can be obtained easily by shifting the graph of $y = \sqrt{x}$ to the *left* 4 units (see Figure 2.4.35). Notice, for example, that $(0, 0) \xrightarrow{XT/-4} (-4, 0)$.

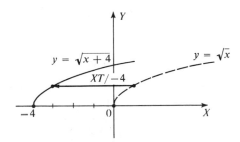

Figure 2.4.35

Problem S8.   Sketch the graph of $y = (x - 3)^3$.

Solution: $y = x^3 \xrightarrow{XT/3} y = (x - 3)^3$ (x is replaced by $x - 3$). Hence, shift the graph of $y = x^3$ to the *right* 3 units, obtaining the graph of $y = (x - 3)^3$. (See Figure 2.4.36.)

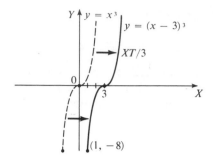

Figure 2.4.36

Problem S9.   Sketch the graph of $y = (x + 2)^2 - 3$.

Solution: $y = x^2 \xrightarrow{XT/-2} y = (x + 2)^2$ [x is replaced by $x - (-2) = x + 2$], and $y = (x + 2)^2 \xrightarrow{YT/-3} y = (x + 2)^2 - 3$ [y is replaced by $y - (-3) = y + 3$, so $y + 3 = (x + 2)^2$ or $y = (x + 2)^2 - 3$]. Thus, first shift the graph of $y = x^2$ to the *left* 2 units $(XT/-2)$, and then shift this graph 3 units downward $(YT/-3)$, as in Figure 2.4.37.

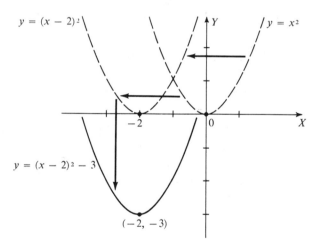

Figure 2.4.37

Problem S10.   Sketch the graph of $y = -1 + \sqrt{x - 3}$.

Solution: $y = \sqrt{x} \xrightarrow{?} y = -1 + \sqrt{x}$
(y is replaced by _____, so
$y + 1 = \sqrt{x}$ or $y = -1 + \sqrt{x}$). Also,
$y = -1 + \sqrt{x} \xrightarrow{?} y = -1 + \sqrt{x - 3}$
(x is replaced by _____). Thus,
to obtain the desired sketch, apply
the translation _____ to the graph of

| $YT/-1$ |
| $y - (-1) = y + 1$ |
| |
| $XT/3$ |
| $x - 3$ |
| |
| $YT/-1$ |

$y = \sqrt{x}$ followed by the translation
_____. That is, shift the graph of
$y = \sqrt{x}$ (*up/down*) 1 unit (*YT*/-1),
and then shift the resulting graph
____ units to the (*right/left*).

XT/3
down

3, right. See Figure 2.4.38.

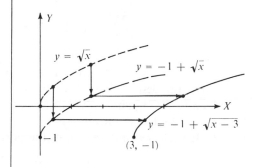

Figure 2.4.38

Problem S11. Using mappings, sketch the graph of $y = x^2 + 4x + 3$.

Solution: Here we need to recognize that the graph of $y = x^2 + 4x + 3$ is closely related to the graph of $y - x^2$ (they are both parabolas). First, we write the original equation in the form $y = (x + a)^2 + b$ and then apply appropriate translations. We can accomplish this by completing the square as follows:

$$y = x^2 + 4x + \left(\frac{4}{2}\right)^2 - \left(\frac{4}{2}\right)^2 + 3 = x^2 + 4x + 4 - 4 + 3$$

Hence, $y = (x + 2)^2 - 1$, so $y = x^2 \xrightarrow{XT/-2} y = (x + 2)^2 \xrightarrow{YT/-1} y = (x + 2)^2 - 1$ imply that the desired graph can be obtained by shifting the graph of $y = x^2$ to the *left* 2 units (*XT*/-2), and then shifting the intermediate graph *down* 1 unit (*YT*/-1). As a partial check of the work, observe that the point $(-2, -1)$ is on the graph of $y = x^2 + 4x + 3$ (see Figure 2.4.39).

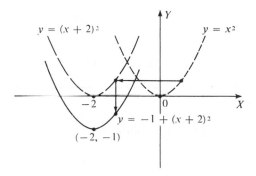

Figure 2.4.39

Problem S12.   Using a sketch of the graph of $g(x) = x^2 - 6x + 7$, solve the inequality $x^2 - 6x + 7 \leq 0$.

Solution: We can obtain a sketch of $y = x^2 - 6x + 7$ by completing the square and applying appropriate mappings to the graph of $y = x^2$, as in Problem S11. Thus,

$$y = x^2 - 6x + \left(\frac{6}{2}\right)^2 - \left(\frac{6}{2}\right)^2 + 7 = x^2 - 6x + 9 - 9 + 7 = (x - 3)^2 - 2$$

Now, $y = x^2 \xrightarrow{XT/3} y = (x - 3)^2 \xrightarrow{YT/-2} y = (x - 3)^2 - 2$ [here x has been replaced by $x - 3$ and y by $y - (-2) = y + 2$, so $y + 2 = (x - 3)^2$ or $y = (x - 3)^2 - 2$] imply that the desired sketch can be found by shifting the graph of $y = x^2$ to the *right* 3 units ($XT/3$), and then shifting the intermediate graph *down* 2 units ($YT/-2$). In Figure 2.4.40, a and b represent the roots of the equation (values of x such that $y = 0$). It is clear from the graph (the portion below the X-axis) that

$$\{x : x^2 - 6x + 7 \leq 0\} = [a, b]$$

To find a and b, let $y = 0$. Since $y = (x - 3)^2 - 2$, $(x - 3)^2 - 2 = 0$ or

$$(x - 3)^2 = 2$$
$$x - 3 = \pm\sqrt{2}$$

Thus, $x = 3 + \sqrt{2}$ or $x = 3 - \sqrt{2}$. Hence, the solution set for $x^2 - 6x + 7 \leq 0$ is $[3 - \sqrt{2}, 3 + \sqrt{2}]$.

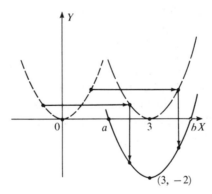

Figure 2.4.40

DRILL PROBLEMS

S13. Sketch the graph of each of the following:
(a) $y = 1/(x + 3)$ (b) $y = (1/x) + 3$
(c) $y = (x - 1)^3 + 2$ (d) $y = x^2 + 2x + 3$
(e) $y = -2 + \sqrt{x - 3}$ (f) $x^2 - 2x + y^2 + 4y - 4 = 0$
S14. Solve (a) $x^2 - 6x + 8 > 0$, (b) $3x - 2x^2 - 1 > 0$

Answers

S13. (a) See Figure 2.4.41.

(b) See Figure 2.4.42.

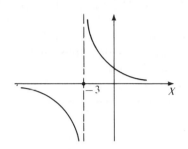

Figure 2.4.41

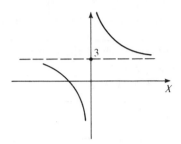

Figure 2.4.42

(c) See Figure 2.4.43.

(d) See Figure 2.4.44.

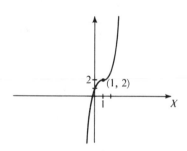

Figure 2.4.43

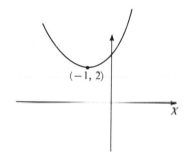

Figure 2.4.44

(e) See Figure 2.4.45.

(f) See Figure 2.4.46.

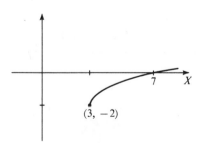

Figure 2.4.45

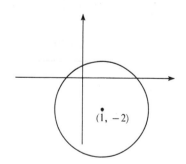

Figure 2.4.46

S14. (a) $(-\infty, 2) \cup (4, \infty)$, (b) $(1/2, 1)$

SUPPLEMENTARY PROBLEMS

S15. Complete each of the following:

(a) $(-2, 3) \xrightarrow{XT/3} (\underline{\quad\quad})$ (b) $(3, -4) \xrightarrow{YT/\frac{1}{2}} (\underline{\quad\quad})$

(c) $x^2 + y^2 = 4 \xrightarrow{\quad ? \quad} (x - 2)^2 + y^2 = 4$

(d) $y = |x| \xrightarrow{\quad ? \quad} y = 1 + |x| \xrightarrow{\quad ? \quad} y = 1 + |x + 2|$.

S16. Sketch each of the following graphs and indicate any mappings used:

(a) $y = \sqrt{x + 3}$ (b) $y = x^2 - 6x + 5$

(c) $y = 3 + |x + 2|$ (d) $y > x^2 + 2$

Answers

S15. (a) $(1, 3)$, (b) $(3, -7/2)$, (c) $XT/2$, (d) $YT/1$, $XT/-2$

S16. (a) $XT/-3$. See Figure 2.4.47.   (b) $XT/3$, $YT/-4$. See Figure 2.4.48.

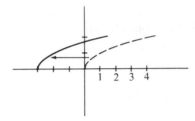

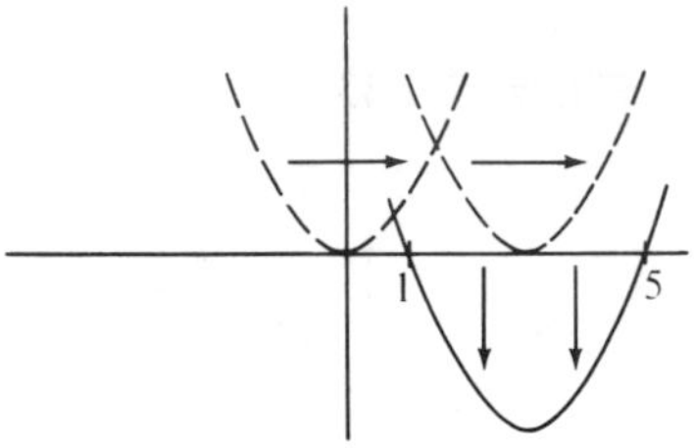

Figure 2.4.47

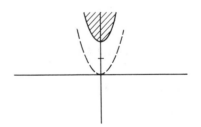

Figure 2.4.48

(c) $XT/-2$, $YT/3$. See Figure 2.4.49.   (d) $YT/2$. See Figure 2.4.50.

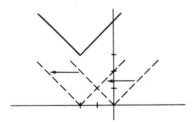

Figure 2.4.49

Figure 2.4.50

2.5 GRAPHING USING MAPPINGS

In the previous section we introduced two kinds of mappings, X- and Y- translations, and learned how to use these mappings to help graph certain equations. Geometrically, these mappings shifted graphs horizontally or vertically without changing the shapes of the graphs. For example, translated circles remained circles. In this section we introduce additional mappings which change graphs in other ways, for example, stretching or shrinking them.

If we transform the circle $x^2 + y^2 = 1$ by doubling the X-coordinate of every point of the circle, yet keeping the Y-coordinate constant, we produce the graph, called an ellipse, shown in Figure 2.5.1. We call this mapping an X-distortion by 2.

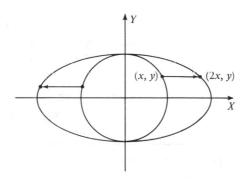

Figure 2.5.1

**Definition
2.5.1**

Let $a > 0$. The distortion XD/a (read "X-distortion by a") is a mapping in which each point (x, y) of a Cartesian plane is mapped onto the corresponding point (ax, y). Similarly, the distortion YD/a maps (x, y) onto (x, ay). We write the above statements symbolically as

$$(x, y) \xrightarrow{XD/a} (ax, y)$$

and

$$(x, y) \xrightarrow{YD/a} (x, ay)$$

respectively.

If $a > 1$, a distortion XD/a stretches a graph horizontally by moving all points of the graph away from the line $x = 0$ (the Y-axis). In the example above, $(1, 0) \xrightarrow{XD/2} (2, 0)$ and $(-1, 0) \xrightarrow{XD/2} (-2, 0)$. If $0 < a < 1$, a distortion XD/a squeezes the graph by horizontally moving all points of the graph closer to the line $x = 0$.

Q1: Sketch the graph of $x^2 + y^2 = 1$ under the distortion $XD/\frac{1}{2}$.

Likewise, a distortion YD/a stretches (if $a > 1$) or squeezes (if $0 < a < 1$) a graph vertically. For example, the distortion $YD/\frac{1}{2}$ applied to the circle $x^2 + (y - 1)^2 = 4$ yields the ellipse shown in Figure 2.5.2. Notice that points on the line $y = 0$ (the X-axis) are invariant (do not change) under a distortion YD/a. Furthermore, since the center of the circle in Figure 2.5.2 is not on the line $y = 0$, it is moved by the distortion $YD/\frac{1}{2}$.

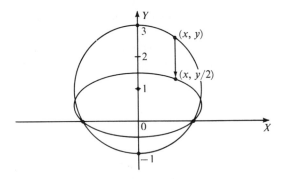

Figure 2.5.2

Q2: Find the coordinates of the centers of the circle and ellipse in Figure 2.5.2.

As is the case with translations, the coordinates of a point transform one way under distortions, while the equations or relations satisfied by the points transform in a different ("inverse") manner. In particular, we have seen that for a translation XT/h, $(x, y) \xrightarrow{XT/h}$ $(x + h, y)$, but $y = f(x) \xrightarrow{XT/h} y = f(x - h)$. That is, to obtain the transformed equation under a translation XT/h, we replace x with $x - h$ (not $x + h$) everywhere in the original equation. The analogous results for distortions are stated as follows.

For *points*:

$$(x, y) \xrightarrow{XD/a} (ax, y)$$
$$(x, y) \xrightarrow{YD/a} (x, ay)$$

For *equations*:

A distortion XD/a replaces x with $\dfrac{x}{a}$.

A distortion YD/a replaces y with $\dfrac{y}{a}$.

For example,

$$x^2 + y^2 = 1 \xrightarrow{XD/2} \left(\frac{x}{2}\right)^2 + y^2 = 1$$

Indeed, notice that the points $(2, 0)$ and $(-2, 0)$ of the ellipse do satisfy this last equation.

Q3: Write the equation of the ellipses shown in Al and in Figure 2.5.2.

The proofs of these "replace" statements are similar to the proofs regarding translations in the previous section. Very briefly, for the distortion XD/a, $(x, y) \xrightarrow{XD/a} (ax, y) = (x', y)$ where $x' = ax$. Thus, $x = \frac{1}{a}x'$, so $(x, y) = \left(\frac{1}{a}x', y\right)$ satisfies the original equation E. Equivalently, (x', y) satisfies the new equation E' obtained by replacing x with $\frac{1}{a}x'$ everywhere in E. Again we drop the prime $(')$ and simply say that the new equation is obtained by replacing x with $\frac{1}{a}x$ everywhere in the original equation.

We can use the distortion mappings together with the translation mappings introduced earlier. However, we must be very careful with the order in which these mappings are applied, as is illustrated in the following example.

A1: See the following figure.

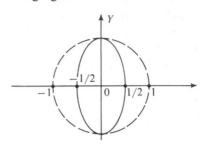

A2: $(0, 1) \xrightarrow{YD/\frac{1}{2}} (0, 1/2)$

A3: In A1, $x^2 + y^2 = 1 \xrightarrow{XD/\frac{1}{2}} (2x)^2 + y^2 = 1$. In Figure 2.5.2, $x^2 + (y - 1)^2 = 4 \xrightarrow{YD/\frac{1}{2}} x^2 + (2y - 1)^2 = 4$.

Example 2.5.1

Sketch the graphs and write the equations of the graphs obtained when the graph of $y = x^2$ is transformed under (a) the translation $YT/-1$ followed by the distortion $YD/2$ and (b) the distortion $YD/2$ followed by the translation $YT/-1$.

Solution

(a) Figure 2.5.3 shows the graph with equations as follows:

$$y = x^2 \xrightarrow{YT/-1} y + 1 = x^2 \xrightarrow{YD/2} \frac{1}{2}y + 1 = x^2 \quad \text{or}$$

$$y = 2(x^2 - 1)$$

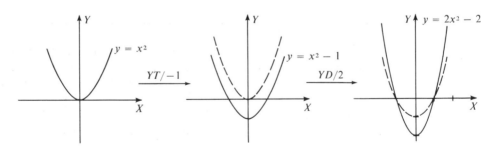

Figure 2.5.3

(b) Figure 2.5.4 shows the graphs with equations as follows:

$$y = x^2 \xrightarrow{YD/2} \tfrac{1}{2}y = x^2 \xrightarrow{YT/-1} (y + 1) = 2x^2 \quad \text{or}$$

$$\text{or} \quad y = 2x^2 \qquad\qquad y = 2x^2 - 1$$

Notice that the results of (a) and (b) are different. That is, these two mappings do not commute.

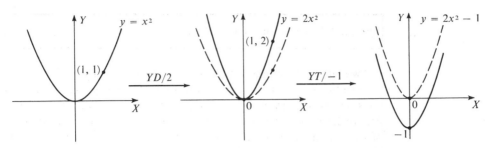

Figure 2.5.4

Since our primary interest in mappings is to enable us to sketch graphs more quickly, we generally shall reverse the procedure of the above example as illustrated in the following example.

Example
2.5.2

Sketch the graph of $y = 2x^2 - 2$.

Solution

Since we are familiar with the graph of $y = x^2$, we can proceed as follows. We observe that the equation $y = 2x^2 - 2$ can be obtained from the equation $y = x^2$ in two steps. First,

$$y = x^2 \xrightarrow{YD/2} \frac{y}{2} = x^2 \quad \text{or} \quad y = 2x^2$$

Next,

$$y = 2x^2 \xrightarrow{YT/-2} y + 2 = 2x^2 \quad \text{or} \quad y = 2x^2 - 2$$

So, in two steps, we have

$$y = x^2 \xrightarrow{YD/2} y = 2x^2 \xrightarrow{YT/-2} y = 2x^2 - 2$$

To find the graph of $y = 2x^2 - 2$, we apply these same mappings to the graph of $y = x^2$ as shown in Figure 2.5.5. Notice that we have used mappings different from those in Example 2.5.1(a), yet we still have achieved the same result.

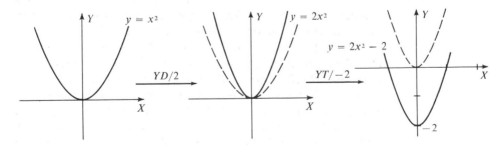

Figure 2.5.5

Q4: Sketch the graph of $y = 3\sqrt{x - 2}$.

Two other mappings which we wish to consider are the X- and Y-reflections. Let us first illustrate one of these geometrically by considering the graph of $y = \sqrt{x}$, shown as the dotted graph in Figure 2.5.6. If we change the sign of the X-coordinate of each point of the graph, we obtain the solid graph of $y = \sqrt{-x}$ as shown in Figure 2.5.6. Geometrically, we say that these two graphs are mirror images, or reflections, of each other. Let us formally define two mappings which produce such mirror images.

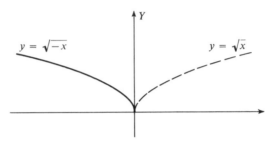

Figure 2.5.6

Definition 2.5.2

The *reflection XR* is a mapping in which each point (x, y) of a Cartesian plane is mapped onto the corresponding point $(-x, y)$. Likewise, the *reflection YR* maps (x, y) onto $(x, -y)$. We write the two statements above symbolically as

$$(x, y) \xrightarrow{XR} (-x, y) \quad \text{and} \quad (x, y) \xrightarrow{YR} (x, -y)$$

respectively.

Q5: Apply the reflection YR to the graph of $y = \sqrt{x}$. Sketch the resulting graph and write an equation for it.

The terminology here can be confusing if you are not careful. By definition, the reflection XR changes the sign of the *X-coordinate* and thus reflects the curve about the line $x = 0$. The confusion arises in that the line $x = 0$ is also called the Y-axis, so the reflection XR reflects about the Y-axis, not the X-axis. Correspondingly, the reflection YR reflects about the line $y = 0$, which is the X-axis.

These reflections have predictable effects on equations of the corresponding graphs. We obtain the equation of the image of the reflection XR by replacing x with $-x$ in the original equation; similarly, for the reflection YR, we replace y with $-y$. The proofs are left for the problems.

**Example
2.5.3**

Sketch the graph of $y = 1 - 2x^2$.

Solution

One way is to proceed as follows:

$$y = x^2 \xrightarrow{YR} y = -x^2 \xrightarrow{YD/2} y = -2x^2 \xrightarrow{YT/1} y = 1 - 2x^2$$

The last two steps are shown in Figure 2.5.7.

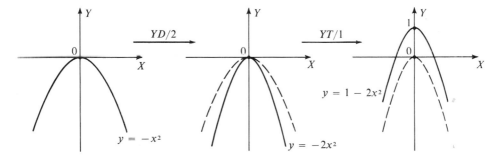

Figure 2.5.7

Sometimes we can save some work in graphing equations if some type of symmetry is involved. For example, Figure 2.5.8 shows the graph of $y = f(x) = x^4 - 2x^2$ for $x \geq 0$. Since $f(-x) = (-x)^4 - 2(-x)^2 = x^4 - 2x^2 = f(x)$, any number and its negative have the same function value. That is, the graph of Figure 2.5.8 together with its image under the reflection XR yields the graph of f as shown in Figure 2.5.9. This graph is said to be symmetric about the Y-axis (or the line $x = 0$) according to Definition 2.5.3.

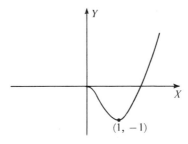

Figure 2.5.8

**Definition
2.5.3**

A subset S of R^2 is *symmetric about the Y-axis* if the reflection XR of every point of S is also a point of S, that is, if the reflection XR maps S onto S. If the reflection YR maps S onto itself, then S is *symmetric about the X-axis*.

A4: $y = \sqrt{x} \xrightarrow{YD/3} \dfrac{y}{3} = \sqrt{x} \xrightarrow{XT/2} \dfrac{y}{3} = \sqrt{x-2}$, or $y = 3\sqrt{x-2}$. The graph is shown in the following figure.

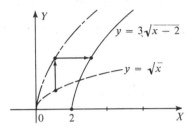

A5: The graph and equation are shown in the following figure.

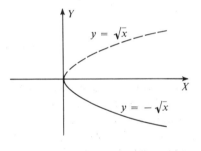

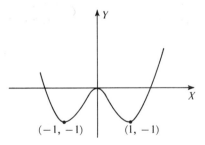

Figure 2.5.9

We can tell from an equation whether its graph is symmetric or not. A point (x_1, y_1) belongs to the graph of the equation $y = f(x)$, precisely when $y_1 = f(x_1)$. The point $(-x_1, y_1)$ belongs to the graph precisely when $y_1 = f(-x_1)$, that is, when $f(x_1) = f(-x_1)$. Thus, the graph of the function f is symmetric about the Y-axis if, and only if, $f(-x) = f(x)$ for each x in the domain of f.

Q6: (a) Which of the following functions have graphs that are symmetric about the Y-axis? (b) Can the graph of a function f be symmetric about the X-axis?

$$h(x) = x^3 \quad g(x) = x^4 \quad F(x) = x^2 + 2 \quad G(x) = (x + 2)^2$$

Most of the graphs in this section have been parabolas, that is, graphs obtained by applying mappings to the graph of $y = x^2$. In the problems you will show that the graph of $y = ax^2 + bx + c$ can always be obtained in such a manner. However, we did not introduce these mappings simply to graph parabolas. In fact, it is often easier to graph parabolas using other techniques. Rather, the mappings and techniques introduced are quite general and apply to any subset of R^2. They will be especially useful in dealing with the transcendental functions of the next two chapters.

PROBLEMS 2.5

1. Complete each of the following:
   (a) $(2, -3) \xrightarrow{XD/2} (\underline{\quad}, \underline{\quad})$
   (b) $(2, -3) \xrightarrow{XR} (\underline{\quad}, \underline{\quad})$
   (c) $(2, -3) \xrightarrow{YD/4} (\underline{\quad}, \underline{\quad})$
   (d) $(2, -3) \xrightarrow{YR} (\underline{\quad}, \underline{\quad})$
   (e) $y = x^2 \xrightarrow{YD/2} y = \underline{\quad}$
   (f) $y = x^2 \xrightarrow{XT/2} y = \underline{\quad}$
   (g) $x^2 + y^2 = 1 \xrightarrow{XD/3} \underline{\quad} = 1$
   (h) $x^2 - 2x + 3y = 4 \xrightarrow{XR} \underline{\quad} = 4$
   (i) $y = 2x^2 + 1 \xrightarrow{YD/4} y = \underline{\quad}$
   (j) $x^2 - 2x + 3y = 4 \xrightarrow{YR} \underline{\quad} = 4$

2. Use mappings to sketch the following sets of graphs:
   (a) $y = x^3, y = 2x^3, y = 2x^3 - 1$
   (b) $y = \sqrt{x}, y = 3\sqrt{x}, y = \sqrt{3x}$
   (c) $y = 1/x^2, y = 1/(x - 1)^2, y = 4/(x - 1)^2$
   (d) $y = x^2, y = 1 - x^2, y = |1 - x^2|$
   (e) $y = \sqrt{x}, y = \sqrt{x - 1}, y = \sqrt{2x - 1}$
   (f) $x^2 + y^2 = 1, x^2 + 4y^2 = 1, x^2 + 4y^2 \leq 1$

3. Sketch the graph of the image of $y = 1/x$ under (a) the distortion $XD/2$ and (b) the distortion $YD/2$.

4. Sketch the graph of each of the following:
   (a) $9x^2 + 25y^2 = 225$ (b) $x^2 + 2y^2 + 6x + 7 = 0$
   (c) $x^2 + 2y + 6x + 7 = 0$ (d) $(x^2 + 2x + 1)y = 1$
   (e) $y^2 + 2x + 6y + 7 = 0$ (f) $x^2 + 2y^2 + 6x + 4y + 12 = 0$

5. Show that the graph of $y = ax^2 + bx + c$ ($a \neq 0$) is always the image of the parabola $y = x^2$ under some set of mappings. (Hint: See Problem 5 in Section 2.4. Consider two cases: $a > 0$ and $a < 0$.) What are the coordinates of the vertex of the parabola, that is, the image of $(0, 0)$?

A6: (a) $g(x)$, $F(x)$. (b) Yes, but only if $f(x) = 0$ for every x in the domain of f.

6. Show that the graph of $Ax^2 + By^2 + Cx + Dy + E = 0$ is either the image of the circle $x^2 + y^2 = 1$ under some set of mappings or a point or empty. Assume that $A, B > 0$. (Hint: See Problem 6 of Section 2.4.) What are the coordinates of the center of the resulting ellipse?

7. Given that the image of any parabola under the mappings introduced thus far is a parabola, show that the graph of $Ax^2 + By^2 + Cx + Dy + E = 0$ is always a parabola *provided that* either A or B, but not both, equals zero.

8. Show that the distortion YD/a applied to the graph of an equation E yields a graph which satisfies the equation obtained by replacing y with y/a everywhere in E.

9. Show that the reflection XR applied to the graph of an equation E yields a graph which satisfies the equation obtained by replacing x with $-x$ everywhere in E.

ANSWERS
2.5

1. (a) $(4, -3)$, (b) $(-2, -3)$, (c) $(2, -3/2)$, (d) $(2, 3)$, (e) $y = 2x^2$, (f) $y = (x - 2)^2$, (g) $(x/3)^2 + y^2 = 1$, (h) $x^2 + 2x + 3y = 4$, (i) $y = x^2 + (1/2)$, (j) $x^2 - 2x - 3y = 4$

2. (a) See Figure 2.5.10.

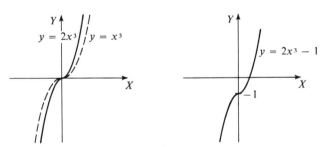

Figure 2.5.10

(b) See Figure 2.5.11.

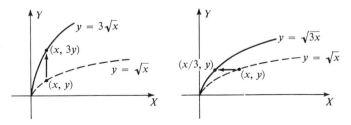

Figure 2.5.11

(c) See Figure 2.5.12.

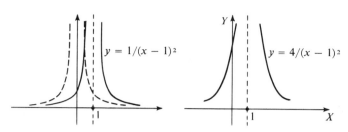

Figure 2.5.12

(d) See Figure 2.5.13.

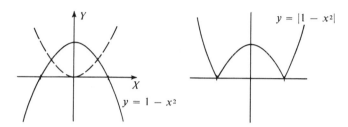

Figure 2.5.13

(e) See Figure 2.5.14.

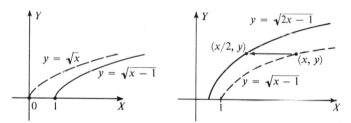

Figure 2.5.14

(f) See Figure 2.5.15.

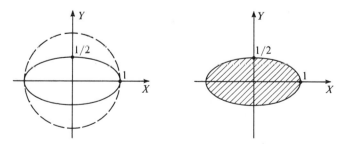

Figure 2.5.15

3. (a) See Figure 2.5.16.

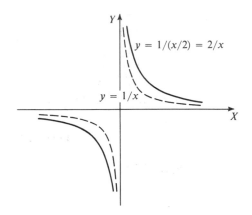

Figure 2.5.16

(b) See Figure 2.5.17.

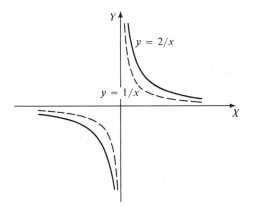

Figure 2.5.17

4. (a) See Figure 2.5.18.   (b) See Figure 2.5.19.

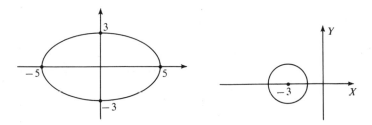

Figure 2.5.18   **Figure 2.5.19**

(c) See Figure 2.5.20.

(d) See Figure 2.5.21.

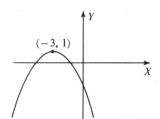

Figure 2.5.20

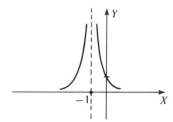

Figure 2.5.21

(e) See Figure 2.5.22.

(f) See Figure 2.5.23.

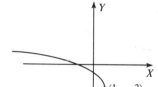

Figure 2.5.22

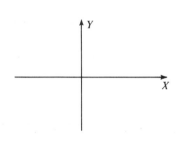

Figure 2.5.23

Supplement 2.5

I. The Effects of X- and Y-Distortions and Reflections

Problem S1. Sketch the image of the point $(-1, 1)$ and the curve $y = x^2$ under (a) the distortion $YD/3$, (b) the distortion $XD/3$, (c) the reflection XR, and (d) the reflection YR.

Solution: (a) The distortion $YD/3$ multiplies each Y-coordinate by 3, so $(-1, 1) \xrightarrow{YD/3}$ $(-1, 3)$. Thus, the distortion $YD/3$ *stretches* the graph of $y = x^2$ in the Y-direction by a factor of 3. Observe that $y = x^2 \xrightarrow{YD/3} y = 3x^2$ and that the point $(0, 0)$ is left unchanged by the $YD/3$ (see Figure 2.5.24). Would this be the case under the translation $YT/3$?

(b) The distortion $XD/3$ multiplies each X-coordinate by 3, so $(-1, 1) \xrightarrow{XD/3} (-3, 1)$. Thus, the distortion $XD/3$ *stretches* the graph of $y = x^2$ by a factor of 3 in the X-direction (see Figure 2.5.25). Notice that $(-1, 1)$ satisfies the original equation

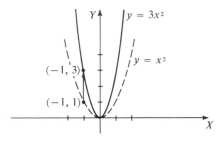

Figure 2.5.24

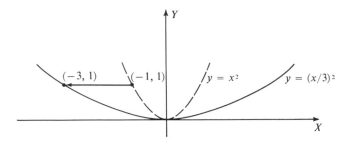

Figure 2.5.25

$y = x^2$ and that $(-3, 1)$ satisfies the equation $y = (x/3)^2$ obtained by replacing x with $x/3$:

$$y = x^2 \xrightarrow{XD/3} y = \left(\frac{x}{3}\right)^2$$

(c) The reflection XR changes the sign of each X-coordinate. Thus, $(-1, 1) \xrightarrow{XR} (1, 1)$, and the reflection XR simply reflects the graph of $y = x^2$ about the Y-axis (the line $x = 0$). Observe that the image graph is *identical* to the original graph (see Figure 2.5.26). That is, the graph of $y = x^2$ is mapped onto itself by the reflection XR. Thus, the graph of $y = x^2$ is symmetric about the Y-axis.

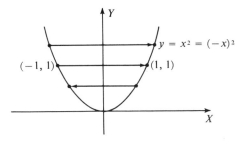

Figure 2.5.26

(d) The reflection YR changes the sign of each Y-coordinate, so $(-1, 1) \xrightarrow{YR} (-1, -1)$. In addition, the reflection YR simply reflects the graph of $y = x^2$ about the line $y = 0$ (the X-axis). Observe that $y = x^2 \xrightarrow{YR} y = -x^2$. (See Figure 2.5.27.)

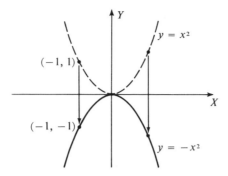

Figure 2.5.27

Problem S2. Sketch the image of the point (4, 2) and the curve $y = \sqrt{x}$ under (a) the distortion $YD/\frac{1}{2}$, (b) the reflection YR, and (c) the distortion $YD/\frac{1}{4}$ followed by the reflection YR.

Solution: (a) The distortion $YD/\frac{1}{2}$ multiplies each Y-coordinate by 1/2, so $(4, 2) \xrightarrow{YD/\frac{1}{2}} (4, 1)$. Thus, the distortion $YD/\frac{1}{2}$ shrinks the graph of $y = \sqrt{x}$ in the Y-direction by a factor of 1/2 (see Figure 2.5.28). Observe that $y = \sqrt{x} \xrightarrow{YD/\frac{1}{2}} y = (1/2)\sqrt{x}$. What effect would the translation $YT/\frac{1}{2}$ have on the graph of $y = \sqrt{x}$?

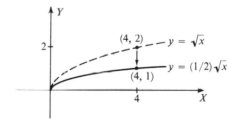

Figure 2.5.28

(b) $(4, 2) \xrightarrow{YR} (4, -2)$. A reflection YR simply reflects the graph of $y = \sqrt{x}$ about the line $y = 0$. Notice that $y = \sqrt{x} \xrightarrow{YR} y = -\sqrt{x}$ (Figure 2.5.29).

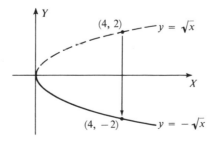

Figure 2.5.29

(c) Using the results of Parts (a) and (b), we obtain $(4, 2) \xrightarrow{YD/\frac{1}{2}} (4, 1) \xrightarrow{YR} (4, -1)$. Thus, we first shrink the graph of $y = \sqrt{x}$ by a factor of $1/2$ in the Y-direction $(YD/\frac{1}{2})$, as shown in Figure 2.5.30, and then reflect this intermediate graph (dashed curve) about the line $y = 0$ (YR). Notice that

$$y = \sqrt{x} \xrightarrow{YD/\frac{1}{2}} y = \frac{1}{2}\sqrt{x} \xrightarrow{YR} y = -\frac{1}{2}\sqrt{x}$$

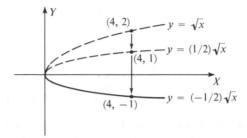

Figure 2.5.30

Problem S3. Sketch the image of $y = f(x)$ under the distortion $XD/\frac{1}{2}$ where $y = f(x)$ is given by the graph in Figure 2.5.31.

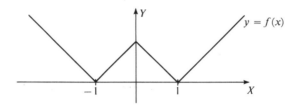

Figure 2.5.31

Solution: Notice that $(-1, 0)$ and $(1, 0) \in f$, so $(-1/2, 0)$ and $(1/2, 0)$ will be points on the image graph because the distortion $XD/\frac{1}{2}$ multiplies each X-coordinate by $1/2$. Thus, the distortion $XD/\frac{1}{2}$ shrinks the graph of $y = f(x)$ by a factor of $1/2$ in the X-direction. Observe that $y = f(x) \xrightarrow{XD/\frac{1}{2}} y = f(2x)$. (See Figure 2.5.32.)

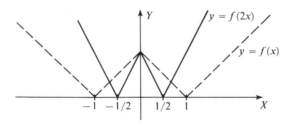

Figure 2.5.32

DRILL PROBLEMS

S4. Sketch the images of the point (1, 1) and the curve $y = x^3$ under the indicated mappings:
(a) $YD/\frac{1}{2}$ (b) XR

S5. Sketch the images of the point $(-1, -1)$ and the curve $y = 1/x$ under the indicated mappings:
(a) $YD/2$ (b) YR

Answers

S4. (a) See Figure 2.5.33. (b) See Figure 2.5.34.

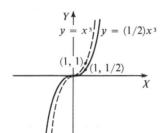

Figure 2.5.33

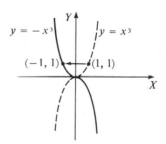

Figure 2.5.34

S5. (a) See Figure 2.5.35. (b) See Figure 2.5.36.

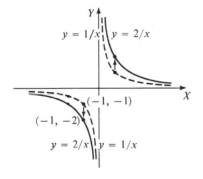

Figure 2.5.35

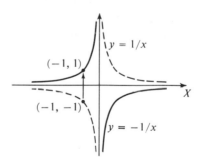

Figure 2.5.36

II. Graphing Using Mappings

Problem S6. Sketch the graph of $y = -(x - 1)^2$.

Solution: Consider the following sequence of mappings:

$$y = x^2 \xrightarrow{XT/1} y = (x - 1)^2 \xrightarrow{YR} y = -(x - 1)^2$$

Here x was replaced by $x - 1$, and then y was replaced by $-y$, yielding $-y = (x - 1)^2$ or $y = -(x - 1)^2$. Thus, the desired sketch is found by shifting the graph of $y = x^2$ to the right 1 unit $(XT/1)$, and then reflecting this intermediate graph about the X-axis (YR). (See Figure 2.5.37).

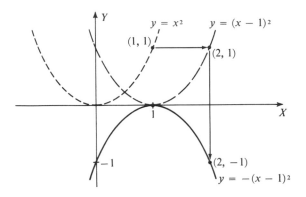

Figure 2.5.37

Problem S7.   Sketch the graph of $y = 3\sqrt{x + 2}$.

Solution: $y = \sqrt{x} \xrightarrow{XT/-2} y = \sqrt{x + 2} \xrightarrow{YD/3} y = 3\sqrt{x + 2}$. [Here x was replaced by $x - (-2) = x + 2$, and y was replaced by $y/3$ in the middle equation. Thus, $y/3 = \sqrt{x + 2}$ or $y = 3\sqrt{x + 2}$.] To sketch $y = 3\sqrt{x + 2}$, apply the translation $XT/-2$ (shift of the graph of $y = \sqrt{x}$ to the left 2 units), and follow that with the distortion $YD/3$ (a stretching 3 units upward of the intermediate graph). See Figure 2.5.38. How do you sketch the graph of $y = 3\sqrt{x} + 2$?

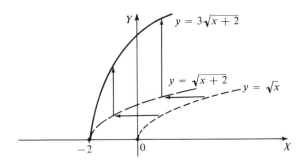

Figure 2.5.38

Problem S8.   Sketch the graph of $y = 3x^2 - 6x + 5$.

Solution: We can write the given equation in the form $y = a(x + b)^2 + c$ by completing the square as follows: $y = 3x^2 - 6x + 5 = 3(x^2 - 2x) + 5 = 3(x^2 - 2x + 1 - 1) + 5 = 3(x - 1)^2 - 3 + 5 = 3(x - 1)^2 + 2$. Thus, $y = x^2 \xrightarrow{XT/1} y = (x - 1)^2 \xrightarrow{YD/3} y = 3(x - 1)^2 \xrightarrow{YT/2} y = 3(x - 1)^2 + 2$. Therefore, to obtain the desired graph, shift the graph of $y = x^2$ to the right 1 unit $(XT/1)$, and then *stretch* it by a factor of 3 in the Y-direction $(YD/3)$. See Figure 2.5.39. Finally, shift this graph 2 units upward $(YT/2)$. As a check of your work, notice that $(0, 0) \rightarrow (1, 2)$ and $(1, 1) \rightarrow (2, 5)$ under this sequence of mappings, and both $(1, 2)$ and $(2, 5)$ satisfy $y = 3x^2 - 6x + 5$.

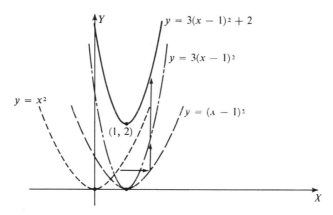

Figure 2.5.39

Problem S9. Sketch the graph of $y = \sqrt{2 - x}$, and label the points where this graph crosses the coordinate axes.

Solution: Notice that $y = \sqrt{2 - x} = \sqrt{-(x - 2)}$. Thus, $y = \sqrt{x} \xrightarrow{XR} y = \sqrt{-x} \xrightarrow{XT/2} y = \sqrt{-(x - 2)}$. (Here x is replaced first by $-x$ and then by $x - 2$.) To sketch $y = \sqrt{2 - x}$, first apply the reflection XR (reflect about the Y-axis) to the graph of $y = \sqrt{x}$, then apply the translation $XT/2$ (shift this intermediate graph 2 units to the right). See Figure 2.5.40. Is the order of the mappings important here? Furthermore when $x = 0$, $y = \sqrt{2}$, and when $y = 0$, $x = 2$.

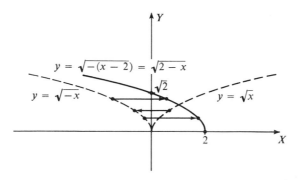

Figure 2.5.40

Problem S10. Sketch the graph of $y = |(x + 1)^2 - 1|$.

Solution: Observe that $y = x^2 \xrightarrow{XT/-1} y = (x + 1)^2 \xrightarrow{YT/-1} y = (x + 1)^2 - 1$. The graph of the last equation is the solid curve in Figure 2.5.41. We find the solution by taking the absolute value of the *Y*-coordinate of each point of this graph. Thus, the portion of the graph *above* the *X*-axis is unchanged (why?), and the portion of the graph below the *X*-axis is reflected about the *X*-axis (why?). The graph of $y = |(x + 1)^2 - 1|$ is shown in Figure 2.5.42.

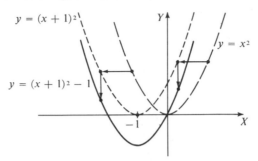

Figure 2.5.41

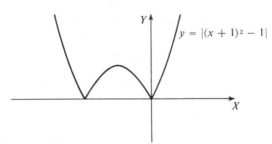

Figure 2.5.42

DRILL PROBLEMS

S11. Sketch the graph of each of the following:
(a) $y = 3/(x - 1)$ (b) $y = 3/(1 - x)$
(c) $y = \sqrt{x - 2} - 1$ (d) $y = 4x^2 + 8x + 9$
(e) $y = 1 - x^3$ (f) $y = |x^2 - 6x + 8|$

Answers

S11. (a) See Figure 2.5.43.   (b) See Figure 2.5.44.

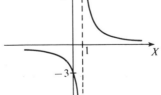

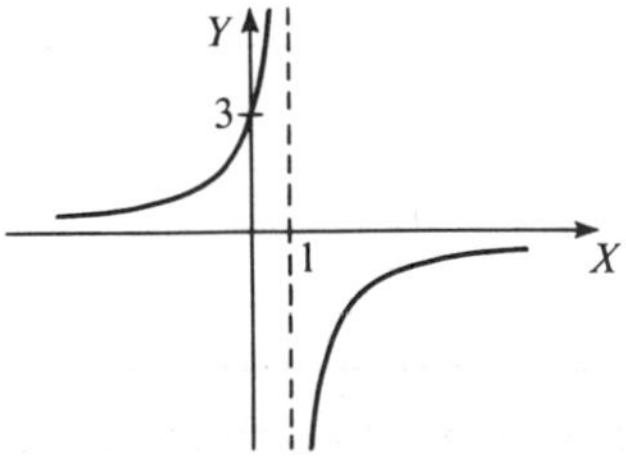

Figure 2.5.43

Figure 2.5.44

(c) See Figure 2.5.45.

(d) See Figure 2.5.46.

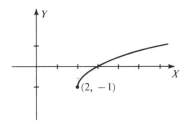

Figure 2.5.45

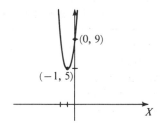

Figure 2.5.46

(e) See Figure 2.5.47.

(f) See Figure 2.5.48.

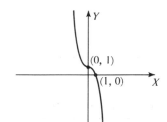

Figure 2.5.47

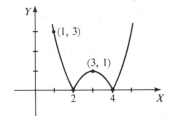

Figure 2.5.48

III. Symmetry

Problem S12. Which of the graphs of the following subsets of R^2 are symmetric about the X-axis: $S = \{(x, y) : y^2 = x^3\}$, $T = \{(x, y) : x^2 = y^3\}$?

Solution: A graph is symmetric about the X-axis (the line $y = 0$) if a Y-reflection maps the graph onto itself. In other words, if a point (x, y) belongs to the graph, then the point $(x, -y)$ must also belong to the graph. The point (x, y) belongs to the graph of S if $y^2 = x^3$; the point $(x, -y)$ belongs to the graph of S if $(-y)^2 = x^3$. Since these two equations are equivalent, S is symmetric about the X-axis. Simply stated, replace y by $-y$ (a YR) in the equation and see if it is *unchanged*. If so, the graph of the equation is *symmetric* about the X-axis. To investigate the symmetry of T, replace y by $-y$ in $x^2 = y^3$, obtaining $x^2 = (-y)^3 = -y^3$. Since these equations are *not* equivalent [for example, (1, 1) satisfies the first but not the second], T is *not* symmetric about the X-axis.

Problem S13. Let $g(x) = x^2 + 2$ and $h(x) = x^2 + 2x$. Are the graphs of g and h symmetric about the Y-axis?

Solution: If a graph is symmetric about the Y-axis and the point (x, y) belongs to the graph, then the point _____ must also belong to the graph.

$(-x, y)$

Equivalently, using function notation, we know that the graph of a function f is symmetric about the Y-axis if, and only if, $f(x) =$ _____.

$f(-x)$

In this problem, $g(x) = x^2 + 2$ and $g(-x) =$ _____. Thus, $g(-x) = g(x)$, and the graph of g (*is/is not*) symmetric about the Y-axis.

$(-x)^2 + 2 = x^2 + 2$

is

   Also, $h(-x) =$ _____.

$(-x)^2 + 2(-x) = x^2 - 2x$

Since $h(-x) \neq h(x)$, the graph of h (*is/is not*) symmetric about the Y-axis.

is not

   Sketch the graphs of each of these functions, observing any symmetry.

DRILL PROBLEMS

S14. Which of the graphs of the following subsets of R^2 are symmetric about (a) the Y-axis, (b) the X-axis: $A = \{(x, y) : x^2 - y^2 = 1\}, B = \{(x, y) : |x - 1| < 3\}$, $C = \{(x, y) : y = x^{2/3}\}$?

Answers

S14. (a) A and C, (b) A and B

SUPPLEMENTARY PROBLEMS

S15. Complete each of the following:
(a) $(-2, 4) \xrightarrow{XD/3} ?$ (b) $(3, -2) \xrightarrow{YD/\frac{1}{3}} ?$
(c) $y = \sqrt{x - 1} \xrightarrow{?} y = \sqrt{-x - 1}$ (d) $y = \sqrt{x} \xrightarrow{?} y = 3\sqrt{x}$
S16. Sketch each of the following graphs and indicate the mappings used:
(a) $y = 3|x|$ (b) $y = 4x^2 - 8x + 1$
(c) $y = 2 - x^2$ (d) $y = \sqrt{1 - x}$
(e) $y = \sqrt{-x - 1}$
S17. Which of the following is (are) symmetric about the (a) Y-axis, (b) X-axis?
(1) $\{(x, y) : y = x^2\}$ (2) $\{(x, y) : y = |x - 2|\}$
(3) $\{(x, y) : y = 3\}$

Answers

S15. (a) $(-6, 4)$, (b) $(3, -2/3)$, (c) XR, (d) $YD/3$
S16. (a) $YD/3$. See Figure 2.5.49.

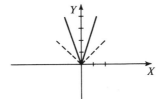

Figure 2.5.49

(b) $XT/1$, $YD/4$, $YT/-3$. See Figure 2.5.50.

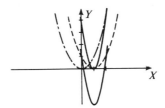

Figure 2.5.50

(c) YR, $YT/2$. See Figure 2.5.51.

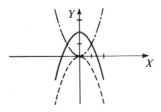

Figure 2.5.51

(d) XR, $XT/1$. See Figure 2.5.52.

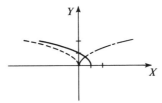

Figure 2.5.52

(e) $XT/1$, XR. See Figure 2.5.53.

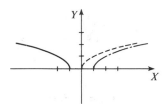

Figure 2.5.53

S17. (a) (1) and (3), (b) none

Achievement Test 2

1. Find $f(-3)$ if $f(x) = |1 + 2x + x^2|$.

2. Find $f(2 + h) - f(2)$ if $f(t) = 2t^2 - t$.

3. Which of the following are functions? (List all.)
   (a) $\{(x, y) : x^2 + y^2 = 1\}$ (b) $\{(x, y) : |x| = 2\}$
   (c) $\{(x, y) : y = 4\}$

4. Find the understood domain of $y = \dfrac{x + 2}{x(x^2 + 5x + 4)}$.

5. Find $f([-2, 1])$ if $f(x) = 2 - |x|$.

6. Find the zeros of $f(x) = 2x^2 + 9x - 5$.

7. Find $f(g(x))$ if $g(x) = 2x + 1$ and $f(x) = x^2 - 2x + 1$.

8. Find the range of f if $f(x) = x + 2$ for $x \leq 3$.

9. Sketch the graph of $\{(x, y) : 0 < x < 2 \text{ or } -1 < y < 1\}$.

10. Sketch the graph of $\{(x, y) : |x| \leq 2\}$.

11. Sketch the graph of $\{(x, y) : 2x + 1 \leq y\}$.

12. Sketch the graph of the image of $y = |x|$ under the translation $XT/2$ followed by the translation $YT/-1$.

13. Sketch the graph of $y = 7 - (x - 3)^2$.

14. Find the image of (a, b) under the distortion $XD/\frac{1}{2}$ followed by the reflection YR.

15. Find a formula for the image of $y = x^2$ under the translation $XT/1$ followed by the reflection YR.

16. Sketch the graph of $y = |x + 3| - 2$.

17. Sketch the graph of $y = -\sqrt{x+1}$.

18. Sketch the graph of $y = 3 - x^3$.

19. Sketch the graph of $y = 1/(x+1)$.

20. Find the zeros of $f(x) = x^4 - x^3 - 5x^2 - x - 6$, and write $f(x)$ in factored form.

21. Let g be symmetric about the Y-axis. Give *another* point on the graph of g if $(2, 5)$ is on the graph of g.

22. Find the solution set for $x^2 + 3x + 5 = 0$.

23. Find an equation for the line with X-intercept 3 and slope $2/5$.

24. Find the center of the circle given by $x^2 + 6x + y^2 = 15$.

25. Find the slope of the line containing $(1/2, 2)$ and $(-3, 3/2)$.

26. Find $\{x : 2|x + 1| + 3 < 0\}$.

27. Find an equation of the line which contains $(2, -4)$ and which is perpendicular to the line $2y - x + 3 = 0$.

28. Find the distance between $(1/2, -2/3)$ and $(-3/2, 7/3)$.

✗**29.** Find a formula for the linear function f if $f(2) = 0$ and $2f(-1) = 5$.

30. Find x such that $(-1, 3)$, $(x, 4)$, and $(2, -1)$ are collinear.

Answers to Achievement Test 2

1. 4

2. $2h^2 + 7h$

3. (c)

4. $\{x \in R : x \neq -4, -1, 0\}$

5. $[0, 2]$

6. $-5, 1/2$

7. $f(g(x)) = 4x^2$

8. $[-\infty, 5]$

9. See Figure 2.A.1.

10. See Figure 2.A.2.

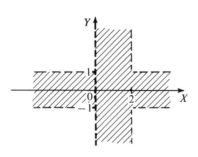

Figure 2.A.1

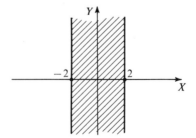

Figure 2.A.2

11. See Figure 2.A.3.

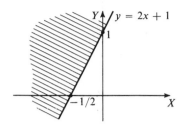

Figure 2.A.3

12. See Figure 2.A.4.

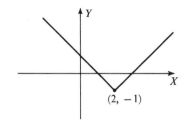

Figure 2.A.4

13. See Figure 2.A.5.

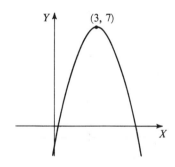

Figure 2.A.5

14. $(a/2, -b)$

15. $y = -(x - 1)^2$ or $y = -x^2 + 2x - 1$

16. See Figure 2.A.6.

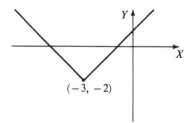

Figure 2.A.6

17. See Figure 2.A.7.

18. See Figure 2.A.8.

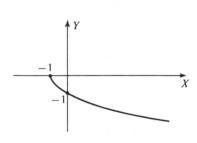

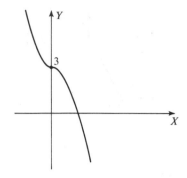

Figure 2.A.7

Figure 2.A.8.

19. See Figure 2.A.9.

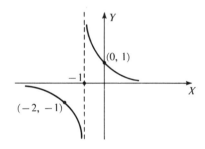

Figure 2.A.9

20. $\{-2, 3\}$; $f(x) = (x + 2)(x - 3)(x^2 + 1)$

21. $(-2, 5)$

22. $\emptyset$

23. $5y - 2x + 6 = 0$

24. $(-3, 0)$

25. $1/7$

26. $\emptyset$

27. $y + 2x = 0$

28. $\sqrt{13}$

29. $f(x) = (10 - 5x)/6$

30. $-7/4$

3 INVERSE FUNCTIONS, LOGARITHMIC AND EXPONENTIAL FUNCTIONS

3.0 Introduction

3.1 Inverse Functions
Supplement

3.2 Exponential Functions
Supplement

3.3 Logarithmic Functions
Supplement

3.4 Computations Using Log_{10}
Supplement

3.5 Applications
Supplement

3.0 INTRODUCTION

In Chapter 2, we introduced the concept of function. Most of our examples dealt with functions whose domains and ranges were subsets of R. In fact, the rules of correspondence for these functions were generally algebraic, that is, defined by equations which involved algebraic operations such as addition, subtraction, multiplication, division, taking powers, and extracting roots. In particular,

$$f(x) = x^2 - 4x - 5$$

$$g(x) = \frac{1}{\sqrt{x^2 + 1}}$$

$$h(x) = \frac{5x^{3/2} - 14}{4x^{23} - \sqrt[3]{x}}$$

are all examples of algebraic functions.

In this chapter we begin the study of *transcendental functions*, functions which cannot be expressed by a finite number of algebraic operations. Some, such as those we shall study here, have fairly straightforward definitions based on certain intuitive or geometrical properties. From these definitions we shall deduce some very important properties which have proved particularly useful in many applications of mathematics. In this chapter we will reinforce many of the concepts of the previous chapter, as well as establish a basic understanding of some of these functions.

3.1 INVERSE FUNCTIONS

In Section 2.1 we introduced the concept of a function. Using an input from the domain, a function produces a single output (an element of its range). The function itself is a set of ordered pairs (see Definition 2.2.2). For example, Figure 3.1.1 is an arrow diagram of a function f with domain $D = \{1, 2, 3\}$ and range $R = \{3, 5, 7\}$. The function is the set

$$f = \{(1, 3), (2, 5), (3, 7)\} \tag{1}$$

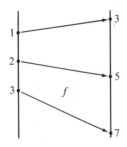

Figure 3.1.1

The correspondences indicated in Figure 3.1.1 can be reversed by changing the direction of the arrows to obtain the diagram shown in Figure 3.1.2. This figure is an arrow diagram of another function,

210

say, *g*. The domain of *g* is the range of *f*, and the range of *g* is the domain of *f*. The function *g* is the set

$$g = \{(3, 1), (5, 2), (7, 3)\} \tag{2}$$

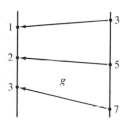

Figure 3.1.2

The functions *f* and *g* are different functions, but they are related in a special way. If an input *x* to *f* yields an output *y*, then using this number *y* as an input to *g* yields the number *x* as an output. For example, from the fact that $f(1) = 3$ we can conclude that $g(3) = 1$. The function *g* is called the *inverse of f*, usually denoted by f^{-1}. Thus, $f^{-1} = \{(3, 1), (5, 2), (7, 3)\}$ and, for example, $f^{-1}(3) = 1$. Notice that $f(f^{-1}(3)) = f(1) = 3$ and that $f^{-1}(f(3)) = f^{-1}(7) = 3$. In fact, $f(f^{-1}(y)) = y$ and $f^{-1}(f(x)) = x$ for each *y* in the domain of f^{-1} and each *x* in the domain of *f*.

> Q1: Verify the preceding statement.

Not all functions have inverse functions. For example, consider the function *h* illustrated in Figure 3.1.3. It does not have an inverse function because reversing all the arrows (see Figure 3.1.4) defines a correspondence in which two numbers, 1 and −1, correspond to 1. Reversing the correspondence defined by a function *f* yields another function precisely when the output of *f* uniquely determines an input. In other words, f^{-1} exists if, and only if, each output of *f* corresponds to exactly one input. In symbols, f^{-1} exists if $f(x_1) =$

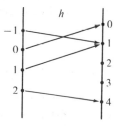

Figure 3.1.3

A1: Clearly, $f^{-1}(f(2)) = f^{-1}(5) = 2$, etc.

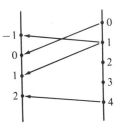

Figure 3.1.4

$f(x_2)$ implies $x_1 = x_2$. We have previously called such a function a one-to-one function (see Definition 2.2.3).

In summary, if f is a one-to-one function, then f has an inverse function f^{-1}. The domain of f^{-1} is the range of f, and the range of f^{-1} is the domain of f.

Q2: Does the function $f(x) = x^2$ have an inverse?

A concise definition of f^{-1} can be given using the idea of ordered pairs. For example, if $f = \{(1, 3), (2, 5), (3, 7)\}$, then $f^{-1} = \{(3, 1), (5, 2), (7, 3)\}$, obtained by interchanging the numbers in each pair.

**Definition
3.1.1**

If f is a one-to-one function, the inverse function is the set

$$f^{-1} = \{(q, p) : (p, q) \in f\}$$

The symbols p and q in Definition 3.1.1 represent elements in the domain and range of f. In most cases that we have considered, p and q are numbers, but recall that the mappings discussed in the preceding two sections are functions whose domains and ranges are R^2. For example, the translation $YT/2$ takes an input point $P = (x, y)$ and produces an output point $Q = (x, y + 2)$. If we denote this mapping by F, we write $F(P) = Q$. Thus, p may be a number, a point in R^2, etc. For example, if F is the translation $YT/2$, then

$$F = \{(P, Q) : P = (x, y) \quad \text{and} \quad Q = (x, y + 2)\}$$

and

$$F^{-1} = \{(Q, P) : Q = (x, y + 2) \quad \text{and} \quad P = (x, y)\}$$

Q3: Show that the function F^{-1}, just defined, is the translation YT/k, and find k.

Summarizing the results illustrated earlier, we have the following:

$$f^{-1}(q) = p \quad \text{if, and only if, } f(p) = q \tag{3}$$

$$\left.\begin{array}{l} \text{Domain of } f = \text{Range of } f^{-1} \\ \text{Domain of } f^{-1} = \text{Range of } f \end{array}\right\} \tag{4}$$

$$\left.\begin{array}{l} f^{-1}(f(x)) = x \quad \text{and} \quad f(f^{-1}(y)) = y \\ \text{for each } x \text{ in the domain of } f \text{ and each } y \text{ in the domain of } f^{-1} \end{array}\right\} \tag{5}$$

If we have the graph of $y = f(x)$ where the domain and range are subsets of R, it is easy to obtain the graph of $y = f^{-1}(x)$. According to Definition 3.1.1, if the point (p, q) lies on the graph of $y = f(x)$, then the point (q, p) lies on the graph of $y = f^{-1}(x)$. Thus, we obtain the graph of $y = f^{-1}(x)$ by interchanging the coordinates of each point of the graph of $y = f(x)$. For example, if $f(x) = 2x + 1$, then $(0, 1)$ lies on the graph of f and $(1, 0)$ lies on the graph of f^{-1}. (See Figure 3.1.5). If we interchange the coordinates of each point on the line $y = 2x + 1$, we obtain the line $x = 2y + 1$ or, equivalently, $y = (1/2)x - 1/2$, shown as $y = f^{-1}(x)$ in Figure 3.1.5. Since this interchange of coordinates can be regarded as mapping (p, q) onto (q, p), let us make the following formal definition.

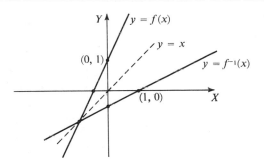

Figure 3.1.5

Definition 3.1.2

The reflection IR is a mapping in which each point (x, y) of a Cartesian plane is mapped onto the corresponding point (y, x). We indicate this symbolically by writing

$$(x, y) \overset{IR}{\longrightarrow} (y, x)$$

It is a direct consequence of Definitions 3.1.1 and 3.1.2 that the graph of $y = f^{-1}(x)$ is the image of the graph of $y = f(x)$ under the reflection IR. Thus, if the graph of f is the solid curve shown in Figure 3.1.6, the graph of f^{-1} is the dashed line curve. We have drawn the line $y = x$ in the figure because the reflection IR is a reflection about this line, as Example 3.1.1 illustrates.

A2: No; for example, if $f(x) = 4$, x could be either 2 or -2.

A3: $F^{-1}(Q) = P = (x, \ (y + 2) - 2)$, so F^{-1} is the translation $YT/-2$.

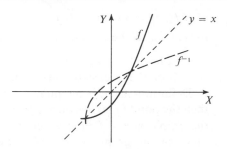

Figure 3.1.6

Example 3.1.1

Suppose that $a \neq b$. Show that the line $y = x$ is the perpendicular bisector of the segment joining $P = (a, b)$ to its image $Q = (b, a)$ under a reflection *IR*. (See Figure 3.1.7.)

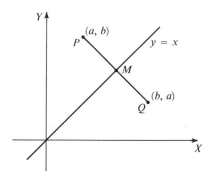

Figure 3.1.7

Solution

The slope of the line PQ is $(b - a)/(a - b) = -1$, so the line $y = x$ with slope 1 is perpendicular to the segment. The equation of the line containing the points P and Q is $y - b = -(x - a)$. This line intersects the line $y = x$ in the point $M = ((a + b)/2, (a + b)/2)$. Using the distance formula, you can verify that

$$\overline{MP} = \sqrt{\frac{(a - b)^2}{4} + \frac{(b - a)^2}{4}} = \overline{MQ}$$

As we have just seen, if we know the graph of $y = f(x)$, it is always easy to obtain the graph of $y = f^{-1}(x)$. However, if we have a formula for f, it may or may not be easy to find an explicit formula for f^{-1}. For example, it was easy to find a formula for f^{-1} using the function $f(x) = 2x + 1$ discussed earlier. The reflection IR applied to $y = f(x) = 2x + 1$ yielded the graph of $x = 2y + 1$, which we easily solved for y, explicitly obtaining $y = f^{-1}(x) = (x - 1)/2$.

Example 3.1.2

Find a formula for f^{-1} if $f(x) = 1/(2x + 3)$.

Solution

Let $y = f(x) = 1/(2x + 3)$. The reflection IR of the graph of $y = f(x) = 1/(2x + 3)$ yields the graph of $y = f^{-1}(x)$ with equation $x = 1/(2y + 3)$. Solving this last equation for y, we obtain

$$2y + 3 = \frac{1}{x} \quad \text{or} \quad 2y = \frac{1}{x} - 3 = \frac{1 - 3x}{x}$$

So $y = f^{-1}(x) = (1 - 3x)/2x$. Notice that the domain of f is $\{x : x \neq -3/2\}$ and the domain of f^{-1} is $\{x : x \neq 0\}$. Thus, according to Equation (4), the range of f must be $\{r : r \neq 0\}$, and the range of f^{-1} is $\{r : r \neq -3/2\}$.

Q4: Let $f(x) - x^3 - 2$. Find a formula for f^{-1}.

We can illustrate Equations (3) and (5) using, for example, the function $f(x) = 2x + 1$ and its inverse $f^{-1}(x) = (x - 1)/2$ found earlier. First, $y = f(x)$ if, and only if, $x = f^{-1}(y)$. Now, $y = f(x) = 2x + 1$, and $x = f^{-1}(y) = (y - 1)/2$. Notice that these two equations are equivalent. We have *not* interchanged x and y here; the same ordered pairs of numbers (x, y) satisfy both equations. That is, both equations describe the graph shown in Figure 3.1.8. Also, notice that

$$f(f^{-1}(y)) = f\left(\frac{y - 1}{2}\right) = 2\left(\frac{y - 1}{2}\right) + 1 = y$$

and

$$f^{-1}(f(x)) = f^{-1}(2x + 1) = \frac{(2x + 1) - 1}{2} = x$$

illustrating Equation (5).

If we apply the reflection IR to $y = f(x) = 2x + 1$, we obtain $x = 2y + 1$ where $y = f^{-1}(x)$. Solving $x = 2y + 1$ for y yields $y = (x - 1)/2 = f^{-1}(x)$. (See Figure 3.1.9.)

A4: The reflection IR applied to the graph of $y = f(x) = x^3 - 2$ yields the graph of $x = y^3 - 2$ which is also the graph of $y = f^{-1}(x)$. Solving for y, we obtain $y = f^{-1}(x) = (x + 2)^{1/3}$.

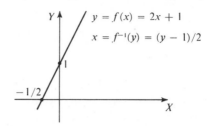

Figure 3.1.8

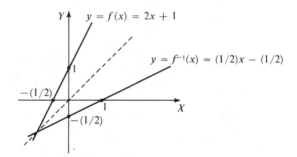

Figure 3.1.9

Example 3.1.3   Let $f(x) = x|x|$. (a) Sketch the graph of f; (b) show that f^{-1} exists; (c) sketch the graph of f^{-1}; (d) find a formula for f^{-1}.

Solution   (a) For $x \geq 0$ the graph of f is the part of the parabola $y = x^2$ to the right of the Y-axis. For $x < 0$, the graph of f is the part of the parabola $y = -x^2$ to the left of the Y-axis. See Figure 3.1.10.

(b) From the sketch of the graph of f we can see that f is one-to-one; so f^{-1} exists. Algebraically, if $0 < x_1 < x_2$, then $0 < x_1^2 < x_2^2$. If $x_1 < x_2 < 0$, then $x_1^2 > x_2^2$ and $-x_1^2 < -x_2^2 < 0$. Thus, in any case, if $x_1 < x_2$, then $f(x_1) < f(x_2)$ and f is one-to-one.

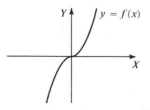

Figure 3.1.10

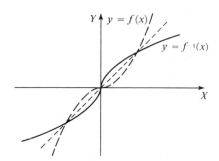

Figure 3.1.11

(c) The reflection IR applied to the graph of f yields the graph of f^{-1} as shown in Figure 3.1.11.

(d) The reflection IR applied to $y = f(x) = x|x|$ yields the graph of $x = y|y|$ as shown in Figure 3.1.11. We can solve this last equation for y by considering the two halves of this graph separately:

> *Case I:* Let $y \geq 0$. Then $|y| = y$, and $x = y|y| = y^2$; so $x \geq 0$. Thus, $y = \sqrt{x}$ (not $-\sqrt{x}$ since $y > 0$).
> *Case II:* Let $y < 0$. Then $|y| = -y$ and $x = y|y| = -y^2$; so $x < 0$. Thus, $y^2 = -x$ and $y = -\sqrt{-x}$ (not $\sqrt{-x}$ since $y < 0$).

Thus,

$$ y = f^{-1}(x) = \begin{cases} \sqrt{x} & \text{if } x \geq 0 \\ -\sqrt{-x} & \text{if } x < 0 \end{cases} $$

PROBLEMS 3.1

1. Which of the following are one-to-one and thus have inverses? Find the inverse function if it exists.

$f = \{(1, 0), (2, 1), (3, 1), (4, 2)\}$ $g = \{(-1, 1), (0, 0), (1, -1)\}$
$F = \{(x, 2x) : x \in N\}$ $G = \{(x, y) : y = |x|\}$

2. Find $f^{-1}(x)$ if $f(x) = 3x - 6$. Sketch the graphs of $y = f(x)$ and $y = f^{-1}(x)$. Verify that $f(f^{-1}(x)) = x$ and $f^{-1}(f(x)) = x$.

3. Find f^{-1} or show that f does not have an inverse for the following:
(a) $f(x) = x^3$ (b) $f(x) = 1/(x - 1)$
(c) $f(x) = 1/(x^2 + 1)$ (d) $f(x) = 1/(x^3 - 2)$
(e) $f(x) = (2x - 3)/(4 + x)$ (f) $f(x) = |2x - 3|$

4. Sketch the graphs of $y = f(x)$ and $y = f^{-1}(x)$, and find a formula for f^{-1} for each of the following:
(a) $f(x) = \sqrt{x - 3}$ (b) $f(x) = x^2 - 4x + 3, x \geq 2$
(c) $f(x) = x^2 - 4x + 3, x \leq 0$

5. Let F be the distortion $YD/2$ and G be the translation $YT/1$. Find the inverse function of F, of G, and of the composition F followed by G.

6. Find f^{-1} if $f(x) = 3(x - 2)|x - 2| + 1$. Sketch the graph of f and f^{-1}.

7. Suppose that functions f and g have inverses and that $h(x) = f(g(x))$. Does h have an inverse? If so, express it in terms of f^{-1} and g^{-1}. Why must we also require that the range of g equal the domain of f?

8. Prove that any linear function $f(x) = mx + b$ with $m \neq 0$ has an inverse.

9. A function is said to be *increasing* if, for each pair of numbers x_1 and x_2 in its domain, $x_1 < x_2$ implies $f(x_1) < f(x_2)$. If $x_1 < x_2$ implies $f(x_1) > f(x_2)$ for any x_1, x_2 in the domain of f, f is said to be *decreasing*.
(a) Give an example of a function which is neither increasing nor decreasing.
(b) Does an increasing function always have an inverse?
(c) Give an example of a function which is neither decreasing nor increasing but which does have an inverse.

ANSWERS
3.1

1. $g^{-1} = \{(1, -1), (0, 0), (-1, 1)\}$, $F^{-1} = \{(x, x/2) : x = 2k, k \in N\}$. f and G do not have inverses.

2. $f^{-1}(x) = 2 + (x/3)$

3. (a) $f^{-1}(x) = x^{1/3}$ (b) $f^{-1}(x) = 1 + (1/x)$
(c) f does not have an inverse.   (d) $f^{-1}(x) = 3\sqrt{(1/x) + 2}$
(e) $f^{-1}(x) = (4x + 3)/(2 - x)$ (f) no inverse

4. (a) $f^{-1}(x) = x^2 + 3$, $x \geq 0$. See Figure 3.1.12.

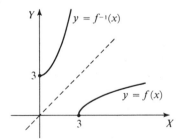

Figure 3.1.12

(b) $f^{-1}(x) = 2 + \sqrt{x + 1}$. See Figure 3.1.13.

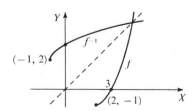

Figure 3.1.13

(c) $f^{-1}(x) = 2 - \sqrt{x + 1}$, $x \geq 3$. See Figure 3.1.14.

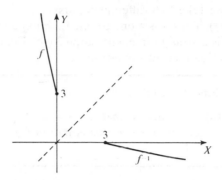

Figure 3.1.14

5. F^{-1} is the distortion $YD/\frac{1}{2}$. G^{-1} is the translation $YT/-1$. The inverse of the composition (F followed by G) is G^{-1} followed by F^{-1}, the translation $YT/-1$ followed by the distortion $YD/\frac{1}{2}$.

Supplement 3.1

Problem S1. Let $f = \{(1, 5), (2, 10), (3, 15)\}$. Does f have an inverse? If so, find f^{-1}, its range, and its domain.

Solution: The function f is a one-to-one function and thus has an inverse, since different inputs always produce different outputs. That is, if $x_1 \neq x_2$, then $f(x_1) \neq f(x_2)$. We obtain the function f^{-1} from f by interchanging the first and second numbers in each pair of f. Thus, $f^{-1} = \{(5, 1), (10, 2), (15, 3)\}$. Notice that the range of f^{-1} is the domain of f, namely, $\{1, 2, 3\}$. The domain of f^{-1} is the range of f, namely, $\{5, 10, 15\}$.

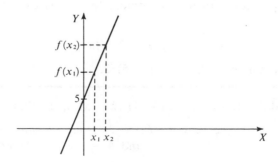

Figure 3.1.15

Problem S2. Does the function $f(x) = 2x + 5$ have an inverse?

Solution: If we take two different inputs $x_1 \neq x_2$ to f, then $2x_1 + 5 \neq 2x_2 + 5$ or $f(x_1) \neq f(x_2)$. Thus, f is a one-to-one function and has an inverse. Alternatively, the graph of f is a straight line with slope 2 (see Figure 3.1.15). Thus, f is clearly an increasing function and thus one-to-one.

Problem S3. Find $f^{-1}(3)$ if $f(x) = 2x + 5$.

Solution: $f^{-1}(3) = x$ means that $f(x) = 3$ [Equation (3)]. Thus, $3 = f(x) = 2x + 5$, and we need only solve this equation for x. Now, $2x = -2$ or $x = -1$, that is, $f^{-1}(3) = -1$.

Problem S4. Find a formula for f^{-1} if $f(x) = 2x + 5$, and sketch the graph of $y = f^{-1}(x)$.

Solution: The graph of $y = f(x) = 2x + 5$ is shown in Figure 3.1.15. The reflection IR yields the equation $x = 2y + 5$ and the graph of $y = f^{-1}(x)$. Solving $x = 2y + 5$ for y yields $2y = x - 5$ or $y = (x - 5)/2 = f^{-1}(x)$. Notice that, for example, $(0, 5) \in f$ and $(5, 0) \in f^{-1}$. (See Figure 3.1.16.)

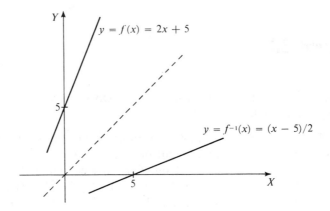

Figure 3.1.16

An alternative procedure to find a formula for f^{-1} is to use Equation (5) as follows:

$$f(f^{-1}(x)) = x = 2f^{-1}(x) + 5$$

Solve for $f^{-1}(x)$, obtaining $f^{-1}(x) = (x - 5)/2$.

Problem S5. Find $f^{-1}(2)$ if $f(x) = (x + 1)/(2 - x)$, given that f^{-1} exists.

Solution: $f^{-1}(2) = x$ if, and only if, $f(x) = 2$. Thus, we can solve $2 = (x + 1)/(2 - x)$ for x. Then $2(2 - x) = x + 1$, and $4 - 1 = x + 2x$ or $x = 1$. That is, $f^{-1}(2) = 1$.

Problem S6. Find a formula for f^{-1} if $f(x) = (x + 1)/(2 - x)$.

Solution: Let $y = f(x) = (x + 1)/(2 - x)$. Then the reflection IR yields $x = (y + 1)/(2 - y)$ where $y = f^{-1}(x)$. Solving $x = (y + 1)/(2 - y)$ for y yields $x(2 - y) = y + 1$ or $2x - 1 = y + xy = (1 + x)y$, so $y = (2x - 1)/(x + 1)$. That is, $y = f^{-1}(x) = (2x - 1)/(x + 1)$.

Alternatively, we can use the equation $f(f^{-1}(x)) = x$ and solve for $f^{-1}(x)$ as follows:

$$f(f^{-1}(x)) = \frac{f^{-1}(x) + 1}{2 - f^{-1}(x)} = x$$

Then we obtain $f^{-1}(x) + 1 = 2x - xf^{-1}(x)$, so $f^{-1}(x) + xf^{-1}(x) = 2x - 1$. Thus, $f^{-1}(x) = (2x - 1)/(x + 1)$, as before.

Problem S7. Find the domain and range of f and of f^{-1} if $f(x) = (x + 1)/(2 - x)$.

Solution: Clearly, the domain of f is $\{x : x \neq 2\}$. From Equation (4) this set is also the range of f^{-1}. Similarly, the domain of f^{-1} is $\{x : x \neq -1\}$ which is automatically the range of f. (The formula for f^{-1} is derived in Problem S6.)

Problem S8. If $f(x) = \sqrt{x - 1}$, find f^{-1}. Sketch the graphs of $y = f(x)$ and $y = f^{-1}(x)$.

Solution: As before, let $y = f(x) = \sqrt{x - 1}$ and perform the reflection IR to obtain $x = \sqrt{y - 1}$ where $y = f^{-1}(x)$. We solve $x = \sqrt{y - 1}$ for y by first squaring both sides of this equation. Thus, $x^2 = y - 1$ or $y = x^2 + 1$. While this formula gives the correct function values of f^{-1}, notice in the preceding equation that $x \geq 0$ since $x = \sqrt{y - 1} \geq 0$. Thus, we must write

$$y = f^{-1}(x) = x^2 + 1, \qquad x \geq 0$$

Furthermore, the graphs are the "half-parabolas" shown in Figure 3.1.17. Notice that each is the reflection IR of the other. Remember, the domain of f^{-1} is the range of f, that is, $[0, \infty]$ in this problem. The range of f^{-1} is the domain of f, that is, $[1, \infty]$ here.

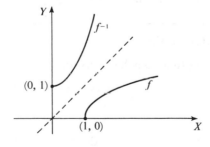

Figure 3.1.17

Problem S9.   Find a formula for f^{-1} if $f(x) = x^3 + 1$.

Solution: Again we let $y = f(x) =$ _____ and apply the reflection *IR* which yields the equation $x =$ _____ where $y =$ _____. We then solve the equation _____ for y, obtaining $y = f^{-1}(x) =$ _____ .	$x^3 + 1$ $y^3 + 1$ $f^{-1}(x)$ $x = y^3 + 1$ $y = (x - 1)^{1/3}$

Problem S10.   Does f have an inverse if $f(x) = 1/(x^2 - 1)$?

Solution: Again let $y = f(x)$ and apply the reflection *IR*, obtaining the equation $x =$ _____ where $y =$ _____. Let us try to solve the equation $x = f(x) = 1/(y^2 - 1)$ for y. Multiplying, we have $x(y^2 - 1) = 1$, or $xy^2 =$ _____. Thus, $y^2 =$ _____. But this permits *two* possible numbers y for each number $x \neq 0$, namely, $y =$ _____ and $y =$ _____. Thus, f (does/does not) have an inverse.	$1/(y^2 - 1)$ $f^{-1}(x)$ $1 + x$ $(1 + x)/x$ $\sqrt{(1 + x)/x}$ $-\sqrt{(1 + x)/x}$ does not

DRILL PROBLEMS

S11. Determine if f has an inverse for each of the following; if it does, find a formula for f^{-1}.

(a) $f(x) = \dfrac{x + 1}{x}$

(b) $f(x) = |x^3|$

(c) $f(x) = 2x^{5/3}$

Answers

S11. (a) $f^{-1}(x) = 1/(x - 1)$, (b) f has no inverse, (c) $f^{-1}(x) = \left(\dfrac{1}{2}x\right)^{3/5}$

SUPPLEMENTARY PROBLEMS

S12. Which of the following functions have inverses:

(a) $f(x) = x^2|x|$

(b) $g(x) = |x^2|x$

(c) $h(x) = \dfrac{|x|}{x}(x^2 + 1)$

S13. Find a formula for f^{-1} or show that f^{-1} does not exist for each of the following:

(a) $f(x) = x - 2x^2$

(b) $f(x) = \dfrac{3 - 2x}{x + 1}$

(c) $f(x) = 1 - 2x^{1/3}$

(d) $f(x) = (x - 3)^3$

(e) $f(x) = x^3 - 3$

(f) $f(x) = \dfrac{4x}{3 + x^2}$

S14. Sketch the graphs $y = f(x)$ and $y = f^{-1}(x)$ for each of the following:

(a) $f(x) = (x - 3)^3$

(b) $f(x) = x^3 - 3$

(c) $f(x) = 1 - 2x^{1/3}$

(d) $f(x) = |x^2|x$

(e) $f(x) = \dfrac{|x|}{x}(x^2 + 1)$

Answers

S12. Only g^{-1} and h^{-1} exist.

S13. (a) $f(1) = f(-1/2) = 1$, so f^{-1} does not exist.

(b) $f^{-1}(x) = \dfrac{3 - x}{x + 2}$

(c) $f^{-1}(x) = \left(\dfrac{1 - x}{2}\right)^3$

(d) $f^{-1}(x) = x^{1/3} + 3$

(e) $f^{-1}(x) = (x + 3)^{1/3}$

(f) $f(1) = f(3) = 1$, so f^{-1} does not exist.

S14. (a) See Figure 3.1.18.

(b) See Figure 3.1.19.

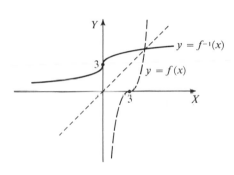

Figure 3.1.18

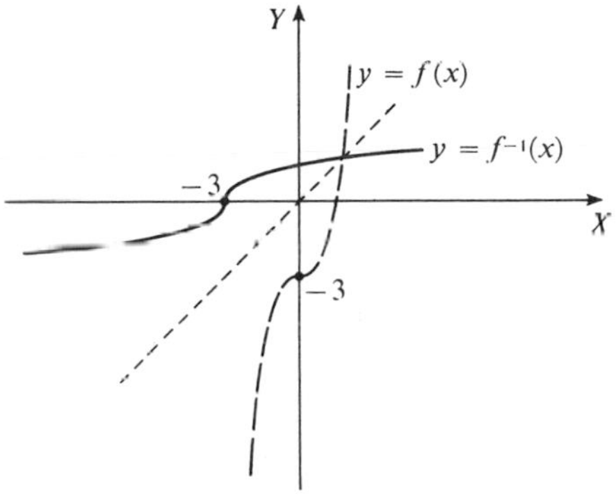

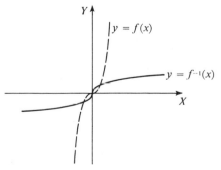

Figure 3.1.19

(c) See Figure 3.1.20.

(d) See Figure 3.1.21.

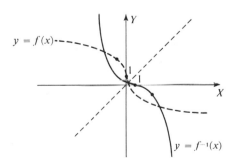

Figure 3.1.20

Figure 3.1.21

(e) See Figure 3.1.22.

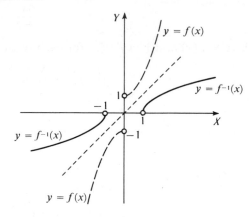

Figure 3.1.22

3.2 EXPONENTIAL FUNCTIONS

The cell of a certain living organism divides into two cells once every hour. If the organism begins with one cell, then at the end of one hour there are two cells. At the end of two hours there are $2 \cdot 2 = 4$ cells, at the end of three hours, $2 \cdot 4 = 8$ cells, etc. If N denotes the number of cells present after x hours, then N is given by the equation $N = 2^x$. (For this physical problem, the equation makes sense only if x is a nonnegative integer. However, the expression 2^x has been defined for any rational number x in Section 1.3.)

Let us sketch the graph of the equation $y = 2^x$. It is easy to plot those points whose X-coordinates are integers. Thus, the graph contains the points $(-3, 1/8)$, $(-2, 1/4)$, $(-1, 1/2)$, $(0, 1)$, $(1, 2)$, etc. On the other hand, it is more difficult to plot points whose X-coordinates are rational numbers but not integers, for then we must compute numbers such as $2^{1/2}$, $2^{2/7}$, $2^{-5/9}$, etc. However, using inequalities, we can show that if r and s are rational numbers and $r < s$, then $2^r < 2^s$ (see Problem 3). Thus, the graph of the equation $y = 2^x$ rises to the right. With this information and some plotted points, we can obtain a sketch of the graph of

$$\{(x, 2^x) : x \in Q\} \qquad \text{(Figure 3.2.1)}$$

Because there are so many rational numbers, it is impossible to tell from Figure 3.2.1 that no point of the graph has an irrational

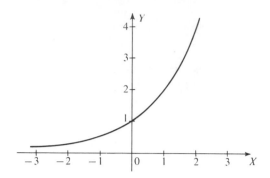

Figure 3.2.1

X-coordinate. It is shown in more advanced mathematics that it is possible to define 2^x for an irrational number x in such a way that the usual properties of exponents hold. In fact, Figure 3.2.1 is a picture of the graph of $\{(x, 2^x) : x \in R\}$. Similarly, b^x can be defined for any positive number b and any real number x in such a way that the usual exponent properties hold. Accepting this fact, we make the following definition.

**Definition
3.2.1**

Let $b > 0$. The exponential function with base b, denoted by $\exp_b$, is the function given by $\exp_b(x) = b^x$. That is, $f(x) = b^x$ represents the exponential function with base b.

Q1: Sketch the graph of the function $\exp_3$, that is, the function $f(x) = 3^x$.

The graph of $y = b^{-x}$ is the reflection XR of the graph of $y = b^x$ (x is replaced by $-x$). Thus, for example, we obtain the graph of $y = (1/2)^x = 2^{-x}$ by applying the reflection XR to the graph of the curve $y = 2^x$ (Figure 3.2.2).

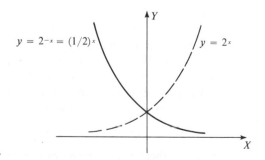

Figure 3.2.2

Q2: Use the result of Q1 to sketch the graph of the function $\exp_{1/3}$.

The general shape of the exponential function $y = b^x$ is shown in Figure 3.2.3 for the case where $b > 1$ and the case where $0 < b < 1$. Notice that in either case the domain of the exponential function is R and the range is the interval $(0, \infty)$.

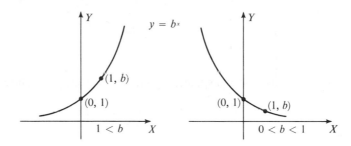

Figure 3.2.3

The usual properties of exponents can be stated as properties of exponential functions. In particular, if $f(x) = b^x$, then

$$f(x + y) = f(x) \cdot f(y) \tag{1}$$

$$[f(x)]^r = f(xr) \tag{2}$$

$$f(x - y) = \frac{f(x)}{f(y)} \tag{3}$$

The proofs of these properties are not difficult. For example, to prove Equation (1), we note that for any x and y

$$f(x + y) = b^{x+y} = b^x b^y$$

by Equation (1) of Section 1.3. Since $f(x) = b^x$ and $f(y) = b^y$, by substitution,

$$f(x + y) = f(x) \cdot f(y).$$

When using the symbol $\exp_b$, we often omit the parentheses and write $\exp_b(x)$ as $\exp_b x$, as in Example 3.2.1.

Example 3.2.1

Evaluate the following: (a) $\exp_3 2$, (b) $\exp_2 3$, (c) $(\exp_3 2)^3$, and (d) $\exp_3 2^3$.

Solution

(a) $\exp_3 2 = 3^2 = 9$ by Definition 3.2.1. (b) $\exp_2 3 = 2^3 = 8$. (c) $[\exp_3 2]^3 = (3^2)^3 = 9^3 = 729$. (d) By convention, $\exp_3 2^3 = \exp_3(2^3) = 3^8 = 6561$.

A1: $y = \exp_3(x) = 3^x$. The graph is shown below.

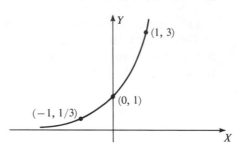

A2: $y = \exp_{1/3}(x) = (1/3)^x$. The graph is shown below.

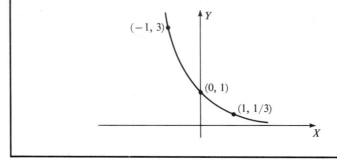

Example
3.2.2

Sketch the graphs of the following equations: (a) $y = 1 + 2^x$, (b) $y = 1 - 2^x$, and (c) $y = 2^{x-1}$.

Solution

(a) The graph of $y = 1 + 2^x$ can be obtained from the graph of $y = 2^x$ by the translation $YT/1$. That is, we move the curve $y = 2^x$ upward one unit (Figure 3.2.4).

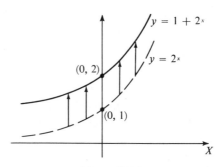

Figure 3.2.4

(b) Since $y = 1 - 2^x = 1 + (-2^x)$, the reflection YR applied to the curve $y = 2^x$ followed by the translation $YT/1$ will yield the graph of $y = 1 - 2^x$ (Figure 3.2.5).

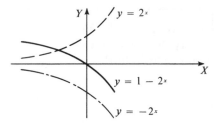

Figure 3.2.5

(c) To obtain $y = 2^{x-1}$ from $y = 2^x$, replace x by $x - 1$. Hence, the curve $y = 2^{x-1}$ can be obtained from $y = 2^x$ by the translation $XT/1$, that is, by moving the graph of $y = 2^x$ one unit to the right (Figure 3.2.6).

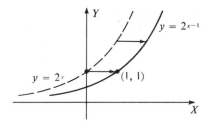

Figure 3.2.6

Alternate Solution to Part (c):

Observe that $y = 2^{x-1} = 2^x 2^{-1} = 2^x(1/2) = (1/2)2^x$. Thus, we can also obtain the graph of $y = 2^{x-1}$ by applying the distortion $YD/\frac{1}{2}$ to the graph of $y = 2^x$. The fact that the translation $XT/1$ and the distortion $YD/\frac{1}{2}$ have the same effect on the graph of the curve $y = 2^x$ is a geometric illustration of Equation (1).

Q3: Use the graph of $y = 2^x$ to sketch the subset of R^2 given by $y < 2^x$ and $y > 0$.

One of the most important properties of exponential functions is that if $b \neq 1$, then $\exp_b$ is a one-to-one function. The fact that $\exp_b(x) = b^x$ is a one-to-one function for $b > 1$ follows from the fact that if $x_1 < x_2$, then $b^{x_1} < b^{x_2}$.

Q4: (a) Explain why $\exp_1$ is not a one-to-one function. (b) Explain why $\exp_b(x) = b^x$ is a one-to-one function if $0 < b < 1$.

From our discussion in Section 3.1 we know that a one-to-one function has an inverse function. Thus, $\exp_b(x) = b^x$, for $b \neq 1$,

A3: All points below the curve $y = 2^x$ and above the X-axis satisfy the inequalities $y < 2^x$ and $y > 0$. Hence, the answer is the region shown in the following figure.

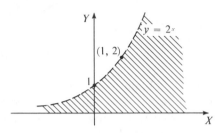

A4: (a) $\exp_1(x) = 1^x = 1$ for every real number x. Its graph is the horizontal line $y = 1$. Clearly, different inputs are associated with the same output (the number 1), so $\exp_1$ is not a one-to-one function. See the figure below.

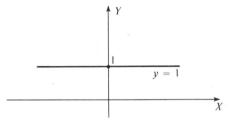

   (b) The fact that $\exp_b(x) = b^x$ is a one-to-one function for $0 < b < 1$ follows from the result that if $x_1 < x_2$, then $b^{x_1} > b^{x_2}$. See the figure below.

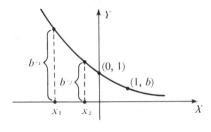

$b > 0$, has an inverse function $\exp_b^{-1}$. The graph of $\exp_b^{-1}$ is the reflection IR of the graph of $\exp_b$. For example, the graph of $\exp_2^{-1}$ is shown in Figure 3.2.7. The inverse function of an exponential function with base b is called a *logarithmic function with base b*. We will study this new function in detail in the next section.

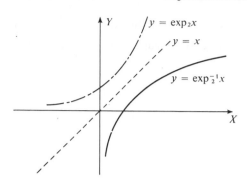

Figure 3.2.7

PROBLEMS
3.2

1. Evaluate each of the following expressions:
(a) $\exp_2 4$ (b) $(\exp_2 3)^2$
(c) $\exp_2 3^2$ (d) $\exp_{1/4} 4$
(e) $\exp_4(1/2)$ (f) $\exp_3(-2)$
(g) $\exp_4(-1/2)$

2. Solve:
(a) $\exp_x 3 = 81$ (b) $\exp_{10} x = 10$
(c) $3^{x^2-1} = 27$

3. Show that for $r < s$, $2^r < 2^s$. (Hint: If $r < s$, then $r + c = s$ for some positive number c. Thus, $2^s = 2^{r+c} = 2^r 2^c$.)

✗ 4. What mapping applied to the graph of $y = b^x$ yields the graph of $y = b^{x+1}$? Of $y = 1 - b^x$? Sketch your results.

5. Sketch the graph of each of the following subsets of R^2:
(a) $y < 3^{x+1}$ (b) $y = 4^x$
(c) $y = 4^{-x}$ (d) $y = 3^{x+2}$
¬(e) $y = -2^{x-1}$ (f) $y = 1 + 3^{x+2}$

6. Give two mappings such that the image of the graph of $y = b^x$ under *either* mapping is the graph of $y = b^{x-1}$. [Hint: See Example 3.2.2(c).]

7. Prove the exponential function properties given by Equations (2) and (3).

ANSWERS
3.2

1. (a) 16, (b) 64, (c) 512, (d) 1/256, (e) 2, (f) 1/9, (g) 1/2

2. (a) $\sqrt[3]{81} = 3\sqrt[3]{3}$, (b) 1, (c) $\{-2, 2\}$

3. $r < s \Rightarrow r + c = s$ for some $c > 0$. Hence, $2^s = 2^{r+c} = 2^r \cdot 2^c > 2^r$ since $2^c > 1$. Thus, $2^r < 2^s$.

4. $XT/-1$, YR followed by $YT/1$, or others.

5. (a) See Figure 3.2.8.   (b) See Figure 3.2.9.

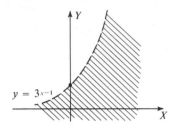

Figure 3.2.8

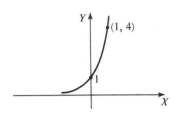

Figure 3.2.9

(c) See Figure 3.2.10.   (d) See Figure 3.2.11.

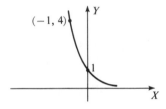

Figure 3.2.10

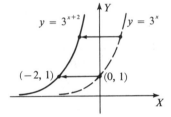

Figure 3.2.11

(e) See Figure 3.2.12.   (f) See Figure 3.2.13.

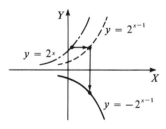

Figure 3.2.12

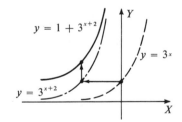

Figure 3.2.13

6. For example, $y = b^x \xrightarrow{XT/1} y = b^{x-1}$ and $y = b^x \xrightarrow{YD/(1/b)} y = (1/b) \cdot b^x = b^{-1} \cdot b^x = b^{x-1}$.

Supplement 3.2

I. Basic Properties

Problem S1.   Find $(\exp_3 2)^3$ and $\exp_{1/2} 4$.

Solution: By definition, $\exp_3 2 = 3^2$, so $(\exp_3 2)^3 = (3^2)^3 = 9^3 = 729$. Also, $\exp_{1/2} 4 = (1/2)^4 = 1/2^4 = 1/16$.

Problem S2.   Solve $\exp_9 3x \cdot \exp_3 x^2 = 1$.

Solution: $\exp_9 3x \cdot \exp_3 x^2 = 9^{3x} \cdot 3^{x^2}$, by definition. Hence, $1 = 9^{3x} \cdot 3^{x^2} = (3^2)^{3x} \cdot 3^{x^2} = 3^{6x} \cdot 3^{x^2} = 3^{x^2+6x}$. Thus, since $3^0 = 1$, $3^0 = 3^{x^2+6x}$ and $x^2 + 6x = 0$ (why?). Hence, $x = 0$ or -6.

DRILL PROBLEMS

S3. Find the following:
(a) $\exp_3(-2)^2 - (\exp_3(-2))^2$
(b) $\exp_4 0 \cdot \exp_5(-1) \cdot \exp_1 6$
(c) $\exp_4 3 \cdot \exp_4(-7) \cdot \exp_4 5$
(d) $\exp_4 3 \cdot \exp_4(-7) \cdot \exp_2 6 \cdot \exp_{16}(1/2)$
S4. Solve the following:
(a) $\exp_4 x = 64$
(b) $\exp_x 4 = 16$
(c) $\exp_x 4 = -16$
(d) $\exp_5 x = 1$
(e) $\exp_8 x = 4$
(f) $\exp_4 x = 2$
(g) $\exp_4(x + 1) \cdot \exp_2(2x - 1) = 8$

Answers

S3. (a) $81 - (1/81)$, (b) .2, (c) 4, (d) 1
S4. (a) 3, (b) 2, (c) $\varnothing$, (d) 0, (e) 2/3, (f) 1/2, (g) 1/2

II. Graphing

Problem S5.   Graph $y = 2^{x-2}$.

Solution: Apply the translation $XT/2$ to the graph of $y = 2^x$. (See Figure 3.2.14.)

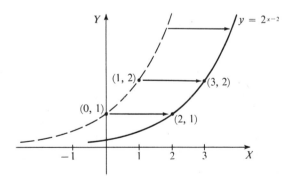

Figure 3.2.14

Problem S6.   Graph $y = 1 - 2^{x-2}$.

Solution: Apply the reflection YR to the graph of $y = 2^{x-2}$ followed by the translation $YT/1$, as shown in Figure 3.2.15.

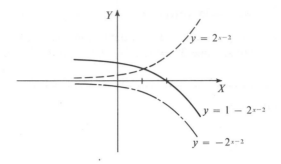

Figure 3.2.15

Problem S7.   Graph $y = |2^{x+1}|$.

Solution: $2^{x+1} > 0$ for any $x \in R$, so $y = |2^{x+1}| = 2^{x+1}$. Thus, we can apply the translation $XT/-1$ to the graph of $y = 2^x$ to obtain the graph of $y = |2^{x+1}|$, shown in Figure 3.2.16.

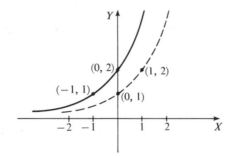

Figure 3.2.16

Problem S8.   Explain how we can sketch $y = 2^{x-2}$ by applying two different mappings to the graph of $y = 2^x$.

Solution: (1) Apply the translation $XT/2$ as in Problem S5. (2) Notice that $y = 2^{x-2} = 2^x \cdot 2^{-2} = (1/4)2^x$. Thus, the distortion $YD/\frac{1}{4}$ will yield the same result. That is, $y = 2^x \xrightarrow{YD/\frac{1}{4}} y = 2^{x-2}$.

Problem S9.   Graph $y = 3^{|x|}$.

Solution: For $x \geq 0$, $y = 3^{|x|} = 3^x$ [see Figure 3.2.17(a)]. For $x < 0$, $|x| = -x$, so $y = 3^{|x|} = 3^{-x}$ for $x < 0$. The graph of $y = 3^{-x} = (1/3)^x$ is shown in Figure 3.2.17(b). Thus, we obtain the sketch of $y = 3^{|x|}$ by combining these results, as shown in Figure 3.2.17(c).

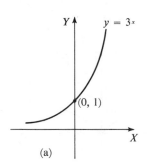

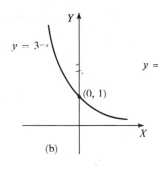

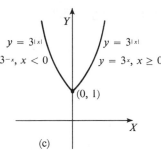

(a) (b) (c)

Figure 3.2.17

DRILL PROBLEMS

S10. Sketch the graph of each of the following:
(a) $y = 3^x + 2$ (b) $y = 3^{x+2}$
(c) $y = 9 \cdot 3^x$ (d) $y = 3^{-|x|}$
(e) $y = \div 3^{-|x|}$ (f) $y = 1 - 3^{-|x|}$

Answers

S10. (a) See Figure 3.2.18.   (b) See Figure 3.2.19.

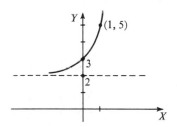

Figure 3.2.18

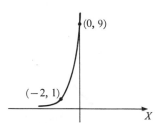

Figure 3.2.19

(c) See Figure 3.2.20.   (d) See Figure 3.2.21.

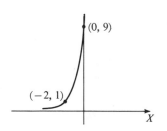

Figure 3.2.20

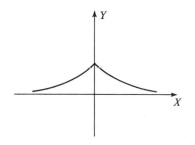

Figure 3.2.21

(e) See Figure 3.2.22.   (f) See Figure 3.2.23.

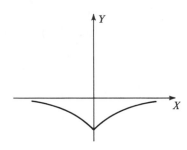

Figure 3.2.22

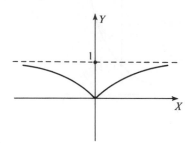

Figure 3.2.23

SUPPLEMENTARY PROBLEMS

S11. Evaluate (a) $\exp_{1/3}(-2)$ and (b) $-\exp_{1/3}(2)$.
S12. Solve (a) $\exp_x 3 = 27$ and (b) $2^{x^2+1} = 4$.
S13. Sketch (a) $y = 1 + 2^x$ and (b) $y > 2^{-x}$.

Answers

S11. (a) 9, (b) $-1/9$
S12. (a) $\{3\}$, (b) $\{-1, 1\}$
S13. (a) See Figure 3.2.24.   (b) See Figure 3.2.25.

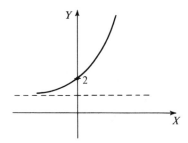

Figure 3.2.24

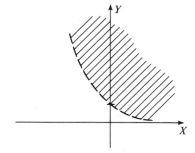

Figure 3.2.25

3.3 LOGARITHMIC FUNCTIONS

The logarithmic function is closely related to the exponential function. For example, both types of functions appear in many mathematical applications, especially when rates of growth are involved. Until recently, logarithms were used extensively in arithmetic computations. However, this use has diminished with the availability of high-speed, digital computers and low-cost, electronic calculators.

From our discussion in Section 3.2 we know that $\exp_b$ for $b > 1$ is one-to-one and has an inverse function $\exp_b^{-1}$. The graph of $y = \exp_b^{-1}(x)$ is the reflection IR of the graph of $y = \exp_b(x)$. The inverse function of the exponential function with base b is called the *logarithmic function with base b*, and is denoted by $\log_b$ (see Figure 3.3.1).

Definition 3.3.1

If $b > 1$, $\log_b(x) = \exp_b^{-1}(x)$.

Q1: (a) Use the graph of $y = 3^x$ to sketch the graph of $y = \log_3(x)$.
(b) Find the domain and the range of the function $\log_3$.

It is customary to write $\log_b x$ for $\log_b(x)$, just as it is to write b^x for $\exp_b(x)$. By convention, $\log_b x^y$ means $\log_b(x^y)$ and *not* $(\log_b x)^y$.

A1: (a) Apply an *I*-reflection to the graph of $y = 3^x$, as shown in the figure below.

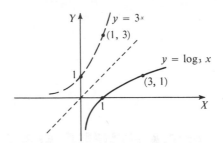

(b) The domain is $(0, \infty)$; the range is R.

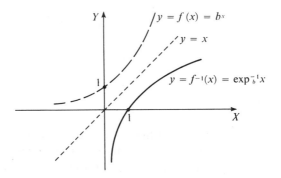

Figure 3.3.1

Since $\log_b$ and $\exp_b$ are inverse functions, $\exp_b x = y$ if and only if $\log_b y = x$. This is just an example of the fundamental inverse function relationship that states that $f(x) = y$ if and only if $f^{-1}(y) = x$ [Equation (3) of Section 3.2] for the particular function $f(x) = \exp_b(x)$. In other words,

$$\log_b y = x \quad \text{if and only if } b^x = y \text{ for } y > 0 \qquad (1)$$

Furthermore, since $f^{-1}(f(x)) = x$, $\log_b(\exp_b(x)) = x$. That is,

$$\log_b(b^x) = x \qquad (2)$$

In particular, $\log_b(b^1) = 1$, so $\log_b b = 1$. Likewise, $f(f^{-1}(x)) = x$, so $\exp_b(\log_b(x)) = x$. Using exponential notation, we have

$$b^{\log_b x} = x \qquad (x > 0) \qquad (3)$$

Example 3.3.1   (a) Find $\log_2 16$. (b) Find $\log_{10}(1/100)$. (c) Find $\log_a a^3$. (d) Show that $\log_b 1 = 0$.

Solution

(a) Let $\log_2 16 = x$. Then $2^x = 16 = 2^4$, so $x = 4$ and $\log_2 16 = 4$. (b) If $\log_{10}(1/100) = N$, then $10^N = 1/100 = (1/10)^2 = 10^{-2}$. So $\log_{10}(1/100) = -2$. (c) $\log_a a^3 = 3$ by Equation (2). (d) If $\log_b 1 = N$, then $b^N = 1$. Thus, $N = 0$ and $\log_b 1 = 0$.

Q2: Find $\log_2 8$, $\log_3(1/81)$, and $\log_{10}(1000)^{-2}$.

Since the function $\log_b$ is the inverse of the function $\exp_b$, we can use Equations (1), (2), and (3) in Section 3.2 to write the following equations:

$$\log_b(xy) = \log_b x + \log_b y \tag{4}$$

$$\log_b(x^r) = r \log_b x \tag{5}$$

$$\log_b\left(\frac{x}{y}\right) = \log_b x - \log_b y \tag{6}$$

The proofs are straightforward. For example, let $\log_b x = u$ and $\log_b y = v$. Then $b^u = x$, $b^v = y$, and $xy = b^u b^v = b^{(u+v)}$. That is, $\log_b(xy) = u + v$ from Equation (1). Substituting for u and v, we obtain $\log_b(xy) = \log_b x + \log_b y$, which is Equation (4).

Q3: Evaluate the following:
(a) $(\log_2 4)^3 - \log_2 4^3$ (b) $\log_2 1024 - \log_2 512$

(c) $\dfrac{\log_2 8}{\log_2 4} - \log_2\left(\dfrac{8}{4}\right)$

Example 3.3.2

Sketch (a) $y = -1 + \log_3 x$, (b) $y = -\log_{10}(x + 1)$.

Solution

(a) The graph of $y = -1 + \log_3 x$ can be obtained from the graph of $y = \log_3 x$ by the translation $YT/-1$. That is, move the curve $y = \log_3 x$ one unit downward, as shown in Figure 3.3.2.

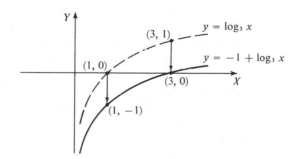

Figure 3.3.2

A2: By Equation (1), if $\log_2 8 = x$, then $2^x = 8 = 2^3$, so $x = 3$. Similarly, $3^{-4} = 1/81$, so $\log_3(1/81) = -4$. Also, $(1000)^{-2} = (10^3)^{-2} = 10^{-6}$, so $\log_{10}(1000)^{-2} = \log_{10}10^{-6} = -6$.

A3: (a) $(\log_2 4)^3 - \log_2 4^3 = 2^3 - 3\log_2 4 = 8 - 3\cdot 2 = 2$
(b) $\log_2 1024 - \log_2 512 = \log_2(1024/512) = \log_2 2 = 1$
(c) $(\log_2 8)/(\log_2 4) = 3/2$, while $\log_2(8/4) = \log_2 2 = 1$. Thus, the answer is $(3/2) - 1 = 1/2$

(b) The graph of $y = -\log_{10}(x + 1)$ can be obtained from the graph of $y = \log_{10}(x + 1)$ by the reflection YR, that is, a reflection about the X-axis. Notice that we can obtain $y = \log_{10}(x + 1)$ from $y = \log_{10}x$ by replacing x by $x - (-1)$. Thus, to obtain the graph of $y = \log_{10}(x + 1)$, apply the translation $XT/-1$ to the graph of $y = \log_{10}x$. That is, move the graph of $y = \log_{10}x$ one unit to the left. Then we find the answer by applying the reflection YR to the graph $y = \log_{10}(x + 1)$. (See Figure 3.3.3.)

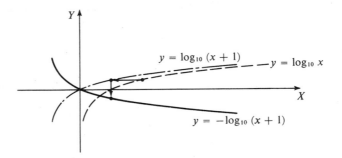

Figure 3.3.3

Given x and b, to find the number $y = \log_b x$ we must be able to solve the equation $b^y = x$ for y. It is easy to solve certain simple equations by inspection. For example, the solution to $2^y = 8$ is $y = 3$, so $\log_2 8 = 3$. The equation $10^y = 2$ cannot be solved as easily. Using numerical methods, we can find that an approximate solution to the equation $10^y = 2$ is $y = .3010$. We write $\log_{10}2 = .3010$ with the understanding that .3010 is the best 4-decimal place approximation to $\log_{10}2$. In order to illustrate some computational work with exponential and logarithmic functions, we have included a table of logarithms to the base 10 of certain numbers between 1 and 10 in Table B.1 in Appendix B.

Q4: Use Table B.1 to find each of the following:
(a) $\log_{10}5.6$ (b) $\log_{10}(3/2)$
(c) $(\log_{10}3)/(\log_{10}2)$. What is $\log_{10}10$?

Using Table B.1, we can approximate $\log_{10}N$ for any number N by writing N in scientific notation as $N = n \times 10^c$, where $1 \le n < 10$ and c is an integer. For example, $56 = 5.6 \times 10^1$. Then, using Equation (4), we have $\log_{10}56 = \log_{10}5.6 + \log_{10}10$. It follows from the preceding equation and Q4 that $\log_{10}56 = 1.7482$. Similarly, since $.0056 = 5.6 \times 10^{-3}$,

$$\log_{10}.0056 = \log_{10}5.6 + \log_{10}10^{-3}$$
$$= .7482 + (-3) = -2.2518$$

Q5: Find $\log_{10}24$, $\log_{10}240$, and $\log_{10}.024$.

**Example
3.3.3**

Solve the following equations for x: (a) $\log_2(x - 3) = 3$, (b) $\log_b x^2 - \log_b \sqrt{x} = 6$, (c) $2^x = 3$.

Solution

(a) If $\log_2(x - 3) = 3$, then $2^3 = x - 3$ by Equation (1). So $8 = x - 3$ or $x = 11$. To check our work, we substitute 11 for x, obtaining $\log_2(11 - 3) = \log_2 8 = 3$. Thus, the solution is $x = 11$.
 (b) Using Equation (5), we have $\log_b x^2 - \log_b x^{1/2} = 2\log_b x - (1/2)\log_b x = (3/2)\log_b x = 6$. So $\log_b x = (2/3)\cdot 6 = 4$. Thus, $x = b^4$. Alternatively, $\log_b x^2 - \log_b \sqrt{x} = \log_b(x^2)/\sqrt{x} = \log_b x^{3/2}$ [by Equation (6)] and we continue as before.
 (c) While it is true that $x = \log_2 3$, we can find a numerical approximation for x by using Table B.1. Since $\log_b$ is a one-to-one function, $3 = 2^x$ is equivalent to $\log_{10}3 = \log_{10}2^x$. By Equation (5), $\log_{10}2^x = x\log_{10}2$. Hence, $\log_{10}3 = x\cdot\log_{10}2$ or $x = (\log_{10}3)/(\log_{10}2)$. Using Table B.1, we find that x is approximately $(.4771)/(.3010)$ or about 1.58.

**PROBLEMS
3.3**

1. Solve the following equations for x:
(a) $\log_4 x = -3$ (b) $\log_4(-16) = x$
(c) $\log_x 4 = 2/3$ (d) $A = A_0 10^{-kx}$
(e) $\log_2 x + \log_2(x + 3) = 2$ (f) $\log_2 x - \log_2(x - 3) = 2$

2. Use Table B.1 to solve the following equations:
(a) $3^x = 5$ (b) $x^3 = 5$
(c) $2^{x+3} = 3^{x+2}$ (d) $x^3 = 75,000$
(e) $x^3 = .0068$

3. Using logarithms, approximate $\sqrt[4]{10}$.

A4: (a) .7482, (b) .1761, (c) $(.4771)/(.3010) \approx 1.58$, $\log_{10} 10 = 1$

A5: (a) 1.3802, (b) 2.3802, (c) -1.6198

4. Prove that $\log_b(xyz) = \log_b x + \log_b y + \log_b z$.

5. What mapping applied to the graph of $y = \log_b x$ yields the graph of each of the following. Illustrate your results with a sketch.
(a) $y = \log_b(x - 2)$ (b) $y = 2\log_b x$
(c) $y = 2 + \log_b x$

6. Sketch the graph of each of the following subsets of R^2:
(a) $y = -\log_5 x$ (b) $y = \log_3(-x)$
(c) $y = -2 + \log_3(x + 2)$ (d) $y = \log_2|x|$
(e) $y = \log_2|x - 1|$ (f) $y = \log_3(1/x)$
(g) $y > 1 - \log_2 x$ (h) $y = |\log_2 x|$

7. Are the functions $f(x) = \log_b x^2$ and $g(x) = 2\log_b x$ identical?

8. A certain minicalculator can manipulate an integer as large as 2^{31}. How many digits are in 2^{31}?

---

9. Solve for x:
(a) $s = (1 - r^x)/(1 - r)$ (b) $\log_b x = \log_x b$
(c) $\log_b x^2 = (\log_b x)^2$ (d) $I = E/R(1 - b^{-Rx/L})$

ANSWERS 3.3

1. (a) $1/64$, (b) no such x, (c) 8, (d) $(-1/k)\log_{10}(A/A_0)$, (e) 1, (f) 4

2. (a) 1.47, (b) 1.7, (c) $-.290$, (d) 42, (e) .19

3. 1.78

4. Using Equation (4) twice, we obtain $\log_b(xyz) = \log_b[(xy)z] = \log_b(xy) + \log_b z = \log_b x + \log_b y + \log_b z$.

5. (a) $XT/2$, (b) $YD/2$, (c) $YT/2$

6. (a) See Figure 3.3.4.   (b) See Figure 3.3.5.

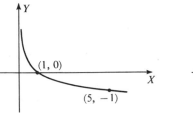

Figure 3.3.4

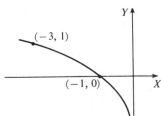

Figure 3.3.5

(c) See Figure 3.3.6.   (d) See Figure 3.3.7.

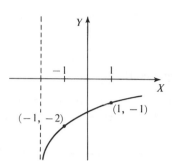

Figure 3.3.6

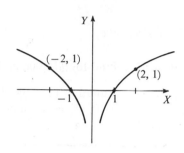

Figure 3.3.7

(e) See Figure 3.3.8.   (f) See Figure 3.3.9.

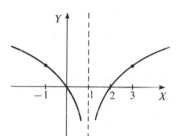

Figure 3.3.8

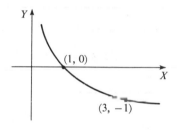

Figure 3.3.9

(g) See Figure 3.3.10.   (h) See Figure 3.3.11.

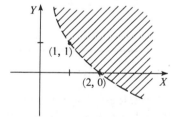

Figure 3.3.10

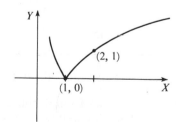

Figure 3.3.11

7. No. The domain of f is $(-\infty, 0) \cup (0, \infty)$, while the domain of g is $(0, \infty)$.

8. $\log_{10} 2^{31} = 31 \cdot \log_{10} 2 = 31(.301) = 9.331$. Thus, $2^{31} = 10^{9.331} = 10^{.331} \cdot 10^9 \approx 2{,}140{,}000{,}000$, which has ten digits.

Supplement 3.3

I. Basic Properties

Problem S1.   Find $\log_{10}10{,}000$ and $\log_{10}.0001$.

Solution: By Equation (1), $\log_{10}10{,}000 = x$ is equivalent to $10^x = 10{,}000 = 10^4$. Thus, $x = \log_{10}10{,}000 = 4$. Similarly, $\log_{10}.0001 = x \Leftrightarrow 10^x = .0001 = 10^{-4}$, so $x = \log_{10}.0001 = -4$.

Problem S2.   Find $\log_2 16$ and $\log_3(1/27)$.

Solution: Again by definition [Equation (1)] $\log_2 16 = x$ is equivalent to $2^x = 16$. But $16 = 2^4$. Thus, $x = \log_2 16 = 4$. Similarly, $\log_3 1/27 = x \Leftrightarrow 3^x = 1/27 = 1/3^3 = 3^{-3}$, so $x = \log_3 1/27 = -3$.

Problem S3.   Solve the following for x: (a) $\log_{10}x = -3$ and (b) $\log_x 9 = 3$.

Solution: (a) $\log_{10}x = -3$ implies $x = 10^{-3} = .001$. (b) $\log_x 9 = 3$ implies $x^3 = 9$, so $x = \sqrt[3]{9}$.

Problem S4.   Solve the following for x: (a) $\log_4(x - 1) = 2$ and (b) $\log_4(x + 1) - \log_4 x = 1$.

Solution: (a) Again by Equation (1), $\log_4(x - 1) = 2$ is equivalent to $4^2 = (x - 1)$, so $x = 4^2 + 1 = 17$. (b) Using Equation (1), we have $\log_4(x + 1) - \log_4 x = \log_4[(x + 1)/x] = 1$ provided $x > 0$. Therefore,

$$\frac{x + 1}{x} = 4^1 = 4 \quad \text{or} \quad 1 + \frac{1}{x} = 4$$

Thus, $1/x = 3$ and $x = 1/3$.

Problem S5.   Solve the following for x: $\log_{10}[\log_{10}(\log_{10}x)] = 0$.

Solution: We work from the outside in. First, $\log_{10}M = 0$ if, and only if, $M = 10^0$. Thus, $\log_{10}[\log_{10}(\log_{10}x)] = 0$ implies $[\log_{10}(\log_{10}x)] = 1$. Now, $\log_{10}N = 1$ if, and only if, $N = 10^1 = 10$. Thus, $\log_{10}(\log_{10}x) = 1$ implies $\log_{10}x = 10$, and hence $x = 10^{10} = 10{,}000{,}000{,}000$.

Problem S6.   If $\log_b 2 = .69$ and $\log_b 3 = 1.10$, find $\log_b 6$, $\log_b 24$, and $\log_b 5$.

Solution: $\log_b 6 = \log_b(2 \cdot 3) = \log_b 2 + \log_b 3$, by Equation (4). Thus, $\log_b 6 = .69 + 1.10 = 1.79$. By Equations (4) and (5),

$$\log_b 24 = \log_b(3 \cdot 2^3) = \log_b 3 + \log_b 2^3$$
$$= \log_b 3 + 3 \cdot \log_b 2 = 1.10 + 3(.69) = 3.17$$

From the information given, we cannot compute $\log_b 5$ directly. (For example, $\log_b 5 = \log_b(2 + 3)$, but $\log_b(2 + 3) \neq \log_b 2 + \log_b 3$.)

DRILL PROBLEMS

S7. Find the following:
(a) $\exp_2(\log_2 8)$ (b) $\log_8 16$
(c) $\log_2(\exp_4(3))$
S8. Solve the following for x:
(a) $\log_2(x^2 - 1) - \log_2(x - 1) = 3$ (b) $\log_x(\exp_3 4) = 2$

Answers

S7. (a) 8, (b) 4/3, (c) 6
S8. (a) 7, (b) 9

II. Graphing

Problem S9. Graph $y = \log_3(x - 1)$.

Solution: Notice that $y = \log_3 x \xrightarrow{XT/1} y = \log_3(x - 1)$. Thus, to obtain the desired sketch, shift the graph of $y = \log_3 x$ to the right 1 unit, as shown in Figure 3.3.12.

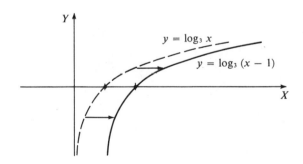

Figure 3.3.12

Problem S10. Graph $y = \log_3 x - 1$.

Solution: Here $y = -1 + \log_3 x$. Now

$$y = \log_3 x \xrightarrow{YT/-1} y = -1 + \log_3 x$$

Thus, to obtain the desired graph, shift the graph of $y = \log_3 x$ one unit downward, as shown in Figure 3.3.13.

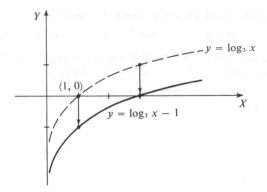

Figure 3.3.13

Problem S11.   Graph $y = -1 - \log_3 x$.

Solution: Observe that

$$y = \log_3 x \xrightarrow{YR} y = -\log_3 x \xrightarrow{YT/-1} y = -1 - \log_3 x$$

Thus, to obtain the desired sketch, first reflect the graph of $y = \log_3 x$ about the X-axis, and then shift this intermediate graph one unit downward. (See Figure 3.3.14).

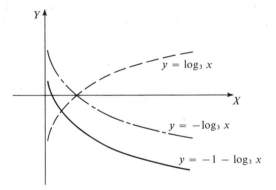

Figure 3.3.14

Problem S12.   Graph $y = |\log_3 x|$.

Solution: For $x \geq 1$, $\log_3 x \geq 0$; so $y = |\log_3 x| = \log_3 x$. For $0 < x < 1$, $\log_3 x < 0$ and thus $y = |\log_3 x| = -\log_3 x$.

The graph of $y = -\log_3 x$ is sketched in Figure 3.3.14. We then find the graph of $y = |\log_3 x|$ by combining the results above, to produce the graph shown in Figure 3.3.15.

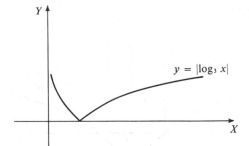

Figure 3.3.15

DRILL PROBLEMS

S13. Graph $y = 2 - \log_3(x + 2)$.
S14. Graph $y = \log_3|x|$.
S15. Graph $y = \log_3|x - 1|$.

Answers

S13. See Figure 3.3.16.

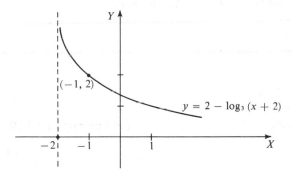

Figure 3.3.16

S14. See Figure 3.3.17.

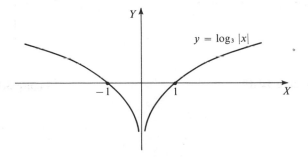

Figure 3.3.17

S15. See Figure 3.3.18.

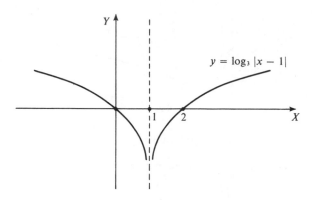

Figure 3.3.18

SUPPLEMENTARY PROBLEMS

S16. Find (a) $\log_3 27$, (b) $\log_4 8$, and (c) $\log_{10}.001$.
S17. Solve (a) $\log_3|1 - x| = 2$ and (b) $\log_3(2x + 1) - \log_3(x - 1) = 1$.
S18. Sketch the graph of each of the following:
(a) $y \leq 1 + \log_5 x$ (b) $y = -\log_5 x^2$
(c) $y = -2 \log_5 x$

Answers

S16. (a) 3, (b) 3/2, (c) -3
S17. (a) $\{-8, 10\}$, (b) $\{4\}$
S18. (a) See Figure 3.3.19.   (b) See Figure 3.3.20.

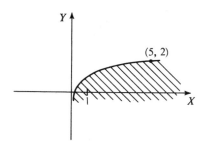

Figure 3.3.19

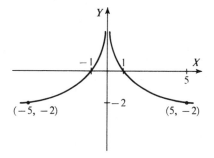

Figure 3.3.20

(c) See Figure 3.3.21.

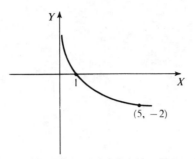

Figure 3.3.21

3.4   COMPUTATIONS USING LOG₁₀

Let us begin by reviewing the basic properties of the logarithmic function from the previous section. Let $N > 0$; then

$$\log_b N = a \quad \text{if, and only if, } b^a = N \tag{1}$$

Thus, the graph of $y = \log_b x$ contains $(1, 0)$ and $(b, 1)$ as shown in Figure 3.4.1.

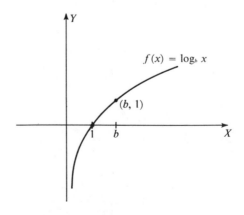

Figure 3.4.1

Also, for M, $N > 0$,

$$\log_b MN = \log_b M + \log_b N \qquad (2)$$

$$\log_b M^r = r \log_b M \qquad (3)$$

$$\log_b\left(\frac{M}{N}\right) = \log_b M - \log_b N \qquad (4)$$

We can use these properties and a table of logarithmic function values to approximate certain numerical expressions.

Example 3.4.1

Use Table B.1 to approximate each of the following numbers: (a) $10^{.3}$, (b) $\sqrt[3]{6}$, (c) $(1.05)^8$, and (d) $847/[(12.6)(43)]$.

Solution

(a) $10^{.3} = x$ is equivalent to $\log_{10} x = .3$ [Equation (1)]. In Table B.1 we see that $\log_{10} 2 = .3010$. Thus, $x = 10^{.3}$ is approximately 2.

(b) Let $x = \sqrt[3]{6}$. Then $\log_{10} x = \log_{10} 6^{1/3} = (1/3)\log_{10} 6 = (1/3)(.7782) = .2594$. Thus, $x = \sqrt[3]{6}$ is about 1.82. We can obtain a more precise estimate by using interpolation as described later in this section.

(c) Let $x = (1.05)^8$. Then $\log_{10} x = \log_{10}(1.05)^8 = 8 \log_{10} 1.05 = 8(.0212) = .1696$. Thus, x is about 1.48.

(d) Let $x = 847/[(12.6)(43)]$. Then $\log_{10} x = \log_{10} 847/[(12.6)(43)]$ $= \log_{10} 847 - (\log_{10} 12.6 + \log_{10} 43)$. Now $\log_{10} 847 = 2.9279$ since $\log_{10} 8.47 = .9279$ and $847 = 8.47 \cdot 10^2$ (see Q5 in Section 3.3). Similarly, $\log_{10} 12.6 = 1.1004$ and $\log_{10} 43 = 1.6335$. Thus,

$$\log_{10} x = 2.9279 - (1.1004 + 1.6335) = .1940$$

So x is about 1.56.

In each part of Example 3.4.1 we have used Table B.1 directly to approximate x given $\log_{10} x$, since $\log_{10} x$ is a number between 0 and 1. If this is not the case, we can proceed as in Example 3.4.2.

Example 3.4.2

Use Table B.1 to approximate each of the following numbers: (a) $10^{2.3}$, (b) $\sqrt[3]{60,000}$, (c) $.0846/[(126)(.43)]$, and (d) $(1.05)^{-8}$.

Solution

(a) $10^{2.3} = 10^{.3} \cdot 10^2 \approx 2 \cdot 100 = 200$ [see Example 3.4.1(a)].

(b) Let $x = \sqrt[3]{60,000}$. Then $\log_{10} x = \log_{10}(60,000)^{1/3} = (1/3)\log_{10} 60,000 = (1/3)(4.7782) = 1.5927 = .5927 + 1$. Equivalently, $x = 10^{.5927+1} = 10^{.5927} \cdot 10^1$. Since $\log_{10} 3.91 = .5922$, x is about $3.91 \cdot 10^1 = 39.1$. Notice that to use Table B.1, we write $\log_{10} x$ as the sum of a number between 0 and 1 (called the *mantissa*) and an integer (called the *characteristic*).

(c) Let $x = .0847/[(126)(.43)]$. Then $\log_{10} x = \log_{10} .0847 - [\log_{10} 126 + \log_{10} .43] = (.9279 - 2) - [(.1004 + 2) + (.6335 - 1)] = .1940 - 3$. Thus, x is about $1.56 \cdot 10^{-3} = .00156$.

(d) Let $x = (1.05)^{-8}$. Then $\log_{10} x = \log_{10}(1.05)^{-8} = -8 \cdot \log_{10} 1.05 = -8(.0212) = -.1696$. To change $\log_{10} x = -.1696$

into the form [Number between 0 and 1] + [Integer], we "add and subtract" 1 and write $\log_{10}x = -.1696 = 1 - .1696 - 1 = .8304 - 1$. Thus, x is about $6.77 \cdot 10^{-1} = .677$.

Q1: Find x if $\log_{10}x =$ (a) 2.8451, (b) $-.0200$, and (c) -1.3002.

Suppose that we are asked to find x such that $\log_{10}x = .3021$. In Table B.1 we see that $\log_{10}2 = .3010$ and $\log_{10}2.01 = .3032$. Since .3021 is halfway between .3010 and .3032, it seems reasonable to assume that x is about halfway between 2 and 2.01, or x is about 2.005. The generalization of this procedure of using "proportional parts" to approximate function values which occur between entries in a table is referred to as *linear interpolation*. Figure 3.4.2 illustrates this technique with a portion of the graph of $y = f(x)$. Here P and Q are two points of the curve given by consecutive tabular entries. We desire to find $f(a)$, whose true value is obtained from point T of the graph. We approximate T with point L which lies on the straight line through points P and Q, since we can find v, the Y-coordinate of L. We can find v by using either the equation of the straight line or similar triangles to calculate d as shown in Figure 3.4.3.

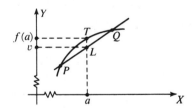

Figure 3.4.2

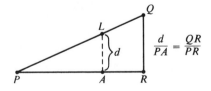

$$\frac{d}{PA} = \frac{QR}{PR}$$

Figure 3.4.3

In Figure 3.4.4, we add d to h, the Y-coordinate of P, to obtain v since this function is increasing. In fact, if we know that the shape of the graph is as shown, we see that the interpolated value v is less than the true value $f(a)$.

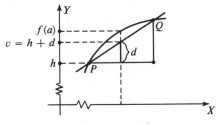

Figure 3.4.4

**Example
3.4.3**

Use linear interpolation and Table B.1 to approximate each of the following numbers: (a) $\log_{10}4.528$, (b) $\log_{10}3.117$, and (c) x if $\log_{10}x = .3456$.

Solution

(a) From Table B.1, $\log_{10}4.52 = .6551$ and $\log_{10} 4.53 = .6561$. In Figures 3.4.2 and 3.4.3, $P = (4.52, .6551)$ and $Q = (4.53, .6561)$. Now 4.528 is 8/10 of the way from 4.52 to 4.53, so $\log_{10}4.528$ is about 8/10 of the way from .6551 to .6561. Thus, $\log_{10}4.528 = .6559$.

(b) From Table B.1, $\log_{10}3.11 = .4928$ and $\log_{10}3.12 = .4942$. Using Figure 3.4.5, we obtain $d/14 = 7/10$, so $d = 98/10 \approx 10$. Thus, $\log_{10}3.117$ is about $.4928 + .0010 = .4938$.

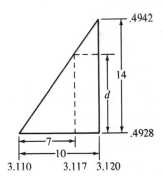

Figure 3.4.5

(c) From Table B.1, $\log_{10}2.21 = .3444$ and $\log_{10}2.22 = .3464$. Using Figure 3.4.6, we have $a/10 = 12/20$, so $a = 6$. Thus, $x = 2.210 + .006 = 2.216$.

Q2: Use linear interpolation and Table B.1 to approximate each of the following numbers:
(a) $\log_{10}1.234$ (b) x if $\log_{10}x = .1234$

A1: (a) 700, (b) .955, (c) .0501

A2: (a) log 1.234 = .0899 + (4/10)(.0934 − .0899) = .0913
    (b) (1234 − 1206)/(1239 − 1206) = 28/33 ≈ 8/10, so x =
1.320 + .008 = 1.328

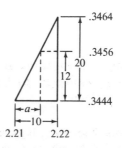

Figure 3.4.6

**PROBLEMS
3.4**

1. Use Table B.1 to find each of the following numbers:
(a) $\log_{10} 23.4$ (b) $\log_{10} .2$
(c) $\log_{10} .00537$ (d) $\log_{10} 10^{.45}$
(e) $\log_{10}(-1.23)$ (f) $\log_{10}(16)^{12}$

2. Find x if $\log_{10} x$ equals
(a) 2.4771   (b) −.3410
(c) 12.73   (d) −1.52
(e) −2.52   (f) −3.1002

3. Use Table B.1 and linear interpolation to approximate each of the following numbers:
(a) $\log_{10} 1776$ (b) $\log_{10} 1492$
(c) $\log_{10} .1492$ (d) $\log_{10} 2001$

4. Use linear interpolation to approximate x if $\log_{10} x$ equals
(a) .7777   (b) 1.2345
(c) .4567

5. Use Table B.1 to approximate each of the following numbers. You need not interpolate.
(a) $\sqrt{7}$ (b) $\sqrt[5]{32}$
(c) $(1.06)^{10}$ (d) $794/63.8$
(e) $(17.6)(.0123)/747$ (f) $\sqrt[4]{27.6} \cdot \sqrt[3]{45}$

**ANSWERS
3.4**

1. (a) 1.3692, (b) −.6990, (c) −2.27, (d) .45, (e) undefined, (f) 12(1.2041) =
14.45

2. (a) 300, (b) .456, (c) 5.37×10^{12}, (d) .0302, (e) .00302, (f) .000794
3. (a) 3.2495, (b) 3.1738, (c) $-.8262$, (d) 3.3012
4. (a) 5.994, (b) 17.16, (c) 2.862
5. (a) 2.65, (b) 2, (c) 1.79, (d) 12.4, (e) .00029, (f) 8.15

Supplement 3.4

I. Computations Involving Properties of Log

Problem S1. Use Table B.1 to find (a) $\log_{10}2.63$, (b) $\log_{10}263$, and (c) $\log_{10}.0263$.

Solution: (a) Looking directly into the table, we see that $\log_{10}2.63 = .4200$. (b) $\log_{10}263 = \log_{10}[(10^2)(2.63)] = \log_{10}10^2 + \log_{10}2.63 = 2\cdot\log_{10}10 + .4200 = 2.4200$ (remember, $\log_{10}10 = 1$). (c) $\log_{10}.0263 = \log_{10}[(10^{-2})(2.63)] = \log_{10}10^{-2} + \log_{10}2.63 = -2\ \log_{10}10 + .4200 = -2 + .4200 = -1.5800$ (Note: $\log_{10}.0263 = -2 + .4200 \neq -2.4200$.)

Problem S2. Use Table B.1 to find N if $\log_{10}N = .3692$.

Solution: Using Table B.1 directly, observe that $\log_{10}2.34 = .3692$. Hence, $N = 2.34$.

Problem S3. Use Table B.1 to find N if $\log_{10}N = 3.3692$.

Solution: $\log_{10}N = 3.3692 \Leftrightarrow N = 10^{3.3692} = (10^3)(10^{.3692})$. Since $10^{.3692} = x$ implies $\log_{10}x = .3692$, $x = 2.34$ from Problem S2. Therefore, $N = (10^3)(2.34) = 2340$.

Problem S4. Use Table B.1 to find N if $\log_{10}N = -2.6308$.

Solution: $\log_{10}N = -2.6308$ implies $N = 10^{-2.6308} = 10^{-3+.3692} = (10^{-3})(10^{.3692})$. Since $10^{.3692} = 2.34$ (see Problem S2), $N = (10^{-3})(2.34) = .00234$.

Note: It should be clear that $10^{-2.6308} \neq (10^{-2})(10^{.6308})$. The key to this problem is to recognize that writing $N = 10^{-2.6308} = 10^{-2+(-.6308)}$ is of *no help* because the numbers in Table B.1 are *positive*. Thus, -2.6308 must be written in the form $k + m$ where $k \in Z$ and $0 < m < 1$. In general, to accomplish this, add and subtract $|k|$. For example, $-2.6308 = 3 - 2.6308 - 3 = .3692 - 3$.

Problem S5. Use Table B.1 to approximate $(56)(127)$.

Solution: By Equation (4) in Section 3.3, $\log_{10}(56\cdot127) = \log_{10}56 + \log_{10}127$. Now $\log_{10}56 = \log_{10}[(10)(5.6)] = \log_{10}10 + \log_{10}5.6 = 1 + .7482 = 1.7482$, and

$\log_{10}127 = \log_{10}[(10^2)(1.27)] = \log_{10}10^2 + \log_{10}1.27 = 2 + .1038 = 2.1038$. Thus, $\log_{10}(56)(127) = \log_{10}56 + \log_{10}127 = 1.7482 + 2.1038 = 3.8520$. So $(56)(127) = 10^{3.8520} = (10^3)(10^{.8520})$. Using Table B.1, notice that $\log_{10}7.11 = .8519$, so $10^{.8519} = 7.11$. Thus, $(56)(127) = 10^3(7.11) = 7110$.* (Note: $(56)(127)$ is exactly 7112.)

Problem S6.   Use Table B.1 to approximate $56/127$.

Solution: $\log_{10}(56/127) = \log_{10}56 - \log_{10}127 = 1.7482 - 2.1038 = -.3556$. (See Problem S5 for $\log_{10}56$ and $\log_{10}127$.) So $\log_{10}(56/127) = -.3556$ or $56/127 = 10^{-.3556} = 10^{-1+.6444} = (10^{-1})(10^{.6444})$. Using Table B.1, notice that $\log_{10}4.41 = .6444$, so $56/127 = (10^{-1})(4.41) = .441$. (Note: $56/127 = .440945$ to six places.)

Problem S7.   Use Table B.1 to approximate 127^{56}.

Solution: $\log_{10}(127)^{56} = 56 \log_{10}127 = 56(2.1038) = 117.8128$ (see Problem S5 for log 127). So $(127)^{56} = 10^{117.8128} = (10^{117})(10^{.8128})$. By Table B.1, $\log_{10}6.50 = .8129$. $(127)^{56} = (10^{117})(6.50) = 6.5 \times 10^{117}$ in scientific notation (a very large number — 118 digits!).

DRILL PROBLEMS

S8. Use Table B.1 to find (a) $\log_{10}1630$ and (b) $\log_{10}.0615$.

S9. Use Table I to find N if (a) $\log_{10}N = 4.2175$ and (b) $\log_{10}N = -2.8013$.

S10. How many digits are in the number 5^{50}?

Answers

S8. (a) 3.2122, (b) -1.2111
S9. (a) 16,500, (b) .00158
S10. 35

II. Linear Interpolation

Problem S11.   Use linear interpolation to find $\log_{10}4523$.

Solution: $\log_{10}4523 = \log_{10}[(10^3)(4.523)] = 3 + \log_{10}4.523$.
Now consider the following difference chart:

$$.010\left\{ \begin{matrix} \log_{10}4.53 \ = .6561 \\ .003\left\{ \begin{matrix}\log_{10}4.523 = N \\ \log_{10}4.52 \ = .6551 \end{matrix}\right\} x \end{matrix} \right\}.0010$$

---

*Remember, it is understood that the equality symbol in numerical logarithmic computations denotes an *approximation*.

Equating corresponding ratios, we obtain $.003/.010 = x/.0010$ where $N = .6551 + x$. So $x = (3/10)(.001) = .0003$. Hence, $N = \log_{10}4.523 = .6551 + .0003 = .6554$ and $\log_{10}4523 = 3.6554$.

Problem S12. Use linear interpolation to find N if $\log_{10}N = 2.5495$.

Solution: $\log_{10}N = 2.5495$ implies $N = 10^{2+.5495} = (10^2)10^{.5494}$. So we want to find M where $\log_{10}M = .5495$. Next we use Table B.1 to construct the difference chart:

$$.01\left\{\begin{matrix} \log_{10}3.55 = .5502 \\ x\begin{cases} \log_{10}M \ \ = .5495 \\ \log_{10}3.54 = .5490 \end{cases}.0005 \end{matrix}\right\}.0012$$

Next, equating corresponding ratios, we obtain $x/.01 = .0005/.0012 = 5/12$ where $3.54 + x = M$. So $x = (.01)(5/12) = .004$. Thus, $M = 3.54 + .004 = 3.544$. Therefore, $N = 354.4$.

DRILL PROBLEMS

S13. Use linear interpolation to find (a) $\log_{10}.02634$ and (b) $\log_{10}13.57$.
S14. Use linear interpolation to find N if (a) $\log_{10}N = 1.9045$ and (b) $\log_{10}N = 2.8410$.

Answers

S13. (a) -1.5794, (b) 1.1325
S14. (a) 80.26, (b) 693.4

SUPPLEMENTARY PROBLEMS

S15. Use Table B.1 (without interpolation) to approximate each of the following:
(a) $\dfrac{(.0248)(35,700)}{4560}$ (b) $\sqrt[4]{1230}$

(c) $(1.01)^{100}$ (d) $\dfrac{1}{\sqrt[3]{123}}$

(e) the number of digits in 2^{100}

S16. Use Table B.1 and linear interpolation to approximate each of the following:
(a) $\dfrac{1776}{1984}$ (b) $\sqrt[3]{23.3}$
(c) $(1.01)^{100}$ (d) $\sqrt[4]{12340}$

Answers

S15. (a) $.194$, (b) 5.91, (c) 2.69, (d) $.201$, (e) 31
S16. (a) $.895$, (b) 2.856, (c) 2.69, (d) 10.58

3.5 APPLICATIONS

One application of logarithms occurs in problems involving compound interest. For example, suppose that a bank pays 5% annual interest on the amount on deposit each year and that an account initially contains $100 (called the *principal*). At the end of one year the account would be credited with $5 interest (5% of $100) and would contain $105 [(100 + 5)% of $100]. If this amount ($105) were left on deposit for another year, the interest added would be 5% of $105, so the total amount on deposit would then be [$100(1.05)]1.05 or $100·(1.05)^2 = $110.25. Similarly, the amount on deposit after 3 years of compounding would be $100· (1.05)^3; after 4 years, it would be $100(1.05)^4. The amount after n years would be $100(1.05)^n = $100(1 + 5/100)^n.

In general, if a principal amount P is deposited in an account which pays i% interest annually, then the compounded amount A after the end of n years is given by the formula

$$A = P\left(1 + \frac{i}{100}\right)^n \tag{1}$$

Q1: Use Example 3.4.1 to approximate the compounded amount on $500 at 5% compounded annually for 8 years.

Example 3.5.1

(a) In how many years will an amount double if left to compound annually at 6%? (b) What interest rate is necessary for compounding to double an amount in 10 years?

Solution

(a) Since the principal P is to be doubled, Equation (1) becomes $2P = P(1.06)^n$, and we must solve for n. Since $P \neq 0$, then $2 = (1.06)^n$ or, equivalently, $\log_{10}2 = \log_{10}(1.06)^n = n \log_{10}1.06$. Thus,

$$n = \frac{\log_{10}2}{\log_{10}1.06} = \frac{.3010}{.0253} \approx 11.9$$

Therefore, the amount will more than double after 12 years.

(b) Here Equation (1) becomes $2P = P(1 + r)^{10}$ or $2 = (1 + r)^{10}$ where r is the interest rate (expressed as a decimal, not a percent). Thus, $\log_{10}2 = \log_{10}(1 + r)^{10} = 10 \log_{10}(1 + r)$ or

$$\log_{10}(1 + r) = \frac{\log_{10}2}{10} = \frac{.301}{10} = .0301$$

Interpolating using Table B.1, we see that $1 + r$ is about 1.0715, so an interest rate of about 7.15% is necessary.

Example 3.5.2

One thousand dollars is deposited in each of five banks which pay interest at an annual rate of 6%. Find the amount accumulated at the end of one year if the banks compound interest (a) yearly, (b) semiannually, (c) quarterly, (d) monthly, and (e) daily.

Solution

(a) $1060.00.

(b) An annual rate of 6% compounded semiannually means that $(6/2)\% = 3\%$ is compounded each 6 months. Thus, at the end of one year the amount is $1000 \cdot (1.03)^2 = 1060.90$.

(c) The rate for each quarter is $(6/4)\% = 1.5\%$, so the amount at the end of one year is $1000(1.015)^4 = 1061.36$.

(d) Following the above examples, the amount equals

$$\$1000 \cdot \left(1 + \frac{.06}{12}\right)^{12} = \$1000(1.005)^{12}$$

This number can be computed to be $1061.68.

(e) The amount is $1000(1 + .06/365)^{365}$ which is about $1061.76. Can you guess the result if interest were compounded every hour? Every minute? What would it mean to say that interest is compounded "continuously"? Using calculus, we can show that this amount would be $1061.84.

Next, let us consider the amount that accumulates when an amount P is deposited at the beginning of *each* period and interest is compounded at the end of each period. For example, suppose that $100 is deposited each year for 4 years and interest compounded at the rate of 5% per year. Then we can find the amount

> A1: $A = \$500(1.05)^8 = \$500(1.48) = \$740$

accumulated at the end of 4 years by considering the "value" of each deposit individually. The initial deposit of $100 now has grown to $100(1.05)^4$ at the end of the fourth year, by Equation (1). The $100 deposited the second year has increased to $100(1.05)^3$ by the end of the fourth year, as it has been on deposit for the last 3 years. The third $100 has grown to $100(1.05)^2$; the fourth $100 has been on deposit one year and is worth $100(1.05)$. The total amount in the account at the end of 4 years is the sum of these 4 numbers, namely,

$$A = 100(1.05) + 100(1.05)^2 + 100(1.05)^3 + 100(1.05)^4$$

If we let $P = 100$ and $x = 1.05$, we can write the above equation as

$$A = Px + Px^2 + Px^3 + Px^4 = P(x + x^2 + x^3 + x^4)$$

In general, if an amount P is deposited each period for n periods, and interest at a rate r per period is compounded at the end of each period, then the total amount A at the end of n periods is given by the equation

$$A = P(x + x^2 + x^3 + \cdots + x^n) \quad \text{where } x = 1 + r \qquad (2)$$

We can shorten the appearance of Equation (2) by the use of the symbol $\sum$ and the summation notation, as shown in Definition 3.5.1.

Definition 3.5.1

Let $m, n \in Z$ with $m < n$, and f be a function. Then

$$\sum_{k=m}^{n} f(k) = f(m) + f(m + 1) + \cdots + f(n - 1) + f(n)$$

For example,

$$\sum_{k=1}^{4} 2^k = 2^1 + 2^2 + 2^3 + 2^4 = 30$$

$$\sum_{i=3}^{5} i^2 = 3^2 + 4^2 + 5^2 = 50$$

Certain properties follow immediately from Definition 3.5.1. For example,

$$\sum_{k=m}^{n} (a_k + b_k) = (a_m + b_m) + (a_{m+1} + b_{m+1}) + \cdots + (a_n + b_n)$$

$$= (a_m + a_{m+1} + \cdots + a_n) + (b_m + b_{m+1} + \cdots + b_n)$$

$$= \left(\sum_{k=m}^{n} a_k\right) + \left(\sum_{k=m}^{n} b_k\right).$$

We usually omit the parentheses on the last terms and write

$$\sum_{k=m}^{n} (a_k + b_k) = \sum_{k=m}^{n} a_k + \sum_{k=m}^{n} b_k \tag{3}$$

In the problems you are asked to show that

$$c \sum_{k=m}^{n} a_k = \sum_{k=m}^{n} c a_k \tag{4}$$

Q2: Compute the following:

(a) $\sum_{i=1}^{4} (i-1)^2$ (b) $\sum_{k=3}^{5} \frac{k^2 - 1}{(-1)^k}$

We can now rewrite Equation (2) as

$$A = P \sum_{k=1}^{n} (1 + r)^k \tag{5}$$

The task of computing $\sum_{k=1}^{n} (1 + r)^k$ can become rather formidable if n is large. However, we can derive a useful result to help us. Again, let us write $x = 1 + r$ and $S = \sum_{k=1}^{n} x^k = x + x^2 + \cdots + x^n$. Then $xS = x(x + x^2 + \cdots + x^n) = x^2 + x^3 + \cdots + x^{n+1}$. Computing the difference between S and xS, we have $S - xS = (x + x^2 + \cdots + x^n) - (x^2 + x^3 + \cdots + x^n + x^{n+1})$ so $S - xS = x - x^{n+1}$, since all but the first and last terms cancel. That is $S - xS = x - x^{n+1}$, or $S(1 - x) = x - x^{n+1}$. Thus, if $x \neq 1$,

$$S = \sum_{k=1}^{n} x^k = \frac{x - x^{n+1}}{1 - x} = \frac{x}{1 - x}(1 - x^n) \tag{6}$$

Q3: Compute $\sum_{k=1}^{10} 2^k$.

Often a slightly different formula is written where the term $x^0 = 1$ is added to the summation. By Equation (6), if $x \neq 1$,

$$x^0 + \sum_{k=1}^{n} x^k = \sum_{k=0}^{n} x^k = 1 + \frac{x - x^{n+1}}{1 - x} = \frac{1 - x + x - x^{n+1}}{1 - x}$$

$$= \frac{1 - x^{n+1}}{1 - x}$$

This sum is usually called the (finite) *geometric series,* and we have just shown that

$$\sum_{k=0}^{n} x^k = \frac{1 - x^{n+1}}{1 - x} \quad \text{if } x \neq 1 \tag{7}$$

A2: (a) 14, (b) -17

A3: $2(2^{10} - 1) = 2046$

For example,

$$\sum_{k=0}^{10} \left(\frac{1}{2}\right)^k = \frac{1 - (1/2)^{11}}{1 - (1/2)} = \frac{1 - (1/2048)}{(1/2)}$$
$$= 2 \cdot \frac{2047}{2048} = \frac{2047}{1024} = 1\frac{1023}{1024}$$

Example 3.5.3

One thousand dollars is deposited at the beginning of each year and compounds annually at a rate of 5%. Find the total amount accumulated after 8 years.

Solution

According to Equation (5),

$$A = 1000 \cdot \sum_{k=1}^{8} (1.05)^k$$

Using Equation (6), we obtain

$$A = 1000 \cdot \left(\frac{1.05}{1 - 1.05}\right)[(1 - (1.05)^8]$$
$$= 1000 \cdot 21 \cdot (1.48 - 1) = 10,080 \text{ (dollars)}$$

since $(1.05)^8 = 1.48$ from Example 3.4.1(c).

PROBLEMS 3.5

1. Find each of the following numbers:

(a) $\sum_{i=2}^{4} \frac{1}{i}$

(b) $\sum_{k=2}^{5} (k^3 - 3^k)$

(c) $\sum_{k=1}^{100} (-1)^k$

(d) $\sum_{k=1}^{100} [k^2 - (k+1)^2]$

2. Find each of the following numbers:

(a) $\sum_{i=1}^{100} i$ (Hint: Group the terms.)

✗ (b) $\sum_{k=1}^{100} (4k + 7)$ [Hint: Use Part (a) with Equations (3) and (4).]

3. Find each of the following numbers. You need not simplify.

(a) $\sum_{k=1}^{12} 3^k$

(b) $\sum_{k=1}^{20} 3^k$

(c) $\displaystyle\sum_{k=13}^{20} 3^k$ 　　　　　　(d) $\displaystyle\sum_{k=0}^{10} 1^k$

(e) $\dfrac{1}{6} + \dfrac{1}{12} + \cdots + \dfrac{1}{3 \cdot 2^k}$ 　　　(f) $\dfrac{1}{3} + \dfrac{1}{6} + \dfrac{1}{12} + \cdots + \dfrac{1}{3 \cdot 2^k}$

4. Write an expression for the amount accumulated if $500 is compounded at an annual rate of 8% for each of the following. You need not simplify.
(a) annually for 4 years 　　　　　(b) semiannually for 4 years
(c) quarterly for 4 years 　　　　　(d) monthly for 4 years

5. What interest rate (compounded annually) is required if an investment is to triple in (a) 20 years, (b) 25 years, and (c) 15 years?

6. How many years are required for an investment to double at an interest rate (compounded annually) of (a) 5%, (b) 7%, and (c) 8%?

7. Five hundred dollars is deposited each year for a period of 10 years and compounded at an annual rate of 6%. Find the amount accumulated at the end of 10 years if compounding is (a) annual, (b) semiannual [Hint: Compute an "annualized" rate, called the *effective annual rate*, which accounts for the semiannual compounding and then use Equation (6) with $n = 10$.], (c) semiannual (assume $250 deposited each 6 months.)

8. How many years are necessary to accumulate $10,000 if annual payments of $1000 are made and interest is compounded annually at 6%?

9. Find the yearly deposit necessary to accumulate $10,000 at the end of 10 years with a rate of 5% compounded annually.

---

10. Approximate the interest rate necessary to accumulate $20,000 in 12 years if $1000 is deposited each year and interest is compounded semiannually.

ANSWERS 3.5

1. (a) 13/12, (b) -136, (c) 0, (d) $1 - (101)^2 = -10,200$

2. (a) $50(101) = 5050$, (b) 20,900

3. (a) $3(3^{12} - 1)/2$, (b) $3(3^{20} - 1)/2$, (c) $3^{13}(3^8 - 1)/2$, (d) 11, (e) $[1 - (1/2)^k]/3$, (f) $[2 - (1/2)^k]/3$

4. (a) $500(1.08)^4$, (b) $500(1.04)^8$, (c) $500(1.02)^{16}$, (d) $500(1.00667)^{48}$

5. (a) about 5.65%, (b) 4.5%, (c) 7.6%

6. (a) 15 years, (b) 11 years, (c) about 9 years (not quite double)

7. (a) $6986, (b) $7021, (c) $6919

8. 8

9. $757

Supplement 3.5

Problem S1. Write out the terms of $\sum_{k=1}^{5} (k^2 + 1)$.

Solution: According to Definition 3.5.1,

$$\sum_{k=1}^{5} k^2 + 1 = (1^2 + 1) + (2^2 + 1) + (3^2 + 1) + (4^2 + 1) + (5^2 + 1)$$

Problem S2. Find (a) $\sum_{i=3}^{7} (2i - 1)$ and (b) $\sum_{i=0}^{3} (i - 1)^2$.

Solution: (a) $\sum_{i=3}^{7} (2i - 1) = (2 \cdot 3 - 1) + (2 \cdot 4 - 1) + (2 \cdot 5 - 1) + (2 \cdot 6 - 1) +$

$(2 \cdot 7 - 1) = 2(3 + 4 + 5 + 6 + 7) - 5 = 45.$

(b) $\sum_{i=0}^{3} (i - 1)^2 = (0 - 1)^2 + (1 - 1)^2 + (2 - 1)^2 + (3 - 1)^2 = 1 + 0 + 1 + 4 = 6.$

Problem S3. Find $\sum_{k=1}^{100} \frac{1}{3^k}$.

Solution: Using Equation (6) with $x = 1/3$ and $n = 100$, we obtain

$$\sum_{k=1}^{100} \frac{1}{3^k} = \sum_{k=1}^{100} \left(\frac{1}{3}\right)^k = \frac{1/3}{1 - (1/3)}\left[1 - \left(\frac{1}{3}\right)^{100}\right] = \frac{1}{2}\left[1 - \left(\frac{1}{3}\right)^{100}\right]$$

Problem S4. Find $3 \cdot 2 + 3 \cdot 2^2 + 3 \cdot 2^3 + \cdots + 3 \cdot 2^{50}$.

Solution: Using summation notation and Equation (6), we obtain

$$3 \cdot 2 + 3 \cdot 2^2 + 3 \cdot 2^3 + \cdots + 3 \cdot 2^{50} = \sum_{k=1}^{50} 3 \cdot 2^k = 3 \sum_{k=1}^{50} 2^k$$

$$= 3\left(\frac{2}{1 - 2}\right)(1 - 2^{50})$$

$$= (-6)(1 - 2^{50}) = 6(2^{50} - 1)$$

Problem S5. Five hundred dollars is deposited in a bank paying interest at the rate of 8% compounded annually. What is the amount in the account after 10 years?

Solution: $A = 500(1 + .08)^{10}$ by Equation (1). Our principal task is to find $(1.08)^{10}$. Thus, let $x = (1.08)^{10}$. Then $\log_{10}x = \log_{10}(108)^{10} = 10 \log_{10}1.08 = 10(.0334) = .3340$. So $x = 2.16$ (approximately) since $\log_{10}2.16 = .3345$. Thus, $A = 500(2.16) = \$1080$.

Problem S6. Five hundred dollars is deposited in a bank paying 8% interest compounded quarterly. What is the amount in the account after 10 years?

Solution: In this problem, one period is 3 months or 1/4 year, so the interest rate per period is 2% and the number of periods is 40 (quarterly periods in 10 years). Again, our basic task is to find $(1.02)^{40}$. Let $x = (1.02)^{40}$. Thus, $\log_{10}x = \log_{10}(1.02)^{40} = 40 \log_{10}1.02 = 40(.0086) = .3440$ and $x = (1.02)^{40} = 2.21$ (approximately) since $\log_{10}2.21 = .3444$. Hence, $A = 500(1.02)^{40} = 500(2.21) = \$1105$.

Problem S7. An investor doubles his initial investment in 4 years. What annual rate of growth has he realized?

Solution: From Equation (1) we obtain $2P = P(1 + r)^4$, so $2 = (1 + r)^4$. Using logs to solve for r yields

$$\log_{10}2 = \log_{10}(1 + r)^4 = 4 \log_{10}(1 + r)$$

$$\log_{10}(1 + r) = \frac{\log_{10}2}{4} = \frac{.3010}{4} = .07525$$

From Table B.1, $\log_{10}1.19 = .0755$. Thus, $1 + r = 1.19$ or $r = .19$ or about 19%.

Problem S8. When will an initial deposit triple in value if left in a bank paying 6% compounded annually?

Solution: In this case, the amount after n years is $A = 3P$ where P is the initial deposit. Using Equation (1), we obtain $3P = P(1 + .06)^n$. Next, we use logs to solve for n:

$$\log_{10}3 = \log_{10}(1.06)^n = n \log_{10}1.06 \quad \text{or} \quad n = \frac{\log_{10}3}{\log_{10}1.06} = \frac{.4771}{.0253} = 18.8571$$

Therefore, an initial sum of P will more than triple if left on deposit for 19 years.

Problem S9. One hundred dollars is deposited at the beginning of each year and compounded at the annual rate of 8%. Find the total amount accumulated after 5 years.

Solution: By Equation (5), the accumulated amount is given by $A = 100 \sum_{k=1}^{5} (1.08)^k$.

By Equation (6),

$$\sum_{k=1}^{5} (1.08)^k = \frac{1.08}{1 - 1.08}[1 - (1.08)^5]$$

(Here we let $x = 1.08$ and $n = 5$.) Our task now is to find $(1.08)^5$. Let $y = (1.08)^5$. Thus, $\log_{10}y = 5 \log_{10}1.08 = 5(.0334) = .1670$. Hence, $y = 1.47$ since $\log_{10}1.47 = .1673$. Therefore, $A = [100(1.08)/(1 - 1.08)](1 - 1.47)$ by substitution. Computing, we obtain $A = \$634.50$.

Problem S10. In how many years will an annual deposit of $100 made at the beginning of each year amount to $1000 if interest is compounded annually at 8%?

Solution: According to Equation (5), $1000 = 100 \sum_{k=1}^{n} (1.08)^k$. Thus,

$$10 = \sum_{k=1}^{n} (1.08)^k = \frac{1.08}{1 - 1.08}[1 - (1.08)^n]$$

from Equation (6). Therefore,

$$1 - (1.08)^n = \frac{10(1 - 1.08)}{1.08} \quad \text{or} \quad (1.08)^n = \frac{10(1.08 - 1)}{1.08} + 1$$

$$= 10(.074) + 1 = 1.740$$

Thus, $\log_{10}(1.08)^n = \log_{10}1.74$ or $n \log_{10}1.08 = \log_{10}1.740$. Hence,

$$n = \frac{\log_{10}1.74}{\log_{10}1.08} = \frac{.2405}{.0334} = 7.2$$

Therefore, after 8 years an annual deposit of $100 made at the beginning of each year will accumulate to over $1000.

SUPPLEMENTARY PROBLEMS

S11. Find

(a) $\sum_{i=1}^{4} (2i - 1)^2$ (b) $\sum_{k=0}^{50} \frac{3}{4^k}$

S12. Three hundred dollars is deposited in a bank paying interest at the following indicated rates. What is the amount of the deposit after 6 years (a) 4% compounded annually, (b) 4% compounded semiannually, and (c) 4% compounded quarterly?
S13. When will an initial deposit double in value when invested at 4% compounded annually?
S14. An investor triples the value of his original investment in 10 years. What annual rate of growth has he realized?

Answers

S11. (a) 84, (b) $4[1 - (1/4)^{51}]$
S12. (a) About $379.50, (b) about $380.40, (c) about $380.40 [The actual value is $380.91. Table B.1 is not precise enough for us to distinguish between (b) and (c).]
S13. In 18 years, the initial deposit will slightly more than double in value.
S14. Approximately 11.7%

Achievement Test 3

1. If $f = \{(1, 2), (2, 3), (4, 0)\}$, then find f^{-1}.
2. If $f(x) = \sqrt{x + 2}$, then find $f^{-1}(2)$.
3. Find a formula for $f^{-1}(x)$ if $f(x) = 2/(x - 1)$.
4. Graph $y = f^{-1}(x)$ if the graph of $y = f(x)$ is as shown in Figure 3.A.1.

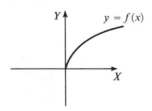

Figure 3.A.1

5. If $f(x) = x^2 + 1$, $x \geq 0$, then find the *domain* of f^{-1}.
6. Compute $\exp_2 5$.
7. Solve $16 = 2^{x^2 - 1}$.
8. Solve $\exp_3(\log_3 7)$.
9. Compute $\log_{16} 4$.
10. Solve $\log_3(1/3) = x$.
11. Solve $\log_2(x - 1) - \log_2 x = -1$.
12. Graph $y = 1 + 2^{-x}$
13. Graph $y = 2^{|x|}$.
14. Graph $y = \log_3(x - 2)$.
15. Graph $y = |\log_3 x|$.
16. Solve $2^x = 1 - x$. (Hint: Graph.)
17. Using logs, compute $(476)(123)$.
18. Using logs, compute 50^{34}.
19. Using logs, compute $346/289$.
20. Using logs, compute $(1.06)^{20}$.
21. Solve $5^x = 8$.

22. Solve for n: $2 = (1.06)^n$.   **23.** Solve for r: $3 = (1 + r)^{20}$.

24. Using interpolation, find N if $\log_{10}N = 2.3621$.

25. Compute $\displaystyle\sum_{i=0}^{5} (i - 1)^2$

26. Find $3 \cdot 4 + 3 \cdot 4^2 + 3 \cdot 4^3 + \cdots + 3 \cdot 4^{15}$.

27. Find the amount accumulated if \$300 is compounded quarterly at an annual rate of 4% for 5 years.

28. How many years are required for an investment to double at an interest rate of 4% compounded annually?

29. What interest rate (compounded annually) is required if an investment is to double in 12 years?

30. If \$100 is deposited at the beginning of each year for ten years and compounded annually at the rate of 6%, then find the amount accumulated at the end of 10 years.

Answers to Achievement Test 3

1. $f^{-1} = \{(2, 1), (3, 2), (0, 4)\}$ **2.** 2

3. $f^{-1}(x) = (2/x) + 1$ **4.** See Figure 3.A.2.

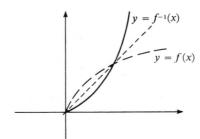

Figure 3.A.2

5. $[1, \infty]$ **6.** 32

7. $x = \pm\sqrt{5}$ **8.** 7

9. 1/2   **10.** $x = -1$

11. $x = 2$ **12.** See Figure 3.A.3.

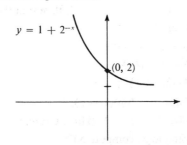

Figure 3.A.3

13. See Figure 3.A.4.

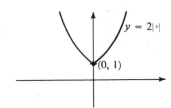

Figure 3.A.4

14. See Figure 3.A.5.

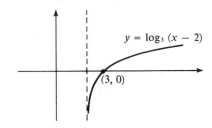

Figure 3.A.5

15. See Figure 3.A.6.

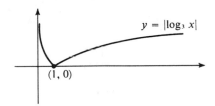

Figure 3.A.6

16. $x = 0$

17. 58,548 (exact answer)

18. $50^{34} \approx 5.83 \times 10^{57}$

19. 1.20

20. 3.2

21. 1.29

22. $11 < n < 12$

23. $.055 < r < .06$

24. 230.2

25. 31

26. $4^{16} - 4$

27. $366

28. 18 years

29. $.055 < r < .06$

30. $1400

4 TRIGONOMETRIC FUNCTIONS

4.0 Introduction

4.1 The Trigonometric Point Function
Supplement

4.2 The Sine and Cosine Functions
Supplement

4.3 The Tangent, Cotangent, Secant, and
Cosecant Functions
Supplement

4.4 Angles and Right Triangles
Supplement

4.5 Inverse Trigonometric Functions
Supplement

4.6 Identities
Supplement

4.0 INTRODUCTION

In Chapter 3 we introduced two basic types of transcendental functions, log and exp. In this chapter we continue our study of transcendental functions by considering the trigonometric functions. While the evolution of trigonometry was initially motivated by the study of plane triangles, the trigonometric functions which resulted have proven to be very useful in many other areas of mathematics and its applications.

4.1 THE TRIGONOMETRIC POINT FUNCTION

We now consider a function P whose domain is the set of real numbers and whose range is the set of points on the *unit circle*, the circle of unit radius centered at the origin (see Figure 4.1.1). That is, the domain of P is R, and the range of P is

$$\{(x, y) : x^2 + y^2 = 1\}.$$

With each nonnegative number t we associate the point $P(t)$ on the unit circle which is located t units counterclockwise from the point $(1, 0)$ along the circle. (See Figure 4.1.2.) Notice that the unit measure equals the radius of the circle. For example, Figure 4.1.2 shows the location of $P(1)$, since the length of the darkened arc equals the radius of the circle.

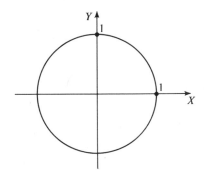

Figure 4.1.1

274

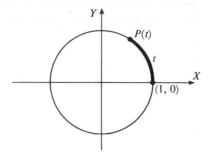

Figure 4.1.2

Q1: Indicate the approximate location of $P(2)$, $P(3)$, and $P(4)$ on the circle shown below.

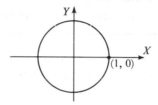

Clearly, $P(0)$ is the point $(1, 0)$. $P(2\pi)$ is also $(1, 0)$ since the circumference of the unit circle is 2π. Since $\pi/2$ is $1/4$ of the circumference, $P(\pi/2)$ is $1/4$ of the way around the circle. Thus, $P(\pi/2) = (0, 1)$. (See Figure 4.1.3.)

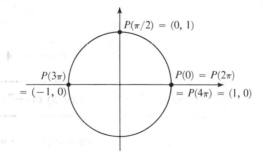

Figure 4.1.3

Q2: Find $P(\pi)$ and $P(3\pi/2)$.

If $t > 2\pi$, we simply continue around the circle until we have traversed a total distance of t units. Thus, $P(3\pi) = (-1, 0)$ and $P(4\pi) = (1, 0)$, as shown in Figure 4.1.3.

A1: The points are shown in the following figure.

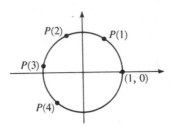

A2: $P(\pi) = (-1, 0)$, $P(3\pi/2) = (0, -1)$

Q3: Find (a) $P(5\pi)$, (b) $P(5\pi/2)$, and (c) $P(50\pi)$.

If $t < 0$, we start at the point $(1, 0)$ but move *clockwise* $|t|$ units along the unit circle. Thus, $P(-\pi/2) = (0, -1)$. (See Figure 4.1.4.)

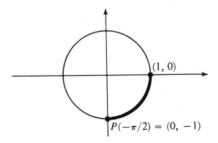

Figure 4.1.4

Q4: Find (a) $P(-3\pi/2)$ and (b) $P(-7\pi/2)$.

We summarize the discussion above with the following definition.

Definition 4.1.1

The *trigonometric point function P* assigns to each real number t the point $P(t)$ on the unit circle which is $|t|$ units along the circle from $(1, 0)$. If $t > 0$, we measure counterclockwise along the circle, and if $t < 0$, we measure clockwise.

(We have assumed that we can precisely measure arc length along a circle. A more detailed discussion of arc length is a topic for more advanced mathematics.)

Although it has been easy to find the coordinates of $P(t)$ for certain values of t, finding precise values of the coordinates for any number t is more difficult. For example, consider the problem of estimating the coordinates of $P(3)$. Since 3 is slightly less than π, $P(3)$ will be slightly above $P(\pi)$ on the circle (see Figure 4.1.5). Thus, the first coordinate of $P(3)$ is almost -1, or perhaps $-.9$, or even $-.99$. The second coordinate of $P(3)$ is a small positive number; perhaps .1 is a good approximation.

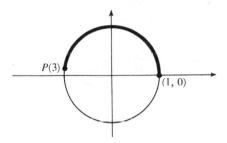

Figure 4.1.5

Q5: Estimate the coordinates of (a) $P(1.5)$ and (b) $P(-1.5)$.

Tables have been prepared to help us estimate more accurately the coordinates of $P(t)$ for any real number t. We shall learn to use these tables in a later section.

Next we shall discuss some useful properties of the function P. Given any real number t, let the coordinates of $P(t)$ be (a, b). Since these coordinates satisfy the equation of the unit circle, we have the following equations:

$$\text{If } P(t) = (a, b), \text{ then } a^2 + b^2 = 1 \tag{1}$$

and

$$-1 \leq a \leq 1, \qquad -1 \leq b \leq 1 \tag{2}$$

We have also observed that moving along the circle 2π units from the point $P(t) = (a, b)$ will take us to the point $P(t + 2\pi)$ which has the coordinates (a, b), the same coordinates as $P(t)$. That is,

$$P(t + 2\pi) = P(t) \quad \text{for any real number } t \tag{3}$$

Similarly, any integer k number of revolutions around the circle from (a, b) traverses a distance $k \cdot 2\pi$, ending at the same point (a, b). In symbols,

$$P(t + k \cdot 2\pi) = P(t) \quad \text{for any } k \in Z \text{ and any } t \in R \tag{4}$$

A3: (a) $(-1, 0)$, (b) $(0, 1)$, (c) $(1, 0)$

A4: (a) $(0, 1)$, (b) $(0, 1)$

A5: (a) $(.07, .997)$, (b) $(.07, -.997)$

A function such as P whose function values repeat with this type of pattern is called periodic.

**Definition
4.1.2**   If there is a positive number p such that $f(x + p) = f(x)$ for each x in the domain of f, then f is *periodic*. The smallest p with this property is called the *period* of f (see Figure 4.1.6).

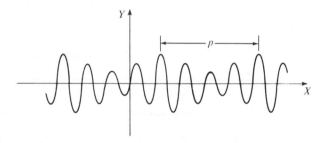

Figure 4.1.6

Q6: Find the period of the function whose graph is shown below.

Next we illustrate another fact which will prove useful later. Since $|t| = |-t|$, the points $P(t)$ and $P(-t)$ are the same distance along the circle from $(1, 0)$. We move clockwise $|t|$ units to reach one point and counterclockwise $|t|$ units to reach the other (Figure 4.1.7). Notice that the first coordinates of both points will always be equal and the second coordinates will be negatives of each other. That is,

$$\text{if } P(t) = (a, b), \text{ then } P(-t) = (a, -b) \qquad (5)$$

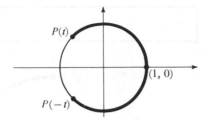

Figure 4.1.7

We conclude this section by determining the coordinates of some other special points on the circle. Since the point $P(\pi/4)$ is equidistant from $P(0)$ and $P(\pi/2)$, it must lie on the line $y = x$ as well as on the circle $x^2 + y^2 = 1$ (see Figure 4.1.8). Therefore, its coordinates are equal, say, (a, a), and satisfy $a^2 + a^2 = 1$ or $a^2 = 1/2$. Since the first coordinate of $P(\pi/4)$ is clearly positive,

$$a = \sqrt{\frac{1}{2}} = \frac{1}{\sqrt{2}} = \frac{\sqrt{2}}{2} \approx .707$$

Thus,

$$P\left(\frac{\pi}{4}\right) = \left(\frac{\sqrt{2}}{2}, \frac{\sqrt{2}}{2}\right)$$

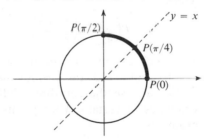

Figure 4.1.8

Next consider the points $P(0)$, $P(\pi/6)$, $P(\pi/3)$ and $P(\pi/2)$, shown in Figure 4.1.9. Since the points are equally spaced along the circle in the first quadrant, $P(\pi/6)$ is the reflection IR of $P(\pi/3)$. Thus, if $P(\pi/3) = (a, b)$, then $P(\pi/6) = (b, a)$. Furthermore, the straight-line distance between $P(\pi/6)$ and $P(0)$ equals the straight-line distance between $P(\pi/3)$ and $P(\pi/6)$. Thus,

$$\sqrt{(b - 1)^2 + a^2} = \sqrt{(a - b)^2 + (b - a)^2}$$

Simplifying yields

$$(b - 1)^2 + a^2 = 2(a - b)^2$$

or

$$b^2 - 2b + 1 + a^2 = 2a^2 - 4ab + 2b^2$$

A6: 2

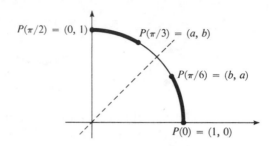

Figure 4.1.9

Using the fact that $a^2 + b^2 = 1$, we obtain

$$2 - 2b = 2 - 4ab \quad \text{or} \quad b = 2ab$$

Since $b > 0$, $a = 1/2$, and (using $a^2 + b^2 = 1$)

$$b = \sqrt{1 - a^2} = \sqrt{1 - \frac{1}{4}} = \sqrt{\frac{3}{4}} = \frac{\sqrt{3}}{2} \approx .866$$

Thus,

$$P\left(\frac{\pi}{3}\right) = \left(\frac{1}{2}, \frac{\sqrt{3}}{2}\right) \quad \text{and} \quad P\left(\frac{\pi}{6}\right) = \left(\frac{\sqrt{3}}{2}, \frac{1}{2}\right)$$

We shall assume familiarity with the coordinates of $P(\pi/4)$, $P(\pi/6)$, and $P(\pi/3)$ throughout this chapter and the next.

We can use the coordinates discovered above to find the coordinates of certain other points, as illustrated in the following examples.

Example 4.1.1   Find the coordinates of $P(5\pi/6)$.

Solution   Since $P(\pi) = (-1, 0)$, $P(5\pi/6)$ is located approximately as shown in Figure 4.1.10. In particular, each of the shaded arcs has length $\pi/6$. This fact and the symmetry of the circle imply that $P(5\pi/6)$ is the reflection XR of $P(\pi/6)$. Since $P(\pi/6) = (\sqrt{3}/2, 1/2)$, $P(5\pi/6) = (-\sqrt{3}/2, 1/2)$.

Q7: Find the coordinates of $P(2\pi/3)$.

Example 4.1.2   Find the coordinates of $P(7\pi/6)$.

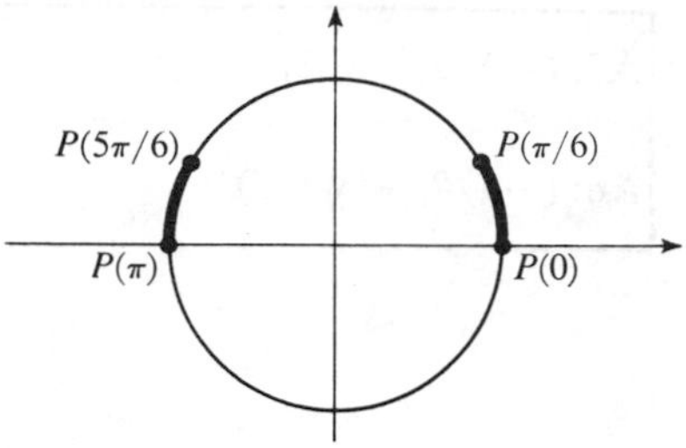

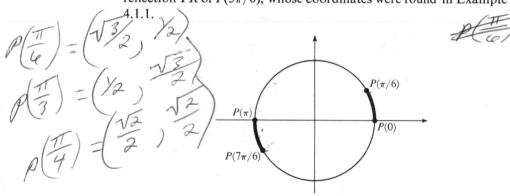

Figure 4.1.10

Solution

Since $P(\pi) = (-1, 0)$, $P(7\pi/6)$ is located in Quadrant III as shown in Figure 4.1.11. We can relate the coordinates of $P(7\pi/6)$ to those of a point in Quadrant I by computing the length of the arc from $P(\pi)$ to $P(7\pi/6)$ to be $\pi/6$ units. Thus, each coordinate of $P(7\pi/6)$ is the negative of the corresponding coordinate of $P(\pi/6)$. Hence,

$$P\left(\frac{7\pi}{6}\right) = \left(-\frac{\sqrt{3}}{2}, -\frac{1}{2}\right)$$

Alternatively, we could have observed that $P(7\pi/6)$ is the reflection YR of $P(5\pi/6)$, whose coordinates were found in Example 4.1.1.

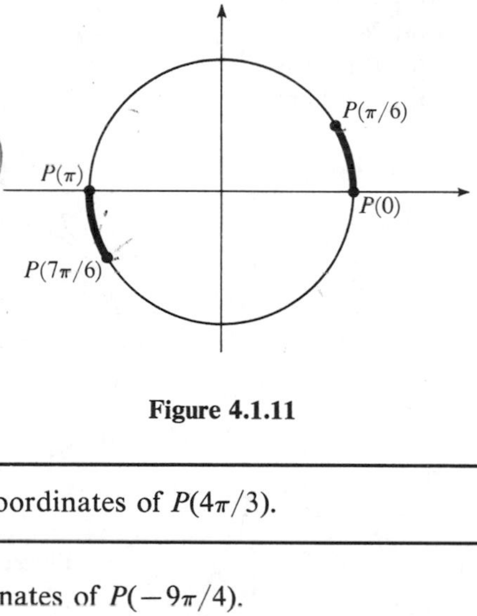

Figure 4.1.11

Q8: Find the coordinates of $P(4\pi/3)$.

Example 4.1.3

Find the coordinates of $P(-9\pi/4)$.

Solution

We see that $P(-9\pi/4) = P(-(\pi/4) - 2\pi) = P(-\pi/4)$. That is, $P(-9\pi/4)$ is located $\pi/4$ units clockwise from the point $(1, 0)$ as shown in Figure 4.1.12. Thus, $P(-9\pi/4)$ is the reflection YR of $P(\pi/4)$. So $P(-9\pi/4) = (\sqrt{2}/2, -\sqrt{2}/2)$.

A7: $(-1/2, \sqrt{3}/2)$

A8: $(-1/2, -\sqrt{3}/2)$

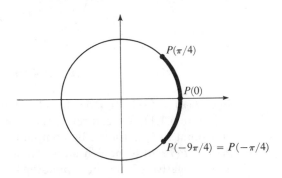

Figure 4.1.12

PROBLEMS   **1.** Locate each of the following on the circle shown in Figure 4.1.13:
4.1   (a) $P(0)$ (b) $P(2)$
(c) $P(\pi)$ (d) $P(-3\pi/2)$
(e) $P(13\pi/4)$ (f) $P(17\pi/2)$
(g) $P(-17\pi/2)$

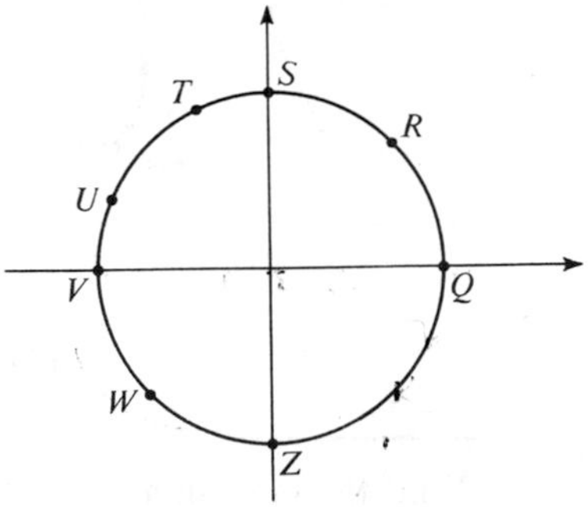

Figure 4.1.13

2. Find the coordinates of each of the following:
(a) $P(3\pi/2)$ (b) $P(-3\pi/2)$
(c) $P(27\pi)$ (d) $P(-27\pi/2)$

3. Estimate the coordinates of each of the following:
(a) $P(1)$ (b) $P(-1)$
(c) $P(4.8)$ (d) $P(6)$

4. Find the coordinates of each of the following:
(a) $P(3\pi/4)$ (b) $P(-3\pi/4)$
(c) $P(-\pi/6)$ (d) $P(11\pi/6)$
(e) $P(-2\pi/3)$

5. Find the period of f if the graph of f is as shown in Figure 4.1.14.

(a)
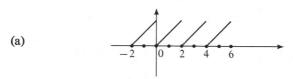

(b)
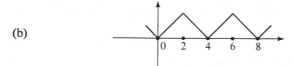

(c)
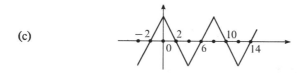

Figure 4.1.14

6. Find all t such that
(a) $P(t) = P(1)$.
(b) $P(t) = P(2\pi - t)$.
(c) the first coordinate of $P(t)$ equals the first coordinate of $P(\pi/6)$.
(d) the second coordinate of $P(t)$ equals the second coordinate of $P(\pi/6)$.
(e) the second coordinate of $P(t)$ equals the first coordinate of $P(\pi/6)$.
(f) $P(t)$ is the reflection YR of $P(-t)$.
(g) $P(t)$ is the reflection XR of $P(-t)$.

7. Which of the following statements are true for all real t:

(a) The first coordinate of $P(t)$ equals the second coordinate of $P\left(t + \dfrac{\pi}{2}\right)$.

(b) The second coodrinate of $P(t)$ equals the first coordinate of $P\left(t - \dfrac{\pi}{2}\right)$.

(c) The first coordinate of $P(t)$ equals the negative of the first coordinate of $P(t + \pi)$.

ANSWERS
4.1

1. (a) Q, (b) T, (c) V, (d) S, (e) W, (f) S, (g) Z

2. (a) $(0, -1)$, (b) $(0, 1)$, (c) $(-1, 0)$, (d) $(0, 1)$

3. (a) $(.5403, .8415)$, (b) $(.5403, -.8415)$, (c) $(.0907, -.9959)$, (d) $(.9611, -.2764)$

4. (a) $(-\sqrt{2}/2, \sqrt{2}/2)$, (b) $(-\sqrt{2}/2, -\sqrt{2}/2)$, (c) $(\sqrt{3}/2, -1/2)$, (d) $(\sqrt{3}/2, -1/2)$, (e) $(-1/2, -\sqrt{3}/2)$

5. (a) 2, (b) 4, (c) 8

6. (a) $\{1 + 2k\pi : k \in Z\}$

(b) $\{k\pi : k \in Z\}$

(c) $\left\{\dfrac{\pi}{6} + 2k\pi : k \in Z\right\} \cup \left\{-\dfrac{\pi}{6} + 2k\pi : k \in Z\right\}$

(d) $\left\{\dfrac{\pi}{6} + 2k\pi : k \in Z\right\} \cup \left\{\dfrac{5\pi}{6} + 2k\pi : k \in Z\right\}$

(e) $\left\{\dfrac{\pi}{3} + 2k\pi : k \in Z\right\} \cup \left\{\dfrac{2\pi}{3} + 2k\pi : k \in Z\right\}$

(f) R

(g) $\varnothing$

7. All are true.

Supplement 4.1

Problem S1.   Find the coordinates of $P(7\pi/2)$.

Solution: Since $7\pi/2 = 2\pi + (3\pi/2)$, $P(7\pi/2) = P(3\pi/2)$ by the periodic property [Equation (4)]. Now $P(3\pi/2)$ is three-fourths of the way along the unit circle from $(1, 0)$, so $P(3\pi/2) = (0, -1)$. Thus, $P(7\pi/2) = (0, -1)$. (See Figure 4.1.15.)

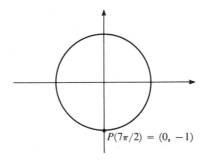

Figure 4.1.15

Problem S2.   Find the coordinates of $P(-3\pi)$.

Solution: Since $-3\pi = -2\pi + (-\pi)$, $P(-3\pi) = P(-\pi)$. Since $-\pi$ is negative, start at the point $(1, 0)$ and move halfway along the unit circle in a clockwise direction. Thus, $P(-\pi) = P(-3\pi) = (-1, 0)$, as shown in Figure 4.1.16.

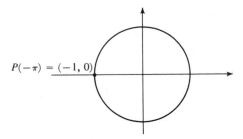

Figure 4.1.16

DRILL PROBLEMS

S3. Find the coordinates of $P(9\pi)$.
S4. Find the coordinates of $P(-13\pi/2)$.

Answers

S3. $(-1, 0)$
S4. $(0, -1)$

Problem S5. Estimate the coordinates of $P(4)$.

Solution: Since $5\pi/4 = \pi + (\pi/4) \approx 3.14 + .78 = 3.92$, $P(4)$ is close to $P(5\pi/4)$. Also, $P(4)$ is between $P(5\pi/4)$ and $P(3\pi/2)$. Thus, locate $P(4)$ as shown in Figure 4.1.17. It appears that $P(4) \approx (-.6, -.8)$.

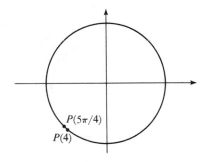

Figure 4.1.17

Problem S6. Estimate the coordinates of $P(7)$.

Solution: Since $7 = 6.28 + .72$, $P(7) \approx P(.72)$ by the periodic property. Since .72 is close to $\pi/4 \approx 3.14/4 = .785$, locate $P(.72)$ as shown in Figure 4.1.18. It appears that $P(.72) \approx (.8, .6)$.

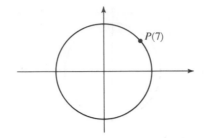

Figure 4.1.18

DRILL PROBLEMS

S7. Estimate the coordinates of $P(2)$.
S8. Estimate the coordinates of $P(-3)$.

Answers

S7. $P(2) \approx (-.4, .9)$
S8. $P(-3) \approx (-.99, -.1)$

Problem S9.   Find the coordinates of $P(4\pi/3)$.

Solution: Recall that if t is any multiple of $\pi/6$ or $\pi/3$ [and $P(t)$ is not on one of the coordinate axes], then each coordinate of $P(t)$ is either $\pm 1/2$ or $\pm\sqrt{3}/2$. Since $4\pi/3 = \pi + (\pi/3)$, $P(4\pi/3)$ is located approximately as shown in Figure 4.1.19. Each shaded arc has length $\pi/3$. From symmetry and because $P(4\pi/3)$ is in the third quadrant, it follows that $P(4\pi/3) = (-1/2, -\sqrt{3}/2.)$

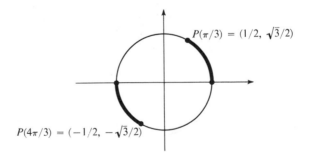

$P(\pi/3) = (1/2, \sqrt{3}/2)$

$P(4\pi/3) = (-1/2, -\sqrt{3}/2)$

Figure 4.1.19

Problem S10.   Find the coordinates of $P(-5\pi/4)$.

Solution: Recall that if t is any odd multiple of $\pi/4$, then each coordinate of $P(t)$ is either $\sqrt{2}/2$ or $-\sqrt{2}/2$. (Even multiples of $\pi/4$ yield points on the coordinate axes.) Since $-5\pi/4 = -\pi - (\pi/4)$, $P(-5\pi/4)$ is located approximately as shown in Figure 4.1.20. The shaded arcs are equal in length. From symmetry and because $P(-5\pi/4)$ is in the second quadrant, $P(-5\pi/4) = (-\sqrt{2}/2, \sqrt{2}/2)$.

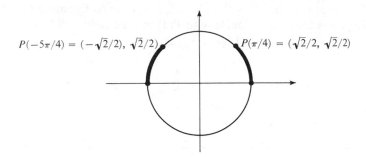

Figure 4.1.20

Problem S11. Find the coordinates of $P(17\pi/6)$.

Solution: Since $17\pi/6 = 2\pi + (5\pi/6)$, $P(17\pi/6) = P(5\pi/6)$ by the periodic property. $P(5\pi/6)$ was previously determined to be $(-\sqrt{3}/2, 1/2)$. (See Example 4.1.1). Thus, $P(17\pi/6) = (-\sqrt{3}/2, 1/2)$ also.

DRILL PROBLEMS

S12. Find the coordinates of $P(11\pi/4)$.
S13. Find the coordinates of $P(-10\pi/3)$.
S14. Find the coordinates of $P(5\pi/3)$.
S15. Find the coordinates of $P(-7\pi/6)$.

Answers

The points are shown in Figure 4.1.21.
S12. $(-\sqrt{2}/2, \sqrt{2}/2) = A$
S13. $(-1/2, \sqrt{3}/2) = B$
S14. $(1/2, -\sqrt{3}/2) = C$
S15. $(-\sqrt{3}/2, 1/2) = D$

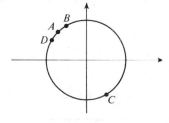

Figure 4.1.21

Problem S16. Find all t such that $P(t) = P(3)$.

Solution: By the periodic property, $P(3) = P(3 + 2\pi) = P(3 + 2 \cdot 2\pi) = \ldots$. In general, $P(3) = P(3 + k \cdot 2\pi)$, $k = 0, \pm 1, \pm 2, \ldots$ [refer to Equation (4)]. It follows that the only points that will coincide with $P(3)$ are the points $P(3 + k \cdot 2\pi)$, $k \in Z$. Thus, $t = 3 + k \cdot 2\pi$, $k \in Z$. See Figure 4.1.22.

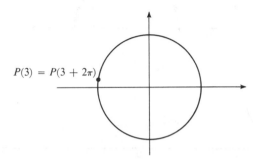

Figure 4.1.22

Problem S17. Find $\{t :$ the first coordinate of $P(\pi/4)$ equals the second coordinate of $P(t)\}$.

Solution: Since $P(\pi/4) = (\sqrt{2}/2, \sqrt{2}/2)$, its first coordinate is $\sqrt{2}/2$. Thus, we are to determine when the second coordinate of $P(t)$ equals $\sqrt{2}/2$. A glance at the unit circle shown in Figure 4.1.23 verifies that exactly two distinct points exist where the second coordinate equals $\sqrt{2}/2$. They are $P(\pi/4)$ and $P(3\pi/4)$. Thus, by the periodic property, the second coordinate of all points $P(\frac{1}{4}\pi + k \cdot 2\pi)$ and $P(\frac{3}{4}\pi + k \cdot 2\pi)$ is $\sqrt{2}/2$. Hence, the solution set is

$$\left\{ \frac{3}{4}\pi + k \cdot 2\pi : k \in Z \right\} \cup \left\{ \frac{1}{4}\pi + k \cdot 2\pi : k \in Z \right\}$$

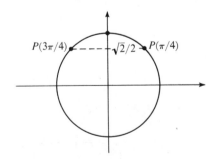

Figure 4.1.23

Problem S18. (A more difficult example.) Show that the first coordinate of $P(t)$ equals the negative of the second coordinate of $P(t - \frac{1}{2}\pi)$ for all $t \in R$.

Solution: Let $P(t)$ be located on the unit circle as shown in Figure 4.1.24. Then $P(t - \frac{1}{2}\pi)$ is located $\frac{1}{2}\pi$ units (one quarter of the circumference) from $P(t)$ on the circle measured in a clockwise direction. Since $\frac{1}{2}\pi - t$ and $t - \frac{1}{2}\pi$ differ by sign only, the shaded arcs are equal in length. Note that $P(\frac{1}{2}\pi - t)$ and $P(t - \frac{1}{2}\pi)$ have the same first coordinates but the second coordinates differ by algebraic sign. Also, by symmetry, the points $P(t)$ and $P(\frac{1}{2}\pi - t)$ are images of each other under the reflection IR, that is, a reflection about the line $y = x$. Hence, if $P(t) = (x, y)$, then $P(\frac{1}{2}\pi - t) = (y, x)$. Then also $P(t - \frac{1}{2}\pi) = (y, -x)$ by an earlier remark. Thus, if $P(t) = (x, y)$, then $P(t - \frac{1}{2}\pi) = (y, -x)$. Hence, the first coordinate of $P(t)$ equals the negative of the second coordinate of $P(t - \frac{1}{2}\pi)$ for all $t \in R$.

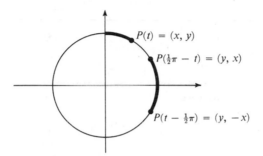

Figure 4.1.24

Problem S19. Indicate whether the following statement is true or false: The second coordinate of $P(t)$ is the negative of the second coordinate of $P(t + \pi)$ for all $t \in R$.

Solution: Let $P(t)$ be located on the unit circle as shown in Figure 4.1.25. $P(t + \pi)$ is located a distance of π units [halfway along the circumference from $P(t)$ measured in a counterclockwise direction]. Observe that the shaded arcs have equal length. Thus, $P(t)$ and $P(t + \pi)$ lie on a line through the origin. By symmetry, if $P(t) = (x, y)$, then $P(t + \pi) = (-x, -y)$. So the second coordinate of $P(t)$ is equal to the negative of the second coordinate of $P(t + \pi)$. Hence, the indicated statement is true.

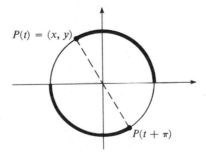

Figure 4.1.25

DRILL PROBLEMS

S20. Find the solution set for the equation $P(t) = P(\pi/6)$.

S21. Find all t such that the first coordinate of $P(t)$ equals the negative of the second coordinate of $P(-\pi/3)$.

S22. Which of the following statements are true for all t:

(a) The first coordinate of $P(t)$ equals the second coordinate of $P(t - \pi)$.

(b) The second coordinate of $P(t)$ equals the negative of the second coordinate of $P(t - \pi)$.

(c) The first coordinate of $P(t)$ equals the second coordinate of $P(\frac{1}{2}\pi - t)$.

Answers

S20. $\{\frac{1}{6}\pi + k \cdot 2\pi : k \in Z\}$

S21. $\{\pm\frac{1}{6}\pi + k \cdot 2\pi : k \in Z\}$

S22. (a) False, (b) true, (c) true

SUPPLEMENTARY PROBLEMS

S23. Find the coordinates of each of the following:

(a) $P(-17\pi/2)$ (b) $P(-5\pi/6)$

(c) $P(17\pi/4)$ (d) $P(14\pi/3)$

S24. Estimate the coordinates of $P(4.5)$.

S25. Find all t such that (a) $P(t) = P(-\pi/3)$ and (b) the second coordinate of $P(t)$ equals the first coordinate of $P(3\pi)$.

S26. Which of the following statements are true for all t:

(a) The first coordinate of $P(t)$ equals the second coordinate of $P(\pi - t)$.

(b) The second coordinate of $P(t)$ equals the first coordinate of $P(t + \frac{3}{2}\pi)$.

Answers

S23. (a) $(0, -1)$, (b) $(-\sqrt{3}/2, -1/2)$, (c) $(\sqrt{2}/2, \sqrt{2}/2)$, (d) $(-1/2, \sqrt{3}/2)$

S24. $(-.2, -.9)$

S25. (a) $\{-\frac{1}{3}\pi + k \cdot 2\pi : k \in Z\}$, (b) $\{\frac{3}{2}\pi + k \cdot 2\pi : k \in Z\}$

S26. Part (b) is true.

4.2 THE SINE AND COSINE FUNCTIONS

In the preceding section we defined the trigonometric point function P whose domain is the set of real numbers and whose range is the set of points on the unit circle. In particular, for any input t the output $P(t)$ is that point on the unit circle which is $|t|$ units measured along the circle from the point $(1, 0)$; we move counterclockwise from $(1, 0)$ if $t > 0$, and clockwise if $t < 0$ (Figure 4.2.1). For example, $P(\pi/2) = (0, 1)$, $P(-\pi/6) = (\sqrt{3}/2, -1/2)$ and $P(3\pi) = (-1, 0)$.

In this section we shall use the function P to define the two basic trigonometric functions, sine and cosine. These functions possess

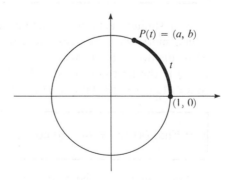

Figure 4.2.1

many interesting properties, some of which we develop in this text. Many of these are useful in a study of calculus and its applications.

Consider the function which, for any real number input t, produces as output the second coordinate of $P(t)$. The name of this function is *sine*, and its output is denoted sin(t). That is, given any $t \in R$, if $P(t) = (a, b)$, then sin(t) is the real number b. For example, sin(0) is the second coordinate of $P(0) = (1, 0)$, so sin(0) = 0 (Figure 4.2.2). Likewise, sin($\pi/2$) = 1, the second coordinate of $P(\pi/2) = (0, 1)$.

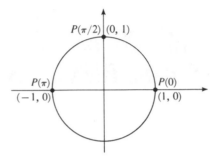

Figure 4.2.2

Q1: Find (a) sin(π) and (b) sin($-\pi/2$).

We often omit the parentheses and write sin t rather than sin(t). By convention, the function operation takes precedence over addition and subtraction, but not over multiplication, division, or exponentiation. Thus, sin $t + 1 = $ (sin t) $+ 1$, not sin($t + 1$). However, sin $2t = $ sin($2t$) and sin $t^2 = $ sin(t^2), not (sin t)^2.

Q2: Find (a) $\sin\left(\dfrac{3\pi}{2} - \pi\right)$ and (b) $3 \sin \dfrac{\pi}{2} - \pi$.

In a similar fashion we define another function named *cosine*, which pairs with each real number t the first coordinate of $P(t)$. This function value is denoted cos(t), or simply cos t. For example, cos 0 is the first coordinate of $P(0)$, so cos 0 = 1 (Figure 4.1.2). Since $P(\pi/2) = (0, 1)$, cos $\pi/2 = 0$.

Q3: Find (a) cos π and (b) cos $\dfrac{3\pi}{2}$.

We summarize some of the discussion above with the following definition.

Definition 4.2.1

Let t be any real number. If the coordinates of the trigonometric point $P(t)$ are (a, b), then $\sin t = b$ and $\cos t = a$. (See Figure 4.2.3.)

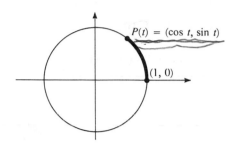

Figure 4.2.3

Q4: Find (a) $\sin(\pi/6)$ and $\cos(\pi/6)$, and (b) $\sin(5\pi/6)$ and $\cos(5\pi/6)$.

From Definition 4.2.1 we see that the domain of each of the functions sine and cosine is R, the set of all real numbers. Since the coordinates of points on the unit circle attain precisely all the numbers in $[-1, 1]$, the range of each of the functions sine and cosine is $[-1, 1]$.

Other properties of these functions follow immediately from their definitions. Since $P(t) = (a, b)$ is on the unit circle, $a^2 + b^2 = 1$. By Definition 4.2.1, $(\cos t)^2 + (\sin t)^2 = 1$ for every real number t. We usually write $(\sin t)^2$ as $\sin^2 t$ and $(\cos t)^2$ as $\cos^2 t$. Thus,

$$\sin^2 t + \cos^2 t = 1 \quad \text{for any } t \in R \tag{1}$$

We often call such a statement an *identity* to emphasize that it is not merely true for some numbers t and false for others, but that it is true for every value of the variable for which all the terms are defined.

In the previous section we observed that if $P(t) = (a, b)$, then $P(-t) = (a, -b)$ (see Figure 4.2.4). Since $\cos t = a$ and $\cos(-t) = a$, $\cos t = \cos(-t)$. Similarly, $\sin t = b$ and $\sin(-t) = -b = -\sin t$. Since these statements are true for any $t \in R$, we have the identities

$$\sin(-t) = -\sin t \quad \text{and} \quad \cos(-t) = \cos t \quad \text{for any } t \in R \tag{2}$$

Example 4.2.1

If $\cos t = 1/3$ and $-\pi/2 < t < 0$, find $\sin t$ and $\sin(-t)$.

A1: (a) 0, (b) -1

A2: (a) 1, (b) $3 - \pi$

A3: (a) -1, (b) 0

A4: (a) $1/2, \sqrt{3}/2$, (b) $1/2, -\sqrt{3}/2$

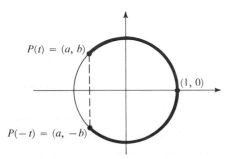

Figure 4.2.4

Solution

We use the fact that $\sin^2 t + \cos^2 t = 1$ to write $\sin^2 t = 1 - (1/3)^2 = 8/9$. Since $-\pi/2 < t < 0$, $P(t)$ lies in Quadrant IV, so $\sin t$, the second coordinate of $P(t)$, is negative. Thus,

$$\sin t = -\sqrt{\frac{8}{9}} = -\frac{2\sqrt{2}}{3}$$

From Equation (2) we obtain $\sin(-t) = -\sin t = 2\sqrt{2}/3$.

Q5: If $\sin t = 2/3$ and $\cos t < 0$, find (a) $\sin(-t)$, (b) $\cos t$, and (c) $\cos(-t)$.

We have also observed the periodic nature of the function P. It follows that each of the functions sine and cosine is also periodic. In fact, the period of each is 2π, since

$$\left. \begin{array}{l} \sin(t + 2k\pi) = \sin t \\ \cos(t + 2k\pi) = \cos t \end{array} \right\} \text{ for } t \in R \text{ and } k \in Z$$

Example 4.2.2

Find all numbers t such that $\sin t = 1/2$.

Solution

Since $\sin t$ is the second coordinate of $P(t)$, we find all points on the unit circle with second coordinate $1/2$, labeled U and V in Figure 4.2.4. From our earlier work we know that $P(\pi/6)$ coincides with U and $P(5\pi/6)$ with V. But from the periodicity of P we know that $P(\pi/6)$ also coincides with $P(\frac{1}{6}\pi + 2\pi)$, $P(\frac{1}{6}\pi + 4\pi)$, etc. Therefore, $P(t)$ coincides with U if $t = \frac{1}{6}\pi + 2k\pi$, $k \in Z$. A similar argument for V yields the solution set

$$\left\{t : t = \frac{\pi}{6} + 2k\pi, \ k \in Z\right\} \cup \left\{t : t = \frac{5\pi}{6} + 2k\pi, \ k \in Z\right\}$$

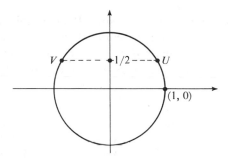

Figure 4.2.5

Q6: Find all t such that $\cos t = 1/2$.

Example 4.2.2 dealt with the familiar number 1/2. We already knew that $\sin(\pi/6) = 1/2$, but suppose that we want to express sin .5 as a decimal or to find t such that $\sin t = .9$. We could draw a large unit circle and try to measure accurately, or we could use a prepared table, such as Table B.2 in Appendix B. From the table we see that sin .5 = .4794, and that one number t such that $\sin t = .9$ is about 1.12. Of course, we write = with the understanding that these numbers are four-place approximations.

Q7: Find (a) sin .9 and (b) two distinct numbers t such that $\sin t = .777$.

Let us now sketch the graph of the sine function. We already know some points of the graph, such as (0, 0), ($\pi/6$, 1/2), and ($\pi/2$, 1). We can obtain others from Table B.2. We have plotted several of these points in Figure 4.2.6. We can describe geometrically the location of any point (t, sin t) by using the unit circle as indicated in Figure 4.2.7. Given t, locate the point $P(t)$ on the unit circle. The second coordinate of $P(t)$, namely, b, is the number sin t. On different coordinate axes, we first move $|t|$ units horizontally from the origin (to the right if $t > 0$) and then $|b| = |\sin t|$ units vertically (up if $b > 0$) to obtain the point (t, sin t) of the

A5: (a) $-2/3$ (b) $-\sqrt{5}/3$ (c) $-\sqrt{5}/3$

A6: $\{\tfrac{1}{3}\pi + 2k\pi : k \in Z\} \cup \{-\tfrac{1}{3}\pi + 2k\pi : k \in Z\}$.

A7: (a) .7833, (b) for example, .89 and $.89 + 2\pi \approx 4.03$.

graph of $f(t) = \sin t$. If we do this for enough numbers $t \in [0, 2\pi]$, we obtain the curve shown at the right of Figure 4.2.8. Since the sine function is periodic with period 2π, we need only repeat this "cycle" every 2π units to obtain the complete graph.

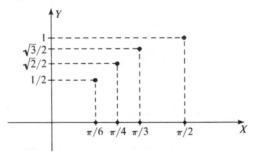

Figure 4.2.6

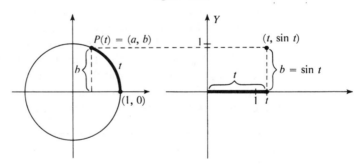

Figure 4.2.7

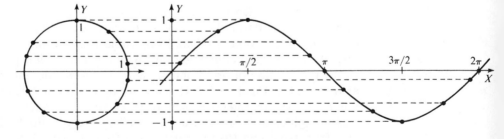

Figure 4.2.8

In this chapter thus far, we avoided using the letter x in our discussion. We have instead used *first coordinate* rather than *X-coordinate*. We have also used t to indicate the input to the functions P, sine, and cosine rather than x. After all, we could not use x as input (arc length) to the function P and then speak of the X-coordinate of $P(x)$ without considerable confusion. However, it is quite common to use x to indicate the input to the functions sine and cosine. With this in mind, we have labeled the horizontal axis as the X-axis in the sketch of the graph of the sine function in Figure 4.2.9.

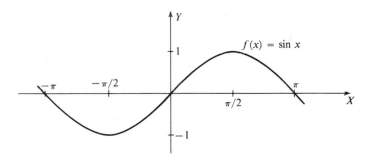

Figure 4.2.9

We can sketch the graph of the cosine function by plotting points such as $(0,\ 1)$, $(\pi/6,\ \sqrt{3}/2)$, $(\pi/4,\ \sqrt{2}/2)$, $(\pi/3,\ 1/2)$, and $(\pi/2,\ 0)$ which belong to the graph and joining them with a smooth curve. The graph of the cosine function appears in Figure 4.2.10. Again, notice the periodicity.

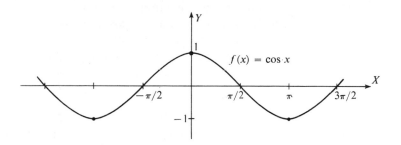

Figure 4.2.10

Example 4.2.3

Graph (a) $y = 2 \sin x$ and (b) $y = \sin 2x$.

Solution

(a) If we start with the graph of $y = \sin x$ (the dotted graph shown in Figure 4.2.11) and double the Y-coordinate of each point of the graph, we will obtain the graph of $y = 2 \sin x$ (dashed graph). More formally, we have used the distortion $YD/2$:

$$y = \sin x \xrightarrow{YD/2} y = 2 \sin x$$

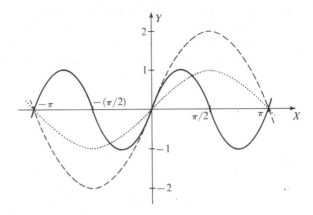

Figure 4.2.11

(b) Again we start with the graph of $y = \sin x$. Since the distortion XD/a replaces x with $(1/a)x$ everywhere in the equation, we apply the distortion $XD/\frac{1}{2}$ to the graph of $y = \sin x$ (dotted graph) to obtain the graph of $y = \sin 2x$ (solid graph):

$$y = \sin x \xrightarrow{XD/\frac{1}{2}} y = \sin 2x$$

Example 4.2.4   Graph $y = \cos(x - \frac{1}{2}\pi)$.

Solution   We obtain the equation $y = \cos(x - \frac{1}{2}\pi)$ by replacing x by $x - \frac{1}{2}\pi$ in the equation $y = \cos x$. Therefore, we obtain the graph of $y = \cos(x - \frac{1}{2}\pi)$ by applying the translation $XT/\frac{1}{2}\pi$ to the graph of $y = \cos x$. (See Figure 4.2.12.) The resulting graph

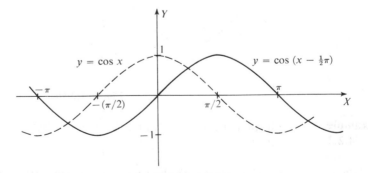

Figure 4.2.12

appears to be the same as the graph of $y = \sin x$. In a later section we shall verify the identities $\cos(x - \frac{1}{2}\pi) = \cos(\frac{1}{2}\pi - x) = \sin x$ and confirm that our observation is correct.

PROBLEMS 4.2

1. Find $\sin t$ and $\cos t$ if $P(t)$ is the point (a) $P(\pi/3)$, (b) $(.6, .8)$, and (c) $P(-7\pi/2)$.

2. Find each of the following numbers:
(a) $\sin 3\pi$ (b) $\cos(-2\pi/3)$
(c) $\sin(11\pi/6)$ (d) $\cos(15\pi/2)$
(e) $\cos(-5\pi/6)$ (f) $\cos(5\pi/4)$

3. Use the unit circle to estimate each of the following numbers:
(a) $\sin 3$ (b) $\cos 2$
(c) $\sin(-2)$

4. Use Table B.2 to find each of the following numbers:
(a) $\sin(1/4)$ (b) $\dfrac{\sin 1}{4}$
(c) $\cos 1 - 2\cos(1/2)$

5. Which of the following are true:
(a) $\sin 2 < \sin 3$ (b) $\cos 2 < \cos 3$
(c) $\sin 10 > 0$ (d) $\sin(-10) > 0$
(e) $\cos(\sin 3) > 0$ (f) $\sin(\sin 4) < 0$

6. Find all $t \in [0, 2\pi]$ such that
(a) $\cos t = 0$ (b) $\sin t = 1$
(c) $\sin t = \cos t$ (d) $\sin t > \cos t$

7. Graph each of the following:
(a) $y = 3 \cos x$ (b) $y = \cos \frac{1}{2}x$
(c) $y = 3 \cos \frac{1}{2}x$ (d) $y = 2 \sin(x - \frac{1}{3}\pi)$
(e) $y = 1 + \sin(x + \frac{1}{4}\pi)$ (f) $y = -1 - \cos(x - \frac{1}{4}\pi)$

8. Find all t such that
(a) $\sin t = -1/2$ (b) $(\sin t)(2 \cos t - 1) = 0$
(c) $\sin t = -\cos t$ (d) $\sin t < -\cos t$
(e) $2 \sin^2 t + \sin t - 1 = 0$ (f) $2 \cos^2 t - \sin t - 1 = 0$

9. Which of the following are true:
(a) $\sin(\frac{1}{2}\pi - t) = \cos t$ (b) $\sin(t + \frac{1}{2}\pi) = \cos t$
(c) $\cos(\frac{1}{2}\pi - t) = \sin t$ (d) $\cos(t + \frac{1}{2}\pi) = \sin t$

10. Sketch the graphs of
(a) $y = \sin x$ and $y = \dfrac{1}{\sin x}$ (b) $y = \cos x$ and $y = \dfrac{1}{\cos x}$

11. Does the reflection IR of $y = \sin x$, $-\pi/2 \le x \le \pi/2$, represent a function? If so, what function?

1. (a) $\sqrt{3}/2$, 1/2, (b) .8, .6, (c) 1, 0

2. (a) 0, (b) $-1/2$, (c) $-1/2$, (d) 0, (e) $-\sqrt{3}/2$, (f) $-\sqrt{2}/2$

3. (a) .1395, (b) $-.4176$, (c) $-.9086$

4. (a) .2474, (b) .2104, (c) -1.2149

5. (d), (e), and (f)

6. (a) $\{\pi/2, 3\pi/2\}$, (b) $\{\pi/2\}$, (c) $\{\pi/4, 5\pi/4\}$, (d) $(\pi/4, 5\pi/4)$

7. (a) See Figure 4.2.13.

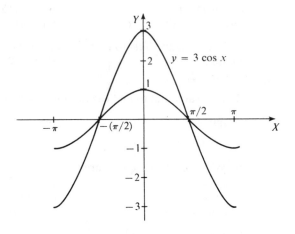

Figure 4.2.13

(b) See Figure 4.2.14.

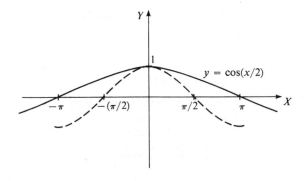

Figure 4.2.14

(c) See Figure 4.2.15.

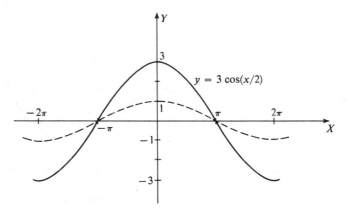

Figure 4.2.15

(d) See Figure 4.2.16.

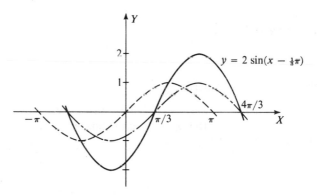

Figure 4.2.16

(e) See Figure 4.2.17.

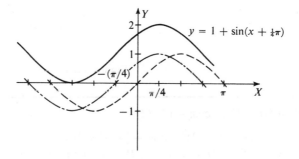

Figure 4.2.17

(f) See Figure 4.2.18.

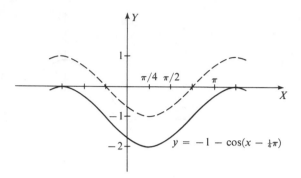

$$y = -1 - \cos(x - \tfrac{1}{4}\pi)$$

Figure 4.2.18

Supplement 4.2

Problem S1.   Find $\sin 15\pi$.

Solution: $\sin 15\pi = \sin(\pi + 7 \cdot 2\pi) = \sin \pi$ by the periodic property. Hence, $\sin 15\pi = \sin \pi = 0$, the *second* coordinate of $P(\pi) = P(15\pi)$. (See Definition 4.2.1 and Figure 4.2.19.)

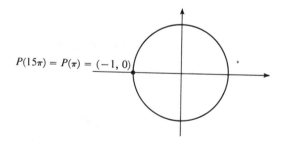

$$P(15\pi) = P(\pi) = (-1, 0)$$

Figure 4.2.19

Problem S2.   Find $\cos(-13\pi/2)$.

Solution: $\cos(-13\pi/2) = \cos(-\tfrac{1}{2}\pi - 3 \cdot 2\pi) = \cos(-\tfrac{1}{2}\pi)$ by the periodic property. Hence, $\cos(-13\pi/2) = \cos(-\pi/2) = 0$, the *first* coordinate of $P(-\pi/2)$. See Figure 4.2.20.

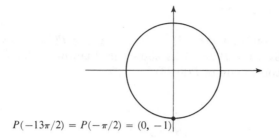

$P(-13\pi/2) = P(-\pi/2) = (0, -1)$

Figure 4.2.20

Problem S3. Find $\cos(7\pi/6)$.

Solution: Locate $P(7\pi/6)$ as shown in Figure 4.2.21. Then $\cos(7\pi/6) = -\sqrt{3}/2$, the *first* coordinate of $P(7\pi/6)$.

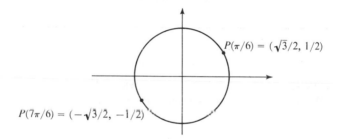

$P(\pi/6) = (\sqrt{3}/2, 1/2)$

$P(7\pi/6) = (-\sqrt{3}/2, -1/2)$

Figure 4.2.21

Problem S4. Find $\sin(-3\pi/4)$.

Solution: Locate $P(-3\pi/4)$ as shown in Figure 4.2.22. Hence, $\sin(-3\pi/4) = -\sqrt{2}/2$, the *second* coordinate of $P(-3\pi/4)$.

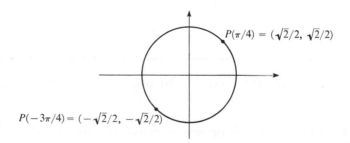

$P(\pi/4) = (\sqrt{2}/2, \sqrt{2}/2)$

$P(-3\pi/4) = (-\sqrt{2}/2, -\sqrt{2}/2)$

Figure 4.2.22

Problem S5.   Find cos $8\pi/3$.

Solution: First note that $P(8\pi/3) = P(2\pi + \frac{2}{3}\pi) = P(2\pi/3)$, by the periodic property. Locate $P(8\pi/3) = P(2\pi/3)$ as shown in Figure 4.2.23. Hence, $\cos(8\pi/3) = -1/2$, the *first* coordinate of $P(8\pi/3)$.

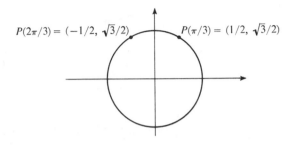

$P(2\pi/3) = (-1/2, \sqrt{3}/2)$ $P(\pi/3) = (1/2, \sqrt{3}/2)$

Figure 4.2.23

Problem S6.   Estimate sin 2.

Solution: Locate (approximately) $P(2)$ as shown in Figure 4.2.24. It appears that $P(2) \approx (-.4, .9)$, so sin $2 \approx .9$, the second coordinate.

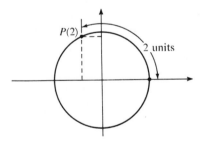

$P(2)$

2 units

Figure 4.2.24

Problem S7.   Which of the following is/are true: (a) sin 2 > sin 3 and (b) sin 1 < sin 4?

Solution: (a) is the only true statement. A glance at Figure 4.2.25 will confirm this result. (a) Clearly, sin 2 > sin 3 [sin 2 is the second coordinate of $P(2)$ — the length of longer vertical segment]. (b) Observe that sin 1 > sin 4 since sin 1 is *positive* and sin 4 is *negative*. [$P(4)$ is in the third quadrant, and both coordinates are negative.]

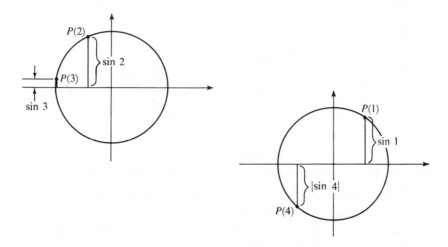

Figure 4.2.25

Problem S8. Find all $t \in [0, 2\pi]$ such that $\sin t = -\sqrt{3}/2$.

Solution: $P(4\pi/3)$ and $P(5\pi/3)$ are the only trig points $P(t)$ for $t \in [0, 2\pi]$ with the second coordinate equal to $-\sqrt{3}/2$. Hence, $\sin(4\pi/3) = \sin(5\pi/3) = -\sqrt{3}/2$ and $t = 4\pi/3$ or $5\pi/3$. (See Figure 4.2.26.)

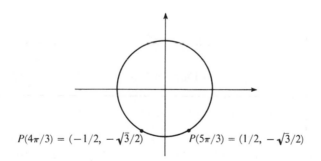

$$P(4\pi/3) = (-1/2, -\sqrt{3}/2) \qquad P(5\pi/3) = (1/2, -\sqrt{3}/2)$$

Figure 4.2.26

Problem S9. Find all $t \in [0, 2\pi]$ such that $\sin t < -\cos t$.

Solution: From a careful inspection of the graphs in Figure 4.2.27, we see that $\sin t$ is clearly larger than $-\cos t$ for $t \in [0, 3\pi/4)$. For $t \in (3\pi/4, 7\pi/4)$, $\sin t$ is less than $-\cos t$, while for $t \in (7\pi/4, 2\pi]$, $\sin t > -\cos t$. Hence, $\sin t < -\cos t$ for $t \in (3\pi/4, 7\pi/4)$.

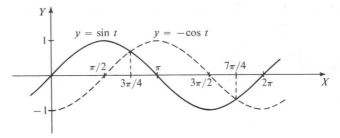

Figure 4.2.27

Problem S10.   Graph $y = 2 \cos(x + 2)$.

Solution: First apply the translation $XT/-2$ to the graph of $y = \cos x$. Then apply the distortion $YD/2$ to the graph of $y = \cos(x + 2)$, obtaining the graph of $y = 2 \cos(x + 2)$ shown in Figure 4.2.28.

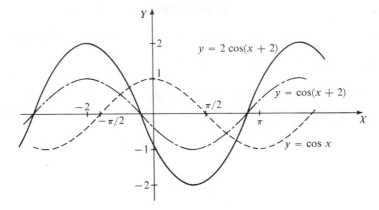

Figure 4.2.28

Problem S11.   Sketch $y = 1 - \cos(x + \tfrac{1}{2}\pi)$.

Solution: Apply the reflection YR to the graph of $y = \cos x$ followed by the translation $XT/-\tfrac{1}{2}\pi$ to obtain the sketch of $y = -\cos(x + \tfrac{1}{2}\pi)$, the solid curve in Figure 4.2.29. Finally, apply the translation $YT/1$ to the graph of $y = -\cos(x + \tfrac{1}{2}\pi)$ to obtain a sketch of $y = 1 - \cos(x + \tfrac{1}{2}\pi)$, shown in Figure 4.2.30.

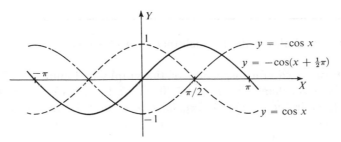

Figure 4.2.29

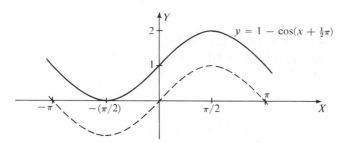

Figure 4.2.30

Problem S12.   Graph $y = |2 \sin(x - \pi)|$

Solution: Apply the translation XT/π followed by the distortion $YD/2$ to the graph of $y = \sin x$ to obtain a sketch of $y = 2 \sin(x - \pi)$, as shown in Figure 4.2.31. We obtain the graph of $y = |2 \sin(x - \pi)|$ by applying the reflection YR to the portions of the graph of $y = 2 \sin(x - \pi)$ below the X-axis (the negative output values), as shown in Figure 4.2.32.

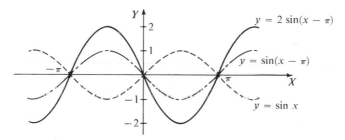

Figure 4.2.31

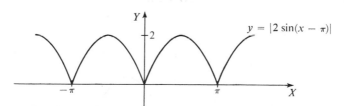

Figure 4.2.32

SUPPLEMENTARY PROBLEMS

S13. Evaluate (a) $\cos(23\pi/2)$, (b) $\sin(-7\pi/4)$, (c) $\cos(-2\pi/3)$, and (d) $\sin 10\pi/6$.
S14. True or false:
(a) $\sin 4 > \sin 6$ (b) $|\sin 4| > |\sin 6|$
(c) $\sin 1 > \cos 1$ (d) $\sin 4 > \cos 4$
S15. Sketch the graph of each of the following:
(a) $y = \sin|x|$ (b) $y = -\cos(x - (\pi/4))$
(c) $y = 2 \sin x - 1$

Answers

S13. (a) 0, (b) $\sqrt{2}/2$, (c) $-1/2$, (d) $-\sqrt{3}/2$
S14. (a) False, (b) true, (c) true, (d) false
S15. (a) See Figure 4.2.33.

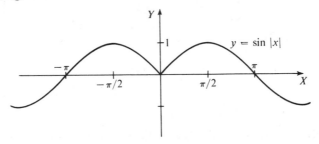

Figure 4.2.33

(b) See Figure 4.2.34.

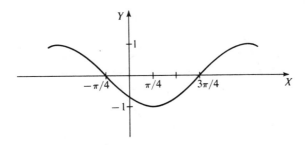

Figure 4.2.34

(c) See Figure 4.2.35.

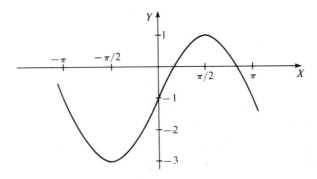

Figure 4.2.35

4.3 THE TANGENT, COTANGENT, SECANT, AND COSECANT FUNCTIONS

Several combinations of the two basic trigonometric functions sine and cosine occur so often in mathematical applications that four of these combinations are given function names. These four are defined from sine and cosine by simply forming quotients and reciprocals. The first of these is the tangent function.

Definition 4.3.1

For every real number t such that $\cos t \neq 0$,

$$\tan t = \frac{\sin t}{\cos t}$$

Using the trigonometric point function P, $\tan t = b/a$ where $P(t) = (a, b)$ and $a \neq 0$. For example, since $P(0) = (1, 0)$, $\sin 0 = 0$ and $\cos 0 = 1$. Therefore, $\tan 0 = 0/1 = 0$. Likewise, since $P(\pi/6) = (\sqrt{3}/2, 1/2)$,

$$\tan \frac{\pi}{6} = \frac{1/2}{\sqrt{3}/2} = \frac{1}{\sqrt{3}} = \frac{\sqrt{3}}{3}$$

Q1: Find (a) $\tan(\pi/4)$, (b) $\tan(\pi/3)$, and (c) $\tan(\pi/2)$.

Since $\cos t$ is the first coordinate of $P(t)$, $\cos t = 0$ if and only if $P(t)$ coincides with either point U or point V in Figure 4.3.1.

A1: (a) 1, (b) $\sqrt{3}$, (c) undefined.

Clearly, $U = P(\pi/2) = P(\frac{1}{2}\pi \pm 2\pi) = \ldots$, and $V = P(3\pi/2) = P(\frac{3}{2}\pi \pm 2\pi) = \ldots$. Thus, $\cos t = 0$ precisely when

$$t \in \left\{\frac{1}{2}\pi + 2k\pi : k \in Z\right\} \cup \left\{\frac{3}{2}\pi + 2k\pi : k \in Z\right\}$$

$$= \left\{\frac{1}{2}\pi + k\pi : k \in Z\right\}$$

Therefore, the domain of the tangent function is

$$\left\{t : t \neq \frac{1}{2}\pi + k\pi : k \in Z\right\}$$

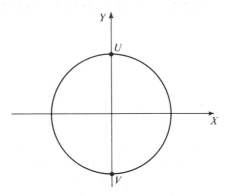

Figure 4.3.1

Q2: For what $t \in [0, 2\pi]$ is $\tan t \geq 0$?

Table B.2 lists approximate numerical values of $\tan t$ for certain numbers t. Notice that for t only slightly smaller than $\pi/2$, the numbers $\tan t$ are quite large since the quotient $(\sin t)/(\cos t)$ has numerator nearly one and denominator positive but close to zero. If t is slightly larger than $\pi/2$, the numerator is still close to one, and the denominator is negative and close to zero. Thus, the number $\tan t$ is a large negative number. In fact, the range of the tangent function is R, the set of all real numbers. We demonstrate this fact as follows.

Given any number r, we shall find a number t such that $\tan t = r$. First plot the points $(1, r)$, $(1, 0)$, and $(0, 0)$, and connect them to form a right triangle as in Figure 4.3.2. The point where the hypotenuse of this triangle intersects the circle is the trigonometric point $P(t) = (a, b)$ corresponding to the length t of the arc darkened in Figure 4.3.2. Next draw the dotted line shown in the figure from $P(t)$ to the X-axis, forming a right triangle which is similar to the first triangle. Thus, the ratios of the lengths of the legs of these two right triangles are equal.

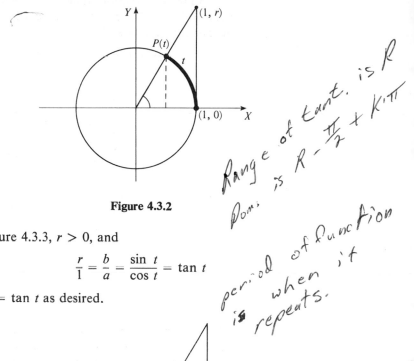

Figure 4.3.2

Range of tant. is R
Dom: is R - π/2 + k·π

period of function is when it repeats.

In Figure 4.3.3, $r > 0$, and

$$\frac{r}{1} = \frac{b}{a} = \frac{\sin t}{\cos t} = \tan t$$

Thus, $r = \tan t$ as desired.

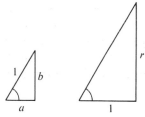

Figure 4.3.3

Clearly, the tangent function is periodic since

$$\tan(t + 2\pi) = \frac{\sin(t + 2\pi)}{\cos(t + 2\pi)}$$

$$= \frac{\sin t}{\cos t} = \tan t$$

However, its period is π, as we can see from the following observations.

For any real number t, the points $P(t)$ and $P(t + \pi)$ are diametrically opposed on the unit circle (Figure 4.3.4). That is, if the coordinates of $P(t)$ are (a, b), then the coordinates of $P(t + \pi)$

A2: $[0, \pi/2) \cup [\pi, 3\pi/2)$.

are $(-a, -b)$. Thus, $\tan(t + \pi) = -b/-a = b/a = \tan t$. This fact and the observations that $\tan t > 0$ if $0 < t < \pi/2$ and $\tan t < 0$ for $\pi/2 < t < \pi$ enable us to conclude that the period of the tangent function is π. Thus,

$$\tan(t + k\pi) = \tan t \quad \text{for } t \in R \text{ and } k \in Z \tag{1}$$

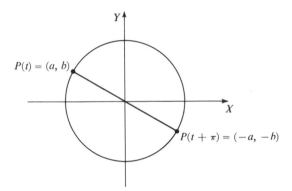

Figure 4.3.4

We can summarize much of the discussion thus far with the graph of $f(x) = \tan x$ sketched in Figure 4.3.5. Notice the domain, range, and periodicity of the tangent function exhibited by the graph. The vertical dashed lines in the figure are called (vertical) *asymptotes*. The graph gets arbitrarily close to an asymptote but

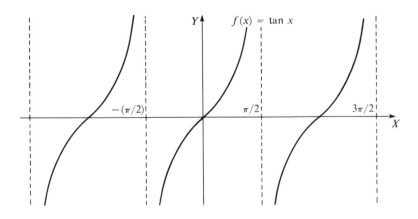

Figure 4.3.5

does not intersect it. The graph also illustrates that, for any t in the domain of tangent,

$$\tan(-t) = -\tan t \qquad (2)$$

Q3: Prove that Equation (2) is true.

**Example
4.3.1**

Sketch the graph of $y = \tan(x - \frac{1}{4}\pi)$.

Solution

We shall obtain the desired graph by applying an appropriate mapping to the graph of $y = \tan x$. Remembering that the translation XT/h replaces x by $x - h$ everywhere in the original equation, we apply the translation $XT/\frac{1}{4}\pi$ to the dotted curve in Figure 4.3.6. That is, the curve $y = \tan x$ is moved $\frac{1}{4}\pi$ units to the right, yielding the solid curve $y = \tan(x - \frac{1}{4}\pi)$.

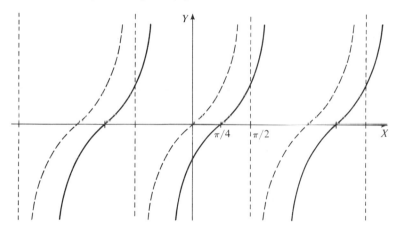

Figure 4.3.6

The reciprocals of the tangent, cosine, and sine functions are used to define the functions cotangent, secant, and cosecant. The following definition expresses the six trigonometric functions using the trigonometric point function P.

**Definition
4.3.2**

Let t be any real number. If the coordinates of the trigonometric point $P(t)$ are (a, b), then

$$\sin t = b \qquad\qquad \cot t = \frac{a}{b} \quad \text{(if } b \neq 0\text{)}$$

$$\cos t = a \qquad\qquad \sec t = \frac{1}{a} \quad \text{(if } a \neq 0\text{)}$$

$$\tan t = \frac{b}{a} \quad \text{(if } a \neq 0\text{)} \qquad \csc t = \frac{1}{b} \quad \text{(if } b \neq 0\text{)}$$

A3: $\tan(-t) = \dfrac{\sin(-t)}{\cos(-t)} = \dfrac{-\sin t}{\cos t} = -\tan t.$

For example, since $P(\pi/2) = (0, 1)$, $\cot(\pi/2) = 0/1 = 0$. Also, $\csc(\pi/2) = 1/1 = 1$, and $\sec(\pi/2)$ is undefined.

Q4: Find the six trigonometric function values for the real number t if $P(t) = (.6, .8)$.

We can use the sine and cosine functions to restate the definitions of the other four trigonometric functions as follows:

$$\tan t = \frac{\sin t}{\cos t}, \qquad \sec t = \frac{1}{\cos t}$$

$$\cot t = \frac{\cos t}{\sin t}, \qquad \csc t = \frac{1}{\sin t} \qquad (3)$$

Of course, the right-hand side of any one of the above identities is undefined for certain real numbers t. For example, although the domain of the sine function is R, the domain of the cosecant function includes only those real numbers t such that $\sin t \neq 0$. Now $\sin t = 0$ if, and only if, $t \in \{k\pi : k \in Z\}$. Thus, the domain of the cosecant function is $\{t \neq k\pi : k \in Z\}$.

Q5: Find the domain of the function (a) cotangent and (b) secant.

Next, let us investigate the range of the cosecant function. Consider any t in the domain of the cosecant function. Since $0 < |\sin t| \leq 1$,

$$\left|\frac{\sin t}{\sin t}\right| \leq \frac{1}{|\sin t|} = \left|\frac{1}{\sin t}\right| \quad \text{or} \quad |\csc t| \geq 1$$

That is, either $\csc t \geq 1$ or $\csc t \leq -1$. Furthermore, since every real number in $[-1, 1]$ is attained by $\sin t$, the range of the cosecant function is $[-\infty, -1] \cup [1, \infty]$.

We can easily observe the domain and range of the cosecant function by constructing the graph of $y = \csc x$. In Figure 4.3.7 we have sketched a part of the graph of $y = \sin x$. For example, the point $(\pi/6, 1/2)$ belongs to that graph. Thus, the point $(\pi/6, 1/(1/2)) = (\pi/6, 2)$ belongs to the graph of $y = \csc x = 1/\sin x$. Similarly, if $\sin x \neq 0$, the point $(x, \sin x)$ can be used to plot the point $(x, 1/\sin x) = (x, \csc x)$, and we obtain the graph shown in Figure 4.3.8. Notice the domain, range, periodicity, and asymptotes in the figure.

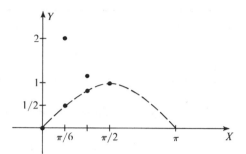

Figure 4.3.7

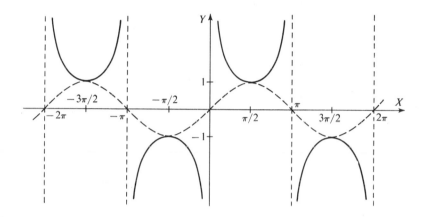

Figure 4.3.8

Q6: Sketch the graph of $y = \sec x$.

It is sometimes necessary to approximate trigonometric function values which are not found directly in a standard table.

**Example
4.3.2**

Using Table B.2, find cos 2.8 and sin 2.8.

Solution

In Figure 4.3.9, locate $U = P(2.8)$ on the unit circle. Next, locate V in the first quadrant so that the shaded arcs have equal length. Thus, the shaded arc has length $\pi - 2.8$ or about $3.14 - 2.8 = .34$. Clearly, the coordinates of V and U are (cos .34, sin .34) and (cos 2.8, sin 2.8), respectively. Since U is the reflection XR of V, (cos 2.8, sin 2.8) = ($-$cos .34, sin .34). Using Table B.2, we obtain cos 2.8 = $-$cos .34 = $-.9428$ and sin 2.8 = $.3335$.

A4: $\sin t = .8$, $\cos t = .6$, $\tan t = 4/3$, $\cot t = 3/4$, $\sec t = 5/3$, $\csc t = 5/4$.

A5: (a) $\{t \neq k\pi : k \in Z\}$, (b) $\{t \neq \frac{1}{2}\pi + k\pi : k \in Z\}$.

A6: The graph is shown in the following figure.

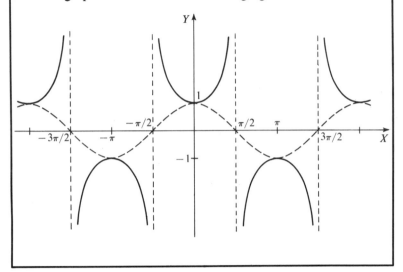

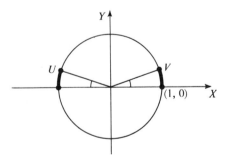

Figure 4.3.9

Example 4.3.3

Find $\cos 4.16$.

Solution

We first locate point P 4.16 units along the unit circle from $(1, 0)$ as shown in Figure 4.3.10. Next, Q is located in the first quadrant, so the shaded arcs have equal length, namely, about $4.16 - 3.14 = 1.02$. The coordinates of P and Q are $(\cos 4.16, \sin 4.16)$ and $(\cos 1.02, \sin 1.02)$, respectively. Clearly, $\cos 4.16 = -\cos 1.02$ or about $-.52$

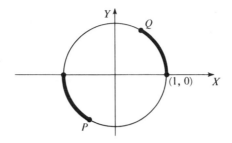

Figure 4.3.10

Example 4.3.4

Use $\pi \approx 3.14$ and Table B.2 to approximate sec 23.

Solution

Since sec $23 = 1/\cos 23$, we shall use Table B.2 to find cos 23. First, using the periodicity of the cosine function, we obtain $\cos 23 = \cos(23 - 2\pi) = \cos(23 - 2 \cdot 2\pi) = \ldots$. Thus, we can repeatedly subtract 2π from 23 until we obtain a "remainder" between 0 and 2π. Since repeated subtraction is just division and since $23 \div 2\pi$ is between 3 and 4, we subtract $3 \cdot 2\pi$ from 23, obtaining $23 - 6\pi \approx 4.16$. Thus, $\cos 23 = \cos 4.16 \approx -.52$ as found in Example 4.3.3. Finally, sec $23 \approx 1/-.52 = -1.92$.

PROBLEMS 4.3

1. Evaluate each of the following:
(a) $\tan \pi$ (b) $\tan(-7\pi/4)$
(c) $\tan(31\pi/2)$ (d) $\tan(2\pi/3)$
(e) $\tan(11\pi/6)$ (f) $\tan(-5\pi/6)$

2. Sketch the graph of each of the following:
(a) $y = \tan 2x$ (b) $y = 2 \tan x$
(c) $y = 2 \tan(x + \frac{1}{2}\pi)$ (d) $y = 2 \tan x + \frac{1}{2}\pi$

3. Find the following:
(a) $\{t : \sin t = \cos t, 0 \le t \le 2\pi\}$
(b) $\{t : \sin t = \sqrt{3} \cos t, 0 \le t \le 2\pi\}$

4. Evaluate each of the following:
(a) $\cot(\pi/4)$ (b) $\sec(2\pi/3)$
(c) $\csc(-\pi/6)$ (d) $\cot \pi$
(e) $\sec(3\pi/2)$ (f) $\csc(7\pi/4)$

5. Evaluate each of the following:
(a) $\sec(-31\pi/6)$ (b) $\cot(28\pi/3)$
(c) $\tan 1.8$ (d) $\sin 7.3$
(e) $\cos(-9)$ (f) $\sec 2$
(g) $\csc 4$ (h) $\cot 5$

6. Sketch the graph of $y = \cot x$.

7. Sketch the graph of each of the following:

(a) $y = \sec x - 1$ (b) $y = \sec(x - 1)$

(c) $y = \cot 2x$ (d) $y = 2 \cot x$

(e) $y = |\csc x|$ (f) $y = \csc |x|$

---

8. Which of the following are true:

(a) $\csc 1 > 0$ (b) $\sec(\sin 3) < 0$

(c) $\sec t \csc t < 0$ for $t \in (\pi/2, \pi)$

(d) $|\csc t + \sec t| \geq 2$

ANSWERS
4.3

1. (a) 0, (b) 1, (c) undefined, (d) $-\sqrt{3}$, (e) $-\sqrt{3}/3$, (f) $\sqrt{3}/3$

2. (a) See Figure 4.3.11.

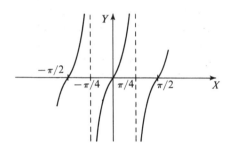

Figure 4.3.11

(b) See Figure 4.3.12.

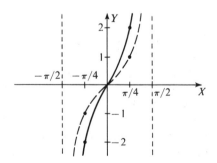

Figure 4.3.12

(c) See Figure 4.3.13.

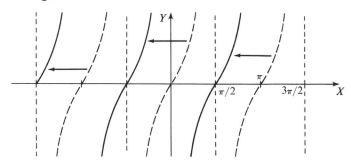

Figure 4.3.13

(d) See Figure 4.3.14.

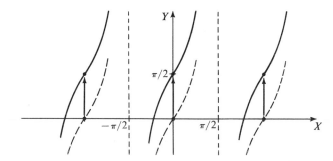

Figure 4.3.14

3. (a) $\{\pi/4, 5\pi/4\}$, (b) $\{\pi/3, 4\pi/3\}$

4. (a) 1, (b) -2, (c) -2, (d) undefined, (e) undefined, (f) $-\sqrt{2}$

5. (a) $-2\sqrt{3}/3$, (b) $\sqrt{3}/3$, (c) -4.256, (d) $.8521$, (e) $-.9131$, (f) -2.395, (g) -1.32, (h) $-.2993$

6. See Figure 4.3.15.

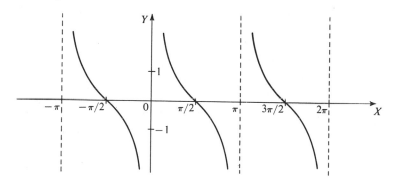

Figure 4.3.15

7. (a) See Figure 4.3.16.   (b) See Figure 4.3.17.

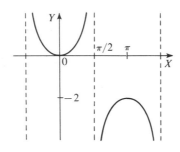

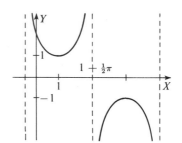

Figure 4.3.16   **Figure 4.3.17**

(c) See Figure 4.3.18.   (d) See Figure 4.3.19.

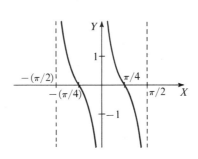

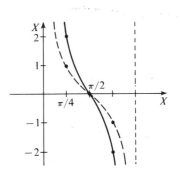

Figure 4.3.18   **Figure 4.3.19**

(e) See Figure 4.3.20.   (f) See Figure 4.3.21.

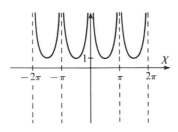

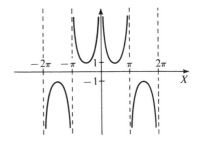

Figure 4.3.20   **Figure 4.3.21**

Supplement 4.3

Problem S1. Complete Table 4.3.1.

Table 4.3.1

t	$P(t)$	$\sin t$	$\cos t$	$\tan t$	$\cot t$	$\sec t$	$\csc t$
0							
$\pi/6$							
$\pi/4$							
$\pi/3$							
$\pi/2$							

Solution: The completed table is shown in Table 4.3.2.

Table 4.3.2

t	$P(t)$	$\sin t$	$\cos t$	$\tan t$	$\cot t$	$\sec t$	$\csc t$
0	$(1, 0)$	0	1	$0/1 = 0$	undef.	$1/1 = 1$	undef.
$\pi/6$	$(\sqrt{3}/2, 1/2)$	$1/2$	$\sqrt{3}/2$	$\sqrt{3}/3$	$\sqrt{3}$	$2\sqrt{3}/3$	2
$\pi/4$	$(\sqrt{2}/2, \sqrt{2}/2)$	$\sqrt{2}/2$	$\sqrt{2}/2$	1	1	$\sqrt{2}$	$\sqrt{2}$
$\pi/3$	$(1/2, \sqrt{3}/2)$	$\sqrt{3}/2$	$1/2$	$\sqrt{3}$	$\sqrt{3}/3$	2	$2\sqrt{3}/3$
$\pi/2$	$(0, 1)$	1	0	undef.	0	undef.	1

Problem S2. Complete Table 4.3.3.

Table 4.3.3

t	$P(t)$	$\sin t$	$\cos t$	$\tan t$	$\cot t$	$\sec t$	$\csc t$
$4\pi/3$							
$7\pi/4$							
$-\pi/6$							
$8\pi/3$							
-3π							
$-3\pi/4$							
$147\pi/2$							

Solution: The completed table is shown in Table 4.3.4.

Table 4.3.4

t	$P(t)$	$\sin t$	$\cos t$	$\tan t$	$\cot t$	$\sec t$	$\csc t$
$4\pi/3$	$(-1/2, -\sqrt{3}/2)$	$-\sqrt{3}/2$	$-1/2$	$\sqrt{3}$	$\sqrt{3}/3$	-2	$-2\sqrt{3}/3$
$7\pi/4$	$(\sqrt{2}/2, -\sqrt{2}/2)$	$-\sqrt{2}/2$	$\sqrt{2}/2$	-1	-1	$\sqrt{2}$	$-\sqrt{2}$
$-\pi/6$	$(\sqrt{3}/2, -1/2)$	$-1/2$	$\sqrt{3}/2$	$-\sqrt{3}/3$	$-\sqrt{3}$	$2\sqrt{3}/3$	-2
$8\pi/3$	$(-1/2, \sqrt{3}/2)$	$\sqrt{3}/2$	$-1/2$	$-\sqrt{3}$	$-\sqrt{3}/3$	-2	$2\sqrt{3}/3$
-3π	$(-1, 0)$	0	-1	0	undef.	-1	undef.
$-3\pi/4$	$(-\sqrt{2}/2, -\sqrt{2}/2)$	$-\sqrt{2}/2$	$-\sqrt{2}/2$	1	1	$-\sqrt{2}$	$-\sqrt{2}$
$147\pi/2$	$(0, -1)$	-1	0	undef.	0	undef.	-1

Problem S3. Evaluate each of the following:

(a) $\tan(\pi/4)$ (b) $\tan(-\pi/4)$
(c) $\tan(\pi/3)$ (d) $\tan(-\pi/3)$
(e) $\tan(5\pi/6)$ (f) $\tan(-5\pi/6)$

Solution: (a) 1, (b) -1, (c) $\sqrt{3}$, (d) $-\sqrt{3}$, (e) $-\sqrt{3}/3$, (f) $\sqrt{3}/3$

Problem S4. Evaluate each of the following:
(a) $\cot(\pi/4)$ (b) $\cot(-\pi/4)$
(c) $\cot(\pi/3)$ (d) $\cot(-\pi/3)$
(e) $\cot(5\pi/6)$ (f) $\cot(-5\pi/6)$

Solution: (a) 1, (b) -1, (c) $\sqrt{3}/3$, (d) $-\sqrt{3}/3$, (e) $-\sqrt{3}$, (f) $\sqrt{3}$

Problem S5. Evaluate each of the following:
(a) $\sec(\pi/4)$ (b) $\sec(-\pi/4)$
(c) $\sec(2\pi/3)$ (d) $\sec(-2\pi/3)$
(e) $\csc(\pi/4)$ (f) $\csc(-\pi/4)$
(g) $\csc(2\pi/3)$ (h) $\csc(-2\pi/3)$

Solution: (a) $\sqrt{2}$, (b) $\sqrt{2}$, (c) 2, (d) 2, (e) $\sqrt{2}$, (f) $-\sqrt{2}$, (g) $2\sqrt{3}/3$, (h) $-2\sqrt{3}/3$

Problem S6. Evaluate each of the following:
(a) $\tan(\pi/6)$ (b) $\tan(7\pi/6)$
(c) $\tan(13\pi/6)$ (d) $\tan(19\pi/6)$
(e) $\tan(-5\pi/6)$ (f) $\tan(-11\pi/6)$
(g) $\tan(-17\pi/6)$ (h) $\tan(-23\pi/6)$

Solution: (a) $\sqrt{3}/3$, (b) $\sqrt{3}/3$, (c) $\sqrt{3}/3$, (d) $\sqrt{3}/3$, (e) $\sqrt{3}/3$, (f) $\sqrt{3}/3$, (g) $\sqrt{3}/3$,
(h) $\sqrt{3}/3$

Problem S7. Evaluate each of the following:

(a) $\csc(\pi/6)$

(b) $\csc(13\pi/6)$

(c) $\csc(25\pi/6)$

(d) $\csc(-11\pi/6)$

(e) $\sec(\pi/4)$

(f) $\sec(9\pi/4)$

(g) $\sec(-7\pi/4)$

(h) $\sec(-15\pi/4)$

(i) $\cot(-5\pi/6)$

(j) $\cot(-11\pi/6)$

(k) $\cot(-17\pi/6)$

(l) $\cot(-23\pi/6)$

Solution: (a) 2, (b) 2, (c) 2, (d) 2, (e) $\sqrt{2}$, (f) $\sqrt{2}$, (g) $\sqrt{2}$, (h) $\sqrt{2}$, (i) $\sqrt{3}$, (j) $\sqrt{3}$, (k) $\sqrt{3}$, (l) $\sqrt{3}$

Problem S8. Graph the following:

(a) $y = \cot x$

(b) $y = |\cot x|$

(c) $y = \tan x + \tfrac{1}{4}\pi$

(d) $y = \tan(x + \tfrac{1}{4}\pi)$

Solution:

(a) See Figure 4.3.22.

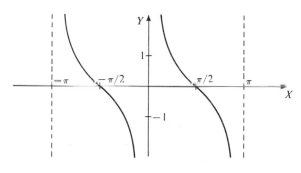

Figure 4.3.22

(b) See Figure 4.3.23.

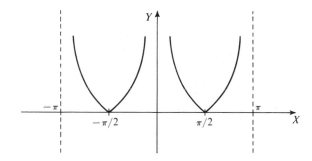

Figure 4.3.23

(c) $y = \tan x \xrightarrow{YT/\frac{1}{4}\pi} y = \frac{1}{4}\pi + \tan x$. The graph is shown in Figure 4.3.24.

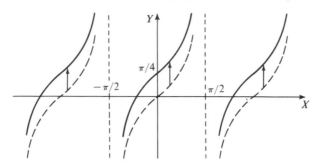

Figure 4.3.24

(d) $y = \tan x \xrightarrow{XT/-\frac{1}{4}\pi} y = \tan(x + \frac{1}{4}\pi)$. The graph is shown in Figure 4.3.25.

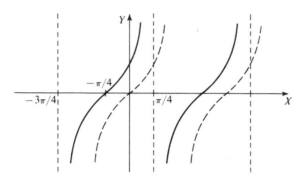

Figure 4.3.25

Problem S9. Graph the following:
(a) $y = -\sec x$ (b) $y = 1 + \csc x$
(c) $y = \csc(x - \pi)$ (d) $y = \sec(x + \frac{1}{2}\pi)$

Solution:
(a) $y = \sec x \xrightarrow{YR} y = -\sec x$. The graph is shown in Figure 4.3.26.

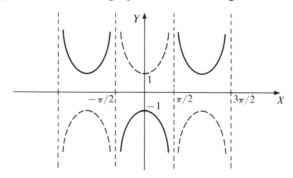

Figure 4.3.26

(b) $y = \csc x \xrightarrow{YT/1} y = 1 + \csc x$. See Figure 4.3.27.

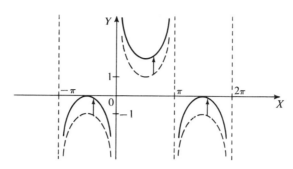

Figure 4.3.27

(c) $y = \csc x \xrightarrow{XT/\pi} y = \csc(x - \pi)$. See Figure 4.3.28.

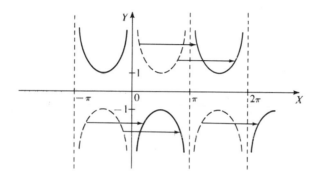

Figure 4.3.28

(d) $y = \sec x \xrightarrow{XT/-\frac{1}{2}\pi} y = \sec(x + \frac{1}{2}\pi)$. The graph is shown in Figure 4.3.29.

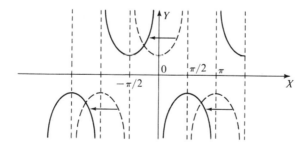

Figure 4.3.29

4.4 ANGLES AND RIGHT TRIANGLES

Historically, trigonometry developed from the study of angles and right triangles. In preceding sections we have taken the "circular function" approach. We now wish to relate the results of our approach to the more traditional, right-triangle aspects of trigonometry.

An angle is formed by two half-lines, or *rays*, intersecting at a common point called the *vertex*. More precisely, we shall say that an angle consists of a ray called the *initial side*, a *rotation*, and a second ray called the *terminal side* (Figure 4.4.1). Thus, an angle consists of a set of points and a rotation.

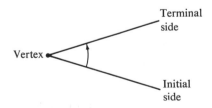

Figure 4.4.1

The angle in Figure 4.4.2 is said to be in *standard position* when its vertex and initial side coincide respectively with the origin and positive X-axis of a Cartesian coordinate system.

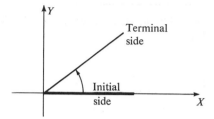

Figure 4.4.2

The "size" of the rotation is called a *measure of the angle*. By convention, counterclockwise rotations yield angles of positive measure, and clockwise rotations yield angles of negative measure. One unit of measure commonly used is the degree. One counterclockwise revolution generates an angle of 360 degrees (written 360°). Thus, one-quarter revolution counterclockwise measures 90°, and $1\frac{1}{4}$ revolution measures 450°. Technically, these angles are different. They consist of the same set of points but have different rotations (see Figure 4.4.3).

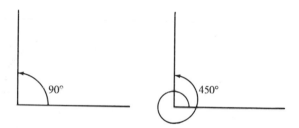

Figure 4.4.3

Q1: Draw, in standard position, angles of measure (a) 180°, (b) −45°, and (c) 630°.

Another way to measure an angle is to use the unit circle with the angle in standard position (Figure 4.4.4). The initial side intersects the circle at (1, 0), and the terminal side intersects it at some point $P(t)$ where $|t|$ is the arc length generated by the rotation of the angle. Of course, t will be a positive number when the rotation is counterclockwise and negative when the rotation is clockwise. Thus, each angle determines a unique real number t, and conversely. Therefore, using the real number t, we can measure the angle. We call the number t the *radian* measure of the angle, or we say that an angle has measure t radians.

A1: The angles are shown in the following figures.
(a)   (b)

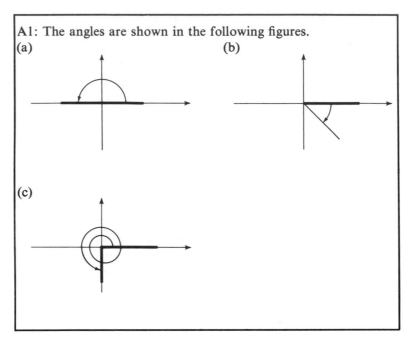

(c)

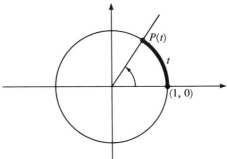

Figure 4.4.4

Q2: Draw, in standard position, angles of radian measure (a) 1, (b) $-\pi/2$, and (c) 3π.

Clearly, one counterclockwise revolution generates an angle of 360° or 2π radians. Thus, an angle of 180° is π radians:

$$180° = \pi \text{ radians} \qquad (1)$$

Using this relationship, we can convert from degree measure to radian measure and conversely. Dividing Equation (1) by 180 yields $1° = \pi/180$ radians. So, for example, $12° = 12(\pi/180)$ radians $= \pi/15$ radians. Also, dividing Equation (1) by π yields

$(180/\pi)° = 1$ radian; so, for example, 1.5 radians $= 1.5(180/\pi)° = (270/\pi)° \approx 86°$.

Q3: Complete the following table of equivalent-angle measure:

radians	0	$\pi/6$			$\pi/2$	$2\pi/3$	$3\pi/4$		
degrees	0		45°	60°				150°	180°

Q4: Find s and t if $20° = s$ radians and 2 radians $= t°$

Recall that the domain of each of the functions sine and cosine is R, the set of all real numbers. Furthermore, the domain of the tangent function is a subset of R, namely, $\{t : t \neq \frac{1}{2}\pi + k\pi, k \in Z\}$. Angle measures can be used as inputs to the trigonometric functions in the following way. If an angle has radian measure t (a real number), the corresponding trigonometric function values are $\sin t$, $\cos t$, $\tan t$, etc. If an angle has degree measure, we use the equivalent radian measure as input. Thus, $\sin 30° = \sin(\pi/6) = 1/2$ since $30° = \pi/6$ radians. Similarly, $\tan 90°$ is undefined since $\pi/2$ is not in the domain of the tangent function.

Q5: Evaluate (a) $\cos 120°$, (b) $\tan(-45°)$, and (c) $\sin 420°$.

Rather than convert from degree measure to radian measure, we shall often make use of Table B.3 which lists decimal approximations of some trigonometric function values for degree inputs.

Q6: Evaluate (a) $\sin 37°$ and (b) $\tan 61°$.

The usefulness of the trigonometric functions in dealing with right triangles can be seen with the aid of Figure 4.4.5. Consider the

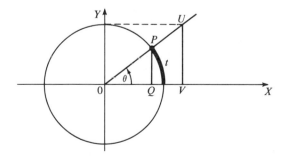

Figure 4.4.5

A2: The angles are shown below.

(a)

(b)

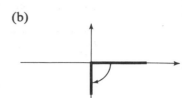

(c)

A3:

radians	0	$\pi/6$	$\pi/4$	$\pi/3$	$\pi/2$	$2\pi/3$	$3\pi/4$	$5\pi/6$	π
degrees	0	30°	45°	60°	90°	120°	135°	150°	180°

A4: $s = \pi/9$, $t = 360/\pi$

A5: (a) $-1/2$, (b) -1, (c) $\sqrt{3}/2$

A6: (a) .6018, (b) 1.8040

right triangle OUV with acute angle θ in standard position as shown. If the length of the hypotenuse is greater than 1, then the hypotenuse will intersect the unit circle at some point P. Next, draw the line through P perpendicular to the X-axis at Q. Then OPQ is a right triangle similar to triangle OUV. The coordinates of P represent the lengths of the legs of the smaller triangle OPQ. If t is the radian measure of θ, then the coordinates of P are $(\cos t, \sin t)$.

In Figure 4.4.6 we have drawn the two right triangles of Figure 4.4.5. Since the triangles are similar,

$$\frac{\sin t}{1} = \frac{\text{length of side opposite } \theta}{\text{length of hypotenuse}}$$

Thus,

$$\sin \theta = \frac{\text{length of side opposite } \theta}{\text{length of hypotenuse}}$$

Figure 4.4.6

Of course, we can measure θ using degrees or radians. In a similar manner, we can find ratios for the other trigonometric functions. We summarize these results in the following abbreviated way:

$$\sin \theta = \frac{\text{opposite}}{\text{hypotenuse}}, \quad \cos \theta = \frac{\text{adjacent}}{\text{hypotenuse}}, \quad \tan \theta = \frac{\text{opposite}}{\text{adjacent}}$$

$$\cot \theta = \frac{\text{adjacent}}{\text{opposite}}, \quad \sec \theta = \frac{\text{hypotenuse}}{\text{adjacent}}, \quad \csc \theta = \frac{\text{hypotenuse}}{\text{opposite}} \tag{2}$$

Q7: For the right triangle shown below, find (a) $\sin \theta$, (b) $\cos \theta$, and (c) $\tan \theta$.

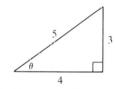

The right triangle relationships in Equation (2) can help you remember the trigonometric function values for certain special angles. In Figure 4.4.7 we have drawn an isosceles right triangle with each leg 1 unit long. By the Pythagorean theorem, the hypot-

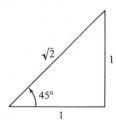

Figure 4.4.7

A7: (a) .6, (b) .8, (c) .75

enuse has length $\sqrt{1 + 1} = \sqrt{2}$. Since each of the acute angles has measure $\pi/4$ radians or 45°,

$$\sin 45° = \cos 45° = \frac{1}{\sqrt{2}} = \frac{\sqrt{2}}{2} \quad \text{and} \quad \tan 45° = \frac{1}{1} = 1$$

Similarly, if we bisect an angle of an equilateral triangle with sides two units long, we obtain the 30°-60° right triangle shown in Figure 4.4.8. Again we use the Pythagorean theorem to find the length of the third side to be $\sqrt{4 - 1} = \sqrt{3}$. Thus,

$$\sin 60° = \frac{\sqrt{3}}{2}$$

$$\cos 60° = \frac{1}{2}$$

$$\tan 60° = \sqrt{3}$$

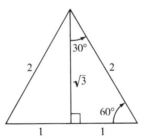

Figure 4.4.8

Q8: Using Figure 4.4.8, find (a) sin 30°, (b) cos 30°, and (c) tan 30°.

We can now use the results we have developed to "solve" right triangles, that is, to find lengths of sides and measures of angles of triangles. These results consist of

1. the trigonometric functions, using Equation 2 and Tables B.2 and B.3;
2. the Pythagorean theorem;
3. the fact that the sum of the measures of the angles in any plane triangle is 180°.

Of course, we must be given sufficient information about the triangle. For example, knowledge of the size of two angles enables us to determine the size of the third, but says nothing about the length of any of the sides.

Example
4.4.1

Find x, y, and α in the right triangle shown in Figure 4.4.9.

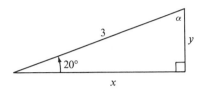

Figure 4.4.9

Solution

Using result 3, we have $\alpha = 180° - 90° - 20° = 70°$. Furthermore, $\sin 20° = y/3$ and $\cos 20° = x/3$. Thus, $y = 3 \sin 20°$ and $x = 3 \cos 20°$. Using Table B.3, we have $y = 3(.3420) = 1.026$ and $x = 3(.9397) = 2.819$.

Example
4.4.2

Find α, β, and r in the right triangle shown in Figure 4.4.10.

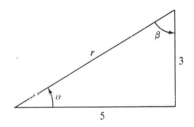

Figure 4.4.10

Solution

Since $r^2 = 3^2 + 5^2 = 34$, $r = \sqrt{34}$. Furthermore, $\tan \alpha = 3/5 = .6$, so we see from Table B.3 that α is about 31°. Thus, $\beta = 59°$ since $\alpha + \beta = 90°$.

Example
4.4.3

A lighthouse L stands 3 miles from the closest point P along a straight shore. Find the distance from P to a point Q along the shore if the angle from P to Q to the lighthouse measures 35°.

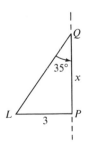

Figure 4.4.11

A8: (a) 1/2, (b) $\sqrt{3}/2$, (c) $1/\sqrt{3} = \sqrt{3}/3$

Solution

Using Figure 4.4.11, we see that $3/x = \tan 35° = .7001$. Thus, $x = 3/.7001$ or x is about 4.3 miles.

PROBLEMS
4.4

1. Find the degree measure of the angle which has radian measure (a) 3, (b) .3, (c) $5\pi/2$, (d) -12, and (e) 3.14.

2. Find the radian measure of the angle which has measure (a) 3°, (b) 300°, (c) $-230°$, and (d) 3780°.

3. Find a, b, c, and α in the triangle shown in Figure 4.4.12 if
(a) $a = 3$, $b = 4$. (b) $\alpha = 41°$, $a = 12$.
(c) $\alpha = 41°$, $c = 10$. (d) $\alpha = 41°$, $b = 15$.
(e) $b = 5$, $c = 7$.

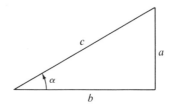

Figure 4.4.12

4. A building casts a shadow 130 feet long when the angle of elevation of the sun (measured from the horizontal) is 38°. How tall is the building?

5. An observer on the ground is one mile from a building 1200 feet tall. What is the angle of elevation from the observer to the top of the building?

6. The angle of elevation from an observer to the top of a building located 200 feet from the observer is 30°. The angle of elevation from the observer to the top of a microwave tower (on top of the building) is 40°. What is the height of the tower?

7. Figure 4.4.13 shows an acute angle θ, part of the unit circle, and six line segments with lengths a through f. Identify each of these lengths with the six trigonometric function values of θ.

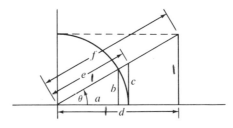

Figure 4.4.13

8. Which of the following are true statements for every α between $0°$ and $90°$:
(a) $\sin(90° - \alpha) = \cos \alpha$ (b) $\cos(90° - \alpha) = \sin \alpha$
(c) $\tan(90° - \alpha) = \tan \alpha$ (d) $\tan(90° - \alpha) = 1/\tan \alpha$
(e) $\tan(180° + \alpha) = \tan \alpha$ (f) $\sin(45° - \alpha) = \cos(\alpha + 45°)$

9. Find the acute angle α ($0° < \alpha < 90°$) if
(a) $\sin(90° - \alpha) = \sin \alpha$ (b) $\sin(\alpha + 45°) = \cos \alpha$
(c) $\tan(90° - \alpha) = \tan \alpha$ (d) $\tan(\alpha + 45°) = 1/\tan \alpha$

10. An observer is an unknown distance from a mountain. Using a sextant, he determines that the angle of elevation from him to the mountain peak is 35°. Moving 1000 feet further away from the mountain, he determines the angle of elevation to be 30°. What is the height of the mountain?

ANSWERS 4.4

1. (a) About 172°, (b) about 17°, (c) 450°, (d) about $-688°$, (e) about 180°

2. (a) About .05 radian, (b) $5\pi/3$ radians, (c) about -4.01 radians, (d) 21π radians

3. (a) $c = 5$, $\alpha \approx 37°$ (b) $c \approx 18.3$, $b \approx 13.8$
(c) $a \approx 6.6$, $b \approx 7.5$ (d) $a \approx 13$, $c \approx 19.9$
(e) $a \approx 4.9$, $\alpha \approx 44.5°$

4. About 101.6 feet

5. About 13°

6. About 52 feet

7. $a = \cos \theta$, $b = \sin \theta$, $c = \tan \theta$, $d = \cot \theta$, $e = \sec \theta$, $f = \csc \theta$

Supplement 4.4

I. Radian and Degree Measure

Problem S1. Since once around the unit circle counterclockwise (CCW) generates an angle of measure 2π radians or 360°, 1/2 revolution CCW subtends an angle of measure _____ radians or _____°.
  Dividing the equation π radians = 180° by π yields _____ radian = _____°.

π
180

1
$180/\pi$ (about 57.3)

Problem S2. Complete the following:
1 radian = ____°, 2 radians = ____°,
2.5 radians = ____°, 5π radians =
____°, -1 radian = ____°, -10
radians = ____°.

$180/\pi$, $360/\pi$
$450/\pi$
900, $-180/\pi$
$-1800/\pi$

Problem S3. Dividing both sides of
Equation (3) by 180 yields ____° =
____ radians.

1
$\pi/180$ (about .017)

Problem S4. Complete the following:
$1° =$ ____ radians, $12° =$ ____ radians,
$120° =$ ____ radians, $-15° =$ ____
radians, $-1080° =$ ____ radians.

$\pi/180$, $\pi/15$
$2\pi/3$, $-\pi/12$
-6π

II. Right Triangle Trigonometry

Recall that $\sin\theta = \dfrac{\text{opp}}{\text{hyp}}$, $\cos\theta = \dfrac{\text{adj}}{\text{hyp}}$, and $\tan\theta = \dfrac{\text{opp}}{\text{adj}}$.

Problem S5. Find $\sin\theta$, $\cos\theta$, and $\tan\theta$
for (a) Figure 4.4.14 and (b) Figure
4.4.15.

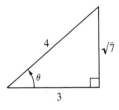

Figure 4.4.14

(a) $\sqrt{7}/4$, $3/4$, $\sqrt{7}/3$

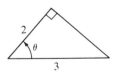

Figure 4.4.15

(b) $\sqrt{5}/3$, $2/3$, $\sqrt{5}/2$

Problem S6. Use the Pythagorean
theorem to find the length of the
third side for (a) Figure 4.4.16,
(b) Figure 4.4.17, and
(c) Figure 4.4.18.

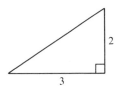

Figure 4.4.16   (a) $\sqrt{2^2 + 3^2} = \sqrt{13}$

Figure 4.4.17   (b) $\sqrt{24}$ or $2\sqrt{6}$

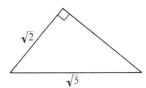

Figure 4.4.18   (c) $\sqrt{3}$

Problem S7: Use the triangles and
results of Problem S6 to evaluate
(a) tan α, (b) cos α,
(c) csc β, (d) sec β,
(e) cos γ, and
(f) cot γ.

(a) 2/3, (b) $3/\sqrt{13} = 3\sqrt{13}/13$
(c) 7/5, (d) $7/2\sqrt{6} = 7\sqrt{6}/12$
(e) $\sqrt{2}/\sqrt{5} = \sqrt{10}/5$
(f) $\sqrt{2}/\sqrt{3} = \sqrt{6}/3$

Problem S8. Use Table B.3 to find
(a) sin 50°, (b) cos 40°, and (c) tan 40°.

(a) .7660   (b) .7660   (c) .8391

Problem S9. Use Table B.3 to
determine the acute angle θ (to the
nearest degree) if (a) cos θ = 3/4 = .75,
(b) tan θ = 2/3, and (c) sin θ = 5/7.

(a) 41°
(b) 34°,   (c) 46°

Problem S10. Use Table B.3 and
Problem S9 to determine θ (to the
nearest degree) in (a) Figure 4.4.19,
(b) Figure 4.4.20, and
(c) Figure 4.4.21.

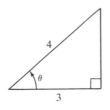

Figure 4.4.19   (a) 41°

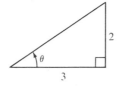

Figure 4.4.20   (b) 34°

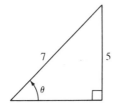

Figure 4.4.21   (c) 46°

Problem S11.   A wire stretches from the top of a vertical pole to level ground. It touches the ground 16 feet from the base of the pole and makes an angle of 62° with the ground. Find the height of the pole and the length of the wire.

Solution: Let h be the height of the pole and l the length of the wire as shown in Figure 4.4.22. Then $h/16 = \tan 62°$, so $h = 16 \tan 62° = 16(1.8807) \approx 30.1$ feet. Also, $16/l = \cos 62°$ or $l = 16/\cos 62° = 16/.4695 \approx 34.1$ feet.

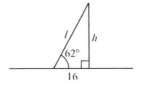

Figure 4.4.22

SUPPLEMENTARY PROBLEMS

S12. Find a, b, c, α, and β in the triangle shown in Figure 4.4.23 if
(a) $a = 2$, $b = \sqrt{5}$ (b) $a = 2$, $c = 5$

(c) $a = 2, \alpha = 34°$ (d) $a = 2, \beta = 34°$
(e) $b = 2, \alpha = 34°$ (f) $b = 2, c = 5$
(g) $c = 5, \alpha = 34°$ (h) $\alpha = 34°, \beta = 56°$

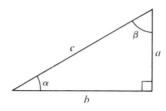

Figure 4.4.23

S13. A rocket climbs for 5 minutes at a constant speed of 1200 mph from ground level at an angle of 20°. Find the altitude of the rocket after five minutes and the distance from a point on the ground directly beneath the rocket to the launch point.

Answers

S12.
(a) $c = 3, \alpha = 42°, \beta = 48°$ (b) $b = \sqrt{21}, \alpha = 24°, \beta = 66°$
(c) $b = 3, c = 3.6, \beta = 56°$ (d) $b = 1.35, c = 2.4, \alpha = 56°$
(e) $a = 1.35, c = 2.4, \beta = 56°$ (f) $a = \sqrt{21}, \alpha = 66°, \beta = 24°$
(g) $a = 2.8, b = 4.1, \beta = 56°$
(h) a, b, and c are not uniquely determined. However, we can conclude that the ratio of a to b to c is .56 to .83 to 1.
S13. About 34 miles, about 94 miles.

4.5 INVERSE TRIGONOMETRIC FUNCTIONS

In Section 3.1 we saw that if f is a one-to-one function, we can reverse or invert it to form a new function called the inverse of f, denoted f^{-1}. For example, if $f(x) = x^3$, then $f^{-1}(x) = \sqrt[3]{x}$ (Figure 4.5.1). Of course, not all functions are one-to-one, so *not all functions have inverses*. Thus, $g(x) = x^2$, $x \in R$, does not have an inverse (Figure 4.5.2). Notice, however, that if we restrict the domain of g to include only the set of positive numbers P, the result is a one-to-one function — let us call it G. Thus, $G(x) = x^2$ for $x > 0$. Furthermore G has an inverse, given by $G^{-1}(u) = \sqrt{u}$ (Figure 4.5.3).

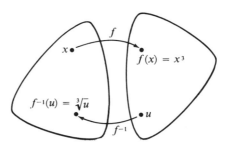

Figure 4.5.1

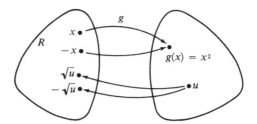

Figure 4.5.2

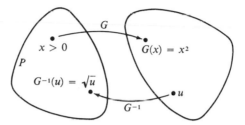

Figure 4.5.3

Let us summarize the discussion above with the aid of Figure 4.5.4 which shows the graphs of f, g, and G. Clearly, f and G are one-to-one functions and g is not one-to-one.

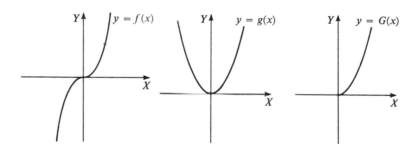

Figure 4.5.4

Recall that the reflection IR of the graphs of $y = f(x)$ and $y = G(x)$ produce the graphs of $y = f^{-1}(x)$ and $y = G^{-1}(x)$ as shown in Figure 4.5.5. In particular, points of the graph of f satisfy $y = x^3 = f(x)$. The reflection IR interchanges coordinates so that points of the graph of f^{-1} satisfy $x = y^3$ or $y = \sqrt[3]{x} = f^{-1}(x)$. Similarly, since $y = G(x) = x^2$ for $x > 0$, points of the graph of G^{-1} satisfy $x = y^2$ for $y > 0$, or equivalently, $y = \sqrt{x} = G^{-1}(x)$. (Notice that $y > 0$ since $\sqrt{x} > 0$.)

We now turn our attention to the functions sine, cosine, and tangent. None of these functions is one-to-one, and therefore none

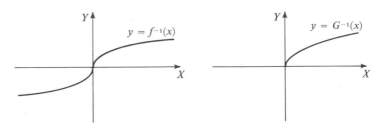

Figure 4.5.5

possesses an inverse. However, we could restrict the domain of each of these functions to define new functions which would have inverses. This is precisely what we did with g in forming the one-to-one function G above. In this section we shall do just that for the sine and tangent functions. We will leave the treatment of the cosine function to the problems. The new functions defined will be useful in your later work in calculus.

We first restrict the domain of the sine function to form a new one-to-one function. We arbitrarily choose the domain to be $[-\pi/2, \pi/2]$ on which the sine function attains each range value precisely once (see Figure 4.5.6). We call this one-to-one function the Sine function. (Notice the capital letter.)

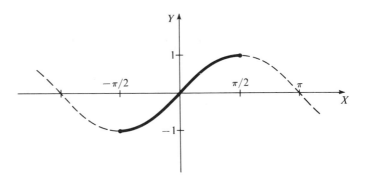

Figure 4.5.6

Definition 4.5.1   Sine $= \{(x, \sin x) : -\pi/2 \leq x \leq \pi/2\}$. That is, $y = \text{Sin } x$ if, and only if, $y = \sin x$ and $-\pi/2 \leq x \leq \pi/2$.

> Q1: Find (a) Sin 0, (b) Sin($\pi/2$), and (c) Sin π.

The graph of $y = \text{Sin } x$ is shown in Figure 4.5.7. Clearly, Sine is a one-to-one function with domain $[-\pi/2, \pi/2]$ and range $[-1, 1]$. We now define the *inverse Sine* or *arcsine* function.

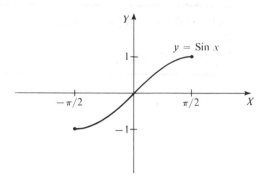

Figure 4.5.7

Definition 4.5.2

$y = \text{Sin}^{-1}x = \arcsin x$ if, and only if, $x = \text{Sin } y$.

Q2: Find (a) $\text{Sin}^{-1}0$ and (b) $\text{Sin}^{-1}1$.

Of course, the domain of the arcsine function is $[-1, 1]$, the range of the Sine function. The range of the inverse Sine function is $[-\pi/2, \pi/2]$, the domain of the Sine function. The graph of $y = \text{Sin}^{-1}x$ is shown in Figure 4.5.8.

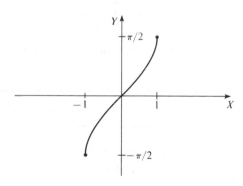

Figure 4.5.8

Example 4.5.1

Find t if (a) $\sin t = 1/2$, (b) $\text{Sin } t = 1/2$.

Solution

(a) In Example 4.2.2 we found that

$$t \in \left\{\frac{\pi}{6} + 2k\pi : k \in Z\right\} \cup \left\{\frac{5\pi}{6} + 2k\pi : k \in Z\right\}$$

See Figure 4.5.9.

A1: (a) 0, (b) 1, (c) undefined

A2: (a) 0, (b) $\pi/2$

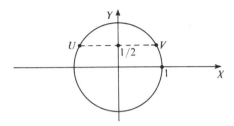

Figure 4.5.9

(b) Here t must lie in $[-\pi/2, \pi/2]$ by definition of Sine. Thus,

$$t = \text{Sin}^{-1} \frac{1}{2} = \frac{\pi}{6}$$

We shall now follow a similar procedure for the tangent function. Again we restrict the domain of the tangent function so that each range value is attained once and only once. We arbitrarily choose the interval $(-\pi/2, \pi/2)$ and make the following definitions of the principal tangent function and its inverse.

**Definition
4.5.3**

$y = \text{Tan } x$ if, and only if, $y = \tan x$ and $-\pi/2 < x < \pi/2$.

Q3: State the domain and the range of the function Tangent.

The graph of $y = \text{Tan } x$ is sketched in Figure 4.5.10. Clearly, Tangent is a one-to-one function and has an inverse.

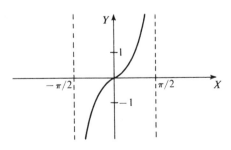

Figure 4.5.10

**Definition
4.5.4**

$y = \text{Tan}^{-1}x = \arctan x$ if, and only if, $x = \text{Tan } y$.

Q4: Find (a) $\text{Tan}^{-1}0$ and (b) $\text{Tan}^{-1}1$.

The graph of $y = \text{Tan}^{-1}x$ is sketched in Figure 4.5.11.

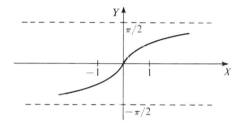

Figure 4.5.11

Q5: State the domain and the range of the inverse Tangent function.

We can read Table B.2 backward to find numerical values of $\text{Sin}^{-1}x$ and $\text{Tan}^{-1}x$.

**Example
4.5.2**

(a) Use Table B.2 to approximate $\text{Sin}^{-1}(2/3)$. (b) Use Table B.2 to approximate

$$q = \frac{\text{Sin}^{-1}(1/2) + \text{Sin}^{-1}(2/3)}{\text{Tan}^{-1}[(1/2) + (2/3)]}$$

Solution

(a) To find $\text{Sin}^{-1}(2/3)$ we use Table B.2 and see that $\sin(.73) = .6669$ or about $2/3$. Thus, $\text{Sin}^{-1}(2/3) = .73$ (decimal approximation), since $.73 \in [-\pi/2, \pi/2]$.

(b) We already know that $\text{Sin}^{-1}(1/2) = \pi/6$ or about .52, since $\sin(\pi/6) = 1/2$ and $\pi/6 \in [-\pi/2, \pi/2]$. Also, $\text{Sin}^{-1}(2/3) = .73$ from Part (a). To find $\text{Tan}^{-1}[(1/2) + (2/3)]$, we first observe that $(1/2) + (2/3)$ is about 1.167. Again reading Table B.2 backward, we see that $\tan .86 = 1.162$. Thus, $\text{Tan}^{-1}[(1/2) + (2/3)] = .86$. Finally, $q = (.52 + .73)/.86 = 1.45$.

In some instances we can use our knowledge of the trigonometric functions to simplify expressions without the use of tables as in Example 4.5.3.

A3: $(-\pi/2, \pi/2), (-\infty, \infty)$

A4: (a) 0, (b) $\pi/4$

A5: $(-\infty, \infty), (-\pi/2, \pi/2)$

Example
4.5.3

Find $\cos(\text{Sin}^{-1}\tfrac{1}{3})$.

Solution

Let $\alpha = \text{Sin}^{-1}(1/3)$. That is, $\sin \alpha = 1/3$ and $\alpha \in [-\pi/2, \pi/2]$. But $\sin^2\alpha + \cos^2\alpha = 1$, so $\cos^2\alpha = 1 - \sin^2\alpha = 1 - (1/3)^2 = 8/9$. Since $\alpha \in [-\pi/2, \pi/2]$, $\cos \alpha \geq 0$ and $\cos \alpha = \sqrt{8}/3 = 2\sqrt{2}/3$. That is, $\cos(\text{Sin}^{-1}\tfrac{1}{3}) = 2\sqrt{2}/3$.

Figure 4.5.12 illustrates the solution above in a more intuitive way. We have drawn a right triangle with one side and with the hypotenuse having lengths 1 and 3, so that $\sin \alpha = 1/3$. We know that α is an acute angle since $\sin \alpha > 0$ and $\alpha \in [-\pi/2, \pi/2]$. The third side of the triangle has length $\sqrt{9-1} = \sqrt{8} = 2\sqrt{2}$ by the Pythagorean theorem. Thus, $\cos \alpha = \cos(\text{Sin}^{-1}\tfrac{1}{3}) = 2\sqrt{2}/3$.

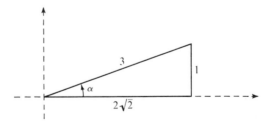

Figure 4.5.12

Warning: One must exercise care in deducing the algebraic signs using this procedure, as shown in the following example.

Example
4.5.4

Find $\sin(\text{Tan}^{-1}(-1/2))$.

Solution

Here let $\beta = \text{Tan}^{-1}(-1/2)$. Then $-\pi/2 < \beta < 0$ since $\text{Tan } \beta = -1/2 < 0$. Thus, $\sin \beta < 0$. We draw a right triangle with sides of length 1 and 2 so that $-\pi/2 < \beta < 0$ and $\tan \beta = -1/2$ (Figure 4.5.13). Thus, the hypotenuse will have length $\sqrt{4+1} = \sqrt{5}$, and $\sin \beta = \sin(\text{Tan}^{-1}(-\tfrac{1}{2})) = -1/\sqrt{5}$ since $-\pi/2 < \beta < 0$.

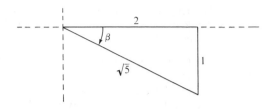

Figure 4.5.13

**Example
4.5.5**

A lighthouse L stands 3 miles from the closest point P along a straight shore. As the light moves along the shore, lighting the shore at Q, the angle θ between the line from Q to L and the line from P to L changes. Express this angle in terms of the distance d from P to Q.

Solution

In Figure 4.5.14 we see that $\tan \theta = d/3$. Taking θ to be an acute angle, we have $\theta = \text{Tan}^{-1}(d/3)$.

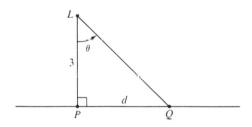

Figure 4.5.14

**PROBLEMS
4.5**

1. Find each of the following numbers:
(a) $\text{Sin}^{-1}1$ (b) $\text{Tan}^{-1}\sqrt{3}$
(c) $\text{Sin}^{-1}(\sqrt{2}/2)$ (d) $\text{Sin}^{-1}(-\sqrt{3}/2)$
(e) $\text{Tan}^{-1}(-\sqrt{3})$ (f) $\text{Sin}^{-1}(-1)$.

2. Use Table B.2 to find each of the following numbers:
(a) $\text{Sin}^{-1}(.324)$ (b) $\text{Tan}^{-1}(1.05)$
(c) $\text{Sin}^{-1}(-.836)$ (d) $\text{Tan}^{-1}(-1.26)$

3. Find each of the following numbers:
(a) $2\cos(\text{Sin}^{-1}\frac{1}{2})$ (b) $\cos(2\,\text{Sin}^{-1}\frac{1}{2})$
(c) $\cos(\text{Tan}^{-1}\sqrt{3} - \text{Tan}^{-1}0)$ (d) $\cos(\text{Tan}^{-1}\sqrt{3}) - \cos(\text{Tan}^{-1}0)$

✗ **4.** Use the technique of Example 4.5.4 to evaluate each of the following:
✓(a) $\cos(\text{Sin}^{-1}.6)$ (b) $\sin(\text{Tan}^{-1}2)$
(c) $\tan(\text{Sin}^{-1}(-.8))$ (d) $\cos(\text{Sin}^{-1}(-.8))$

5. The length L of the shadow cast by a 50-ft.-tall tower depends on θ, the angle of elevation of the sun (measured from the horizontal). Express θ in terms of L.

6. The Cosine function is defined by restricting the domain of the cosine function to $[0, \pi]$.
(a) Sketch the graph of the Cosine function.
(b) Sketch the graph of the inverse Cosine function $y = \text{Cos}^{-1}x$.
(c) Find $\text{Cos}^{-1}0$, $\text{Cos}^{-1}1$, $\text{Cos}^{-1}(-1)$, and $\text{Cos}^{-1}(1/2)$.

7. Which of the following are true:
(a) $\text{Sin}^{-1}(-x) = -\text{Sin}^{-1}x$ (b) $\text{Sin}^{-1}(2x) = 2\,\text{Sin}^{-1}x$
(c) $\sin(\text{Sin}^{-1}x) = x$ (d) $\text{Sin}^{-1}(\sin x) = x$
(e) $\text{Tan}^{-1}(a/b) = \text{Sin}^{-1}(a/\sqrt{a^2 + b^2})$
(f) $\tan(\text{Tan}^{-1}x) = x$ (g) $\text{Tan}^{-1}(\tan x) = x$

8. (a) For each of the functions cotangent, secant, and cosecant, restrict the domain so that each function is one-to-one and attains each range value precisely once.
(b) Sketch the graph of each of these functions and the inverse of each.

ANSWERS
4.5

1. (a) $\pi/2$, (b) $\pi/3$, (c) $\pi/4$, (d) $-\pi/3$, (e) $-\pi/3$, (f) $-\pi/2$

2. (a) .33, (b) .81, (c) $-.99$, (d) $-.90$

3. (a) $\sqrt{3}$, (b) $1/2$, (c) $1/2$, (d) $-1/2$

4. (a) .8, (b) $2\sqrt{5}/5$, (c) $-4/3$, (d) .6

5. $\theta = \text{Tan}^{-1}(50/L)$

6. (a) The graph is shown in Figure 4.5.15.

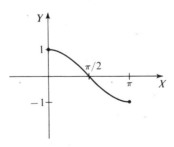

Figure 4.5.15

(b) See Figure 4.5.16.

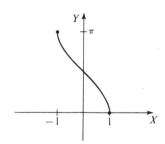

Figure 4.5.16

(c) $\pi/2$, 0, π, $\pi/3$

Supplement 4.5

I. Inverse Sine

Recall that $b = \text{Sin}^{-1}a$ if, and only if, $a = \sin b$ and $b \in [-\pi/2, \pi/2]$.

Problem S1. Given that $t \in [-\pi/2, \pi/2]$, complete the following:
(a) If $\sin t = 1$, then $t = \pi/2$; so $\text{Sin}^{-1}1 = \pi/2$.
(b) If $\sin t = 1/2$, then $t =$ _____; so $\text{Sin}^{-1}(1/2) =$ _____.
(c) If $\sin t = 0$, then $t =$ _____; so $\text{Sin}^{-1}0 =$ _____.
(d) If $\sin t = \sqrt{3}/2$, then $t =$ _____; so $\text{Sin}^{-1}(\sqrt{3}/2) =$ _____.
(e) If $\sin t = -1$, then $t =$ _____; so $\text{Sin}^{-1}(-1) =$ _____.
(f) If $\sin t = -1/2$, then $t =$ _____; so $\text{Sin}^{-1}(-1/2) =$ _____.

Solution: (b) $\pi/6$, $\pi/6$, (c) 0, 0, (d) $\pi/3$, $\pi/3$, (e) $-\pi/2$, $-\pi/2$, (f) $-\pi/6$, $-\pi/6$

Problem S2. Complete the following:
(a) $\text{Sin}^{-1}(\sqrt{2}/2) =$ _____   (b) $\text{Sin}^{-1}(-\sqrt{2}/2) =$ _____
(c) $\text{Sin}^{-1}(-\sqrt{3}/2) =$ _____   (d) $\text{Sin}^{-1}(-\pi/2) =$ _____

Solution: (a) $\pi/4$, (b) $-\pi/4$, (c) $-\pi/3$, (d) undefined since $-\pi/2 \notin [-1, 1]$

Problem S3. Complete the following:

(a) $\sin\left(\text{Sin}^{-1}\dfrac{\sqrt{2}}{2}\right) =$ _____   (b) $\sin\left(\text{Sin}^{-1}\left(-\dfrac{\sqrt{2}}{2}\right)\right) =$ _____

(c) $\sin\left(\text{Sin}^{-1}\left(-\frac{\sqrt{3}}{2}\right)\right) = $ _____   (d) $\sin\left(\text{Sin}^{-1}\left(-\frac{\pi}{2}\right)\right) = $ _____

(e) Is it always true that $\sin(\text{Sin}^{-1}a) = a$?

Solution: (a) $\sqrt{2}/2$, (b) $-\sqrt{2}/2$, (c) $-\sqrt{3}/2$, (d) undefined, (e) yes, if $\text{Sin}^{-1}a$ is defined, that is, if $-1 \leq a \leq 1$.

Problem S4. Use Table B.2 to approximate (a) $\text{Sin}^{-1}.548$, (b) $\text{Sin}^{-1}.77$, (c) $\text{Sin}^{-1}(-.77)$, and (d) $\text{Sin}^{-1}3.14$.

Solution: (a) .58, (b) about .88, (c) $-.88$, (d) undefined

Problem S5. Remember, the number $\text{Sin}^{-1}a$ must lie in $[-\pi/2, \pi/2]$. Complete the following, given that $t \in [\pi/2, 3\pi/2]$. (Compare with Problem S1.)
(a) If $\sin t = 0$, then $t = \pi$; but $\text{Sin}^{-1}0 = 0$.
(b) If $\sin t = 1/2$, then $t = 5\pi/6$; but $\text{Sin}^{-1}(1/2) = \pi/6$.
(c) If $\sin t = -1$, then $t = $ _____; but $\text{Sin}^{-1}(-1) = $ _____.
(d) If $\sin t = \sqrt{2}/2$, then $t = $ _____; but $\text{Sin}^{-1}(\sqrt{2}/2) = $ _____.
(e) If $\sin t = -\sqrt{3}/2$, then $t = $ _____; but $\text{Sin}^{-1}(-\sqrt{3}/2) = $ _____

Solution: (c) $3\pi/2$, $-\pi/2$, (d) $3\pi/4$, $\pi/4$, (e) $4\pi/3$, $-\pi/3$.

Problem S6. Complete the following;
(a) $\sin 0 = 0$, so $\text{Sin}^{-1}(\sin 0) = \text{Sin}^{-1}0 = 0$.
(b) $\sin \pi = $ _____, so $\text{Sin}^{-1}(\sin \pi) = \text{Sin}^{-1}$___ $= $ ___.
(c) $\sin\left(-\frac{\pi}{2}\right) = $ _____, so $\text{Sin}^{-1}\left(\sin\left(-\frac{\pi}{2}\right)\right) = \text{Sin}^{-1}$___ $= $ ___.
(d) $\sin \frac{3\pi}{2} = $ _____, so $\text{Sin}^{-1}\left(\sin \frac{3\pi}{2}\right) = \text{Sin}^{-1}$___ $= $ ___.
(e) $\sin\left(-\frac{\pi}{6}\right) = $ _____, so $\text{Sin}^{-1}\left(\sin\left(-\frac{\pi}{6}\right)\right) = \text{Sin}^{-1}$___ $= $ ___.
(f) $\sin \frac{11\pi}{6} = $ _____, so $\text{Sin}^{-1}\left(\sin \frac{11\pi}{6}\right) = \text{Sin}^{-1}$___ $= $ ___.
(g) Does $\text{Sin}^{-1}(\sin t) = t$?

Solution: (b) 0, 0, 0, (c) -1, -1, $-\pi/2$, (d) -1, -1, $-\pi/2$, (e) $-1/2$, $-1/2$, $-\pi/6$, (f) $-1/2$, $-1/2$, $-\pi/6$, (g) No, (b), (d), and (f) are counter examples.

Problem S7. Graph each of the following. Carefully show the domain and the range of each.
(a) $y = f(x) = \sin x$ (b) $y = F(x) = \text{Sin } x$
(c) $y = F^{-1}(x) = \text{Sin}^{-1}x$

Solution: See Figures 4.2.9, 4.5.7, and 4.5.8.

Problem S8.   Graph each of the following:

(a) $y = \frac{1}{2} \operatorname{Sin}^{-1}x$

(b) $y = \operatorname{Sin}^{-1}\frac{1}{2}x$

(c) $y = \operatorname{Sin}^{-1}\frac{1}{2} + x$

(d) $y = \frac{\pi}{2} + \operatorname{Sin}^{-1}(x + 1)$

Solution:

(a) See Figure 4.5.17.

(b) See Figure 4.5.18.

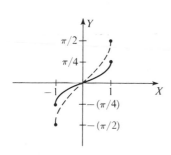

Figure 4.5.17

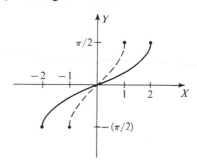

Figure 4.5.18

(c) See Figure 4.5.19.

(d) See Figure 4.5.20.

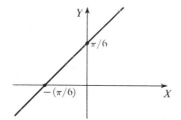

Figure 4.5.19

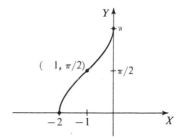

Figure 4.5.20

II. Inverse Tangent

Problem S9.   Complete the following table:

t	0	$\pi/6$	$\pi/4$	$\pi/3$	$\pi/2$	$-\pi/6$	$-\pi/4$	$-\pi/3$
$\tan t$								

Solution:

t	0	$\pi/6$	$\pi/4$	$\pi/3$	$\pi/2$	$-\pi/6$	$-\pi/4$	$-\pi/3$
$\tan t$	0	$\sqrt{3}/3$	1	$\sqrt{3}$	undef.	$-\sqrt{3}/3$	-1	$-\sqrt{3}$

Problem S10.   Complete the following table:

x	0	$\sqrt{3}/3$	1	$\sqrt{3}$	$-\sqrt{3}/3$	-1	$-\sqrt{3}$
$\text{Tan}^{-1}x$							

Solution: Using Problem S9, we obtain the following table:

x	0	$\sqrt{3}/3$	1	$\sqrt{3}$	$-\sqrt{3}/3$	-1	$-\sqrt{3}$
$\text{Tan}^{-1}x$	0	$\pi/6$	$\pi/4$	$\pi/3$	$-\pi/6$	$-\pi/4$	$-\pi/3$

Problem S11.   Complete the following table:

t	$2\pi/3$	$3\pi/4$	$5\pi/6$	π	$5\pi/4$	$4\pi/3$
$\tan t$						

Solution:

t	$2\pi/3$	$3\pi/4$	$5\pi/6$	π	$5\pi/4$	$4\pi/3$
$\tan t$	$-\sqrt{3}$	-1	$-\sqrt{3}/3$	0	1	$\sqrt{3}$

Problem S12.   Complete the following table:

x	$-\sqrt{3}$	-1	$-\sqrt{3}/3$	0	1	$\sqrt{3}$
$\text{Tan}^{-1}x$						

Solution: (See Problem S10.) Compare Problems S11 and S12 with S9 and S10.

x	$-\sqrt{3}$	-1	$-\sqrt{3}/3$	0	1	$\sqrt{3}$
$\text{Tan}^{-1}x$	$-\pi/3$	$-\pi/4$	$-\pi/6$	0	$\pi/4$	$\pi/3$

Problem S13.   Use Table B.2 to find (a) $\text{Tan}^{-1}.631$, (b) $\text{Tan}^{-1}.4$, and (c) $\text{Tan}^{-1}(-.4)$.
Solution: (a) .57, (b) about .38, (c) about $-.38$

Problem S14.   Graph (a) $y = \tan x$, (b) $y = \text{Tan } x$, and (c) $y = \text{Tan}^{-1}x$.
Solution: See Figures 4.3.5, 4.5.10, and 4.5.11.

Problem S15.   Graph (a) $y = \text{Tan}^{-1}(x - 1)$, (b) $y = -\text{Tan}^{-1}(x - 1)$, and (c) $y = (\pi/2) - \text{Tan}^{-1}(x - 1)$.

Solution:
(a) See Figure 4.5.21.   (b) See Figure 4.5.22.

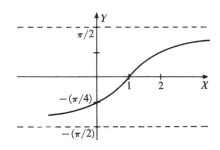

Figure 4.5.21   Figure 4.5.22

(c) See Figure 4.5.23.

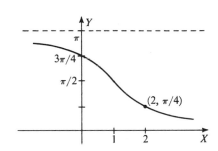

Figure 4.5.23

III. Inverse Cosine

(See Problem 6.)

Problem S16.   Evaluate the following:
(a) $\cos \frac{1}{2}\pi$ (b) $\cos(-\frac{1}{2}\pi)$
(c) $\text{Cos}^{-1}0$ (d) $\cos \frac{1}{4}\pi$
(e) $\cos \frac{9}{4}\pi$ (f) $\text{Cos}^{-1}(\sqrt{2}/2)$
Solution: (a) 0, (b) 0, (c) $\pi/2$, (d) $\sqrt{2}/2$, (e) $\sqrt{2}/2$, (f) $\pi/4$

Problem S17.   Evaluate the following:
(a) $\text{Cos}^{-1}1$ (b) $\text{Cos}^{-1}(\text{Cos}^{-1}1)$
(c) $(\text{Cos}^{-1}0)^{-1}$ (d) $(\text{Cos }0)^{-1}$
(e) $\text{Cos}^{-1}2^{-1}$ (f) $(\text{Cos}^{-1}2^{-1})^{-1}$
Solution: (a) 0, (b) $\pi/2$, (c) $2/\pi$, (d) 1, (e) $\pi/3$, (f) $3/\pi$

Problem S18.   Graph the following:
(a) $y = |\text{Cos}^{-1}x|$ (b) $y = \text{Cos}^{-1}|x|$
(c) $y = \frac{1}{2}\pi - \text{Cos}^{-1}x$

Solution:

(a) See Figure 4.5.24.   (b) See Figure 4.5.25.

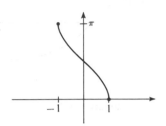

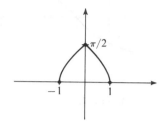

Figure 4.5.24   **Figure 4.5.25**

(c) See Figure 4.5.26.

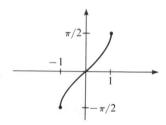

Figure 4.5.26

IV. Right Triangles

In the right triangle shown in Figure 4.5.27, $\sin \theta = 3/5$ and $\tan \theta = 3/4$. Since θ is an acute angle, $\theta = \text{Sin}^{-1}(3/5)$ or $\theta = \text{Tan}^{-1}(3/4)$.

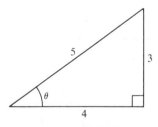

Figure 4.5.27

Problem S19.   Use the function Sin^{-1} to express θ in terms of the length of the sides of the right triangles in Figures 4.5.28, 4.5.29, 4.5.30, and 4.5.31.

(a) (b)

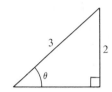 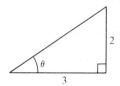

Figure 4.5.28 Figure 4.5.29

(c) (d)

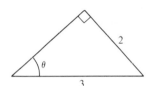 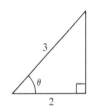

Figure 4.5.30 Figure 4.5.31

Solution:
(a) $\theta = \text{Sin}^{-1}(2/3)$
(c) $\theta = \text{Sin}^{-1}(2/3)$

(b) $\theta = \text{Sin}^{-1}(2\sqrt{13}/13)$
(d) $\theta = \text{Sin}^{-1}(\sqrt{5}/3)$

Problem S20. Repeat Problem S19, but use the function Tan^{-1}.

Solution:
(a) $\text{Tan}^{-1}(2\sqrt{5}/5)$
(c) $\text{Tan}^{-1}(2\sqrt{5}/5)$

(b) $\text{Tan}^{-1}(2/3)$
(d) $\text{Tan}^{-1}(\sqrt{5}/2)$

Problem S21. Repeat Problem S19, but use the function Cos^{-1}.

Solution:
(a) $\text{Cos}^{-1}(\sqrt{5}/3)$
(c) $\text{Cos}^{-1}(\sqrt{5}/3)$

(b) $\text{Cos}^{-1}(3\sqrt{13}/13)$
(d) $\text{Cos}^{-1}(2/3)$

Problem S22. Evaluate each of the following (see Problems S19, S20, and S21):

(a) $\sin\left(\text{Sin}^{-1}\frac{2}{3}\right)$

(b) $\cos\left(\text{Sin}^{-1}\frac{2}{3}\right)$

(c) $\tan\left(\text{Sin}^{-1}\frac{2}{3}\right)$

(d) $\sin\left(\text{Tan}^{-1}\frac{2}{3}\right)$

(e) $\cos\left(\operatorname{Tan}^{-1}\frac{2}{3}\right)$ (f) $\tan\left(\operatorname{Tan}^{-1}\frac{2}{3}\right)$

(g) $\sec\left(\operatorname{Tan}^{-1}\frac{2}{3}\right)$ (h) $\sin\left(\operatorname{Cos}^{-1}\frac{2}{3}\right)$

(i) $\cot\left(\operatorname{Cos}^{-1}\frac{2}{3}\right)$ (j) $\sec\left(\operatorname{Cos}^{-1}\frac{2}{3}\right)$

(k) $\cos\left(\operatorname{Sin}^{-1}\frac{\sqrt{5}}{3}\right)$ (l) $\csc\left(\operatorname{Sin}^{-1}\frac{\sqrt{5}}{3}\right)$

Solution: (See Examples 4.5.3 and 4.5.4.) (a) $2/3$, (b) $\sqrt{5}/3$, (c) $2/\sqrt{5}$, (d) $2/\sqrt{13}$, (e) $3/\sqrt{13}$, (f) $2/3$, (g) $\sqrt{13}/3$, (h) $\sqrt{5}/3$, (i) $2\sqrt{5}/5$, (j) $3/2$, (k) $2/3$, (l) $3\sqrt{5}/5$

Problem S23.   Evaluate each of the following:

(a) $\sin(\operatorname{Tan}^{-1}(-1))$ (b) $\sin(\operatorname{Tan}^{-1}(-1/3))$

(c) $\sin(\operatorname{Cos}^{-1}(-1/2))$ (d) $\sin(\operatorname{Cos}^{-1}(-2/3))$

Solution: (a) $-\sqrt{2}/2$, (b) $-\sqrt{10}/10$, (c) $\sqrt{3}/2$, (d) $\sqrt{5}/3$

4.6 IDENTITIES

Since we have used the sine and cosine functions to define four other trigonometric functions, any problem solved using these six functions could certainly be solved using only sine and cosine. In fact, only one of these two is needed (why?). Nevertheless, in certain applications it is convenient to have any of the six at our disposal. However, this redundancy and the many properties of these functions give rise to a veritable deluge of trigonometric identities. In this section we shall review and develop the most commonly used identities. Although the resulting list of identities will be far from complete, it and the techniques used in development will be adequate for most applications.

Recall that an identity is a statement of equality that it is true for every value of the variable for which all terms are defined. From our definitions we have the following basic identities:

$$\tan t = \frac{\sin t}{\cos t}, \qquad \cot t = \frac{\cos t}{\sin t}, \qquad \cot t = \frac{1}{\tan t} \tag{1}$$
$$\sec t = \frac{1}{\cos t}, \qquad \csc t = \frac{1}{\sin t}, \qquad \sin^2 t + \cos^2 t = 1$$

Notice that the equation $\cot t = 1/\tan t$ is undefined for $t = \pi/2$, for example, even though $\cot \frac{1}{2}\pi$ is defined to be 0.

We can easily generate two more identities by dividing the equation $\sin^2 t + \cos^2 t = 1$ by either $\sin^2 t$ or $\cos^2 t$. For example,

$$\frac{\sin^2 t}{\cos^2 t} + \frac{\cos^2 t}{\cos^2 t} = \frac{1}{\cos^2 t},$$

or

$$\tan^2 t + 1 = \sec^2 t \qquad (2)$$

Q1: Obtain a similar result for the cotangent and cosecant functions by dividing by $\sin^2 t$ rather than $\cos^2 t$.

Since the sine and cosine functions are used to define the other trigonometric functions, many expressions can be simplified using Equations (1), as in Example 4.6.1.

Example 4.6.1   Simplify $\csc t - \dfrac{\cot t}{\sec t}$.

Solution

$$\csc t - \frac{\cot t}{\sec t} = \frac{1}{\sin t} - \frac{\cos t}{\sin t} \cdot \cos t$$
$$= \frac{1 - \cos^2 t}{\sin t} = \frac{\sin^2 t}{\sin t}$$
$$= \sin t$$

Of course, we write $\csc t - (\cot t)/(\sec t) = \sin t$ with the understanding that the left-hand side is undefined (why?) for any integral multiple of $\pi/2$, so the equation is not valid for these values of t.

In Section 4.2 we derived the *negative-angle* formulas:

$$\sin(-t) = -\sin t \quad \text{and} \quad \cos(-t) = \cos t \qquad (3)$$

Q2: Use Equations (1) and (3) to derive negative-angle formulas for the remaining four trigonometric functions.

One set of identities which shall prove useful to us are the so-called addition formulas. We shall now derive one of these, namely,

$$\cos(u - v) = \cos u \cdot \cos v + \sin u \cdot \sin v$$

To verify this equation, consider the points $P(u)$, $P(v)$, $P(u - v)$ and $P(0)$ on the unit circle. By definition, $P(u) = (\cos u, \sin u)$, $P(v) = (\cos v, \sin v)$, $P(u - v) = (\cos(u - v), \sin(u - v))$ and $P(0) = (1, 0)$. If $u > v$, the length of the arc from $P(v)$ to $P(u)$ is $u - v$, the same as the length of the arc from $P(0)$ to $P(u - v)$. Thus, the length of the chord from $P(v)$ to $P(u)$ must equal the length of the chord from $P(0)$ to $P(u - v)$ (see Figure 4.6.1).

Thus,

$$\overline{P(u-v)P(0)} = \overline{P(u)P(v)}$$

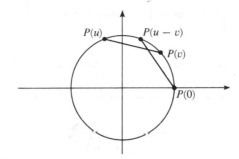

Figure 4.6.1

Using the distance formula and squaring, we obtain

$$[\cos(u-v) - 1]^2 + [\sin(u-v)]^2$$
$$= (\cos u - \cos v)^2 + (\sin u - \sin v)^2$$

or

$$\cos^2(u-v) - 2\cos(u-v) + 1 + \sin^2(u-v)$$
$$= \cos^2 u - 2\cos u \cdot \cos v + \cos^2 v + \sin^2 u - 2\sin u \cdot \sin v + \sin^2 v$$

Using the fact that $\sin^2 t + \cos^2 t = 1$, we simplify the above equation to

$$2 - 2\cos(u-v) = 2 - 2\cos u \cdot \cos v - 2\sin u \cdot \sin v$$

Multiplying by $-1/2$ and then adding 1 to both sides of this last equation yields $\cos(u-v) = \cos u \cdot \cos v + \sin u \cdot \sin v$, as desired.

Q3: Use the formula for $\cos(u-v)$ to show that $\cos\left(x - \dfrac{\pi}{2}\right) = \cos\left(\dfrac{\pi}{2} - x\right) = \sin x$ as suggested by Example 4.2.4.

We obtain a formula for $\cos(u+v)$ by replacing v with $-v$ everywhere in the formula for $\cos(u-v)$. Thus, $\cos(u+v) = \cos(u - (-v)) = \cos u \cdot \cos(-v) + \sin u \cdot \sin(-v)$. Since $\cos(-v) = \cos v$ and $\sin(-v) = -\sin v$,

$$\cos(u+v) = \cos u \cdot \cos v - \sin u \cdot \sin v$$

If we write $t = (t - \tfrac{1}{2}\pi) + \tfrac{1}{2}\pi$, we have

A1: $\dfrac{\sin^2 t}{\sin^2 t} + \dfrac{\cos^2 t}{\sin^2 t} = \dfrac{1}{\sin^2 t}$, or $1 + \cot^2 t = \csc^2 t$

A2: $\tan(-t) = -\tan t$, $\cot(-t) = -\cot t$, $\sec(-t) = \sec t$, and $\csc(-t) = -\csc t$

A3: $\cos(x - \tfrac{1}{2}\pi) = \cos x \cdot \cos(\pi/2) + \sin x \cdot \sin(\pi/2) = (\cos x) \cdot 0 + (\sin x) \cdot 1 = \sin x$

$$\cos t = \cos\left[\left(t - \frac{\pi}{2}\right) + \frac{\pi}{2}\right]$$
$$= \cos\left(t - \frac{\pi}{2}\right) \cdot \cos\frac{\pi}{2} - \sin\left(t - \frac{\pi}{2}\right) \cdot \sin\frac{\pi}{2}$$
$$= \cos\left(t - \frac{\pi}{2}\right) \cdot 0 - \sin\left(t - \frac{\pi}{2}\right) \cdot 1$$
$$= -\sin\left(t - \frac{\pi}{2}\right) = \sin\left(\frac{\pi}{2} - t\right).$$

Repeating this last result and the result of Q3, we have

$$\sin\left(\frac{\pi}{2} - t\right) = \cos t$$

$$\cos\left(\frac{\pi}{2} - t\right) = \sin t$$

(4)

This last pair of equations has the geometrical interpretation that the points $P(t)$ and $P(\tfrac{1}{2}\pi - t)$ are reflections *IR* of each other (see Figure 4.6.2). That is, the first coordinate of $P(t)$ equals the second coordinate of $P(\tfrac{1}{2}\pi - t)$ and vice versa.

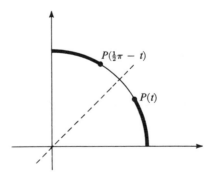

Figure 4.6.2

We can also use the equations derived earlier to obtain formulas for $\sin(u + v)$ and $\sin(u - v)$. For example,

$$\sin(u + v) = \cos\left[\frac{\pi}{2} - (u + v)\right] = \cos\left[\left(\frac{\pi}{2} - u\right) - v\right]$$
$$= \cos\left(\frac{\pi}{2} - u\right)\cdot\cos v + \sin\left(\frac{\pi}{2} - u\right)\cdot\sin v$$

Thus,

$$\sin(u + v) = \sin u\cdot\cos v + \cos u\cdot\sin v$$

Q4: Derive a formula for $\sin(u - v)$ by replacing v with $-v$ everywhere in the formula for $\sin(u + v)$ and simplifying.

To summarize, we have now derived the following addition formulas:

$$\sin(u + v) = \sin u\cdot\cos v + \cos u\cdot\sin v$$
$$\sin(u - v) = \sin u\cdot\cos v - \cos u\cdot\sin v$$
$$\cos(u + v) = \cos u\cdot\cos v - \sin u\cdot\sin v$$
$$\cos(u - v) = \cos u\cdot\cos v + \sin u\cdot\sin v$$

(5)

Q5: Using the second of Equations (5) with $u = 45°$ and $v = 30°$, find $\sin 15°$.

Of course, we can use Equations (5) to generate addition formulas for the remaining trigonometric functions.

Example 4.6.2

Derive an addition formula for the tangent function.

Solution

$$\tan(a + b) = \frac{\sin(a + b)}{\cos(a + b)}$$
$$= \frac{\sin a\cdot\cos b + \cos a\cdot\sin b}{\cos a\cdot\cos b - \sin a\cdot\sin b}$$

We often prefer to write the formula above using only the tangent function. To accomplish this, we divide both numerator and denominator by $\cos a\cdot\cos b$, obtaining

$$\tan(a + b) = \frac{\dfrac{\sin a\cdot\cos b}{\cos a\cdot\cos b} + \dfrac{\cos a\cdot\sin b}{\cos a\cdot\cos b}}{\dfrac{\cos a\cdot\cos b}{\cos a\cdot\cos b} - \dfrac{\sin a\cdot\sin b}{\cos a\cdot\cos b}}$$
$$= \frac{\tan a + \tan b}{1 - \tan a\cdot\tan b}$$

A4: $\sin(u - v) = \sin u \cdot \cos v - \cos u \cdot \sin v$

A5: $\sin(15°) = \sin(45° - 30°) = \sin 45° \cos 30° - \cos 45° \sin 30°$

$$= \frac{\sqrt{2}}{2} \cdot \frac{\sqrt{3}}{2} - \frac{\sqrt{2}}{2} \cdot \frac{1}{2} = \frac{\sqrt{2}}{4}(\sqrt{3} - 1)$$

If we set $u = t$ and $v = t$ in the addition formulas (5), we obtain the following *double-angle* formulas:

$$\sin(t + t) = \sin t \cdot \cos t + \cos t \cdot \sin t$$
$$\cos(t + t) = \cos t \cdot \cos t - \sin t \cdot \sin t$$

That is,

$$\sin 2t = 2 \sin t \cdot \cos t$$
$$\cos 2t = \cos^2 t - \sin^2 t \tag{6}$$

Q6: Find $\sin 2t$ if $\sin t = 2/3$ and $0 < t < \pi/2$.

We can write the identity for $\cos 2t$ in other ways by using the fact that $\sin^2 t + \cos^2 t = 1$. For example,

$$\cos 2t = \cos^2 t - \sin^2 t = (1 - \sin^2 t) - \sin^2 t$$
$$= 1 - 2 \sin^2 t$$

If we now solve this equation for $\sin^2 t$, we obtain $2 \sin^2 t = 1 - \cos 2t$ or

$$\sin^2 t = \frac{1 - \cos 2t}{2} \tag{7}$$

Q7: Starting with the equation $\cos 2t = \cos^2 t - \sin^2 t$, derive the identity $\cos 2t = 2 \cos^2 t - 1$. Now use this identity to obtain a formula for $\cos^2 t$ similar to Equation (7).

We often rewrite Equation (7) with $\theta = 2t$ (so that $t = \theta/2$) to generate the *half-angle* formulas:

$$\sin^2 \frac{\theta}{2} = \frac{1 - \cos \theta}{2}$$
$$\cos^2 \frac{\theta}{2} = \frac{1 + \cos \theta}{2} \tag{8}$$

Q8: Using Equation (8), find $\sin 15°$ and $\cos 15°$.

Example
4.6.3

Derive a half-angle formula for the tangent function.

Solution

Using Equations (8), we obtain

$$\tan^2\frac{\theta}{2} = \frac{\sin^2(\tfrac{1}{2}\theta)}{\cos^2(\tfrac{1}{2}\theta)} = \frac{1 - \cos\theta}{1 + \cos\theta}$$

Thus,

$$\tan\frac{\theta}{2} = \pm\sqrt{\frac{1 - \cos\theta}{1 + \cos\theta}}$$

where the choice of sign is determined by the number θ. Actually we can derive alternate formulas which avoid this sign ambiguity. For example, multiply top and bottom of $\tan\tfrac{1}{2}\theta = (\sin\tfrac{1}{2}\theta)/(\cos\tfrac{1}{2}\theta)$ by $2\sin\tfrac{1}{2}\theta$, obtaining

$$\tan\frac{\theta}{2} = \frac{2\sin^2\tfrac{1}{2}\theta}{2\sin\tfrac{1}{2}\theta\cdot\cos\tfrac{1}{2}\theta} = \frac{1 - \cos\theta}{\sin\theta}$$

by using Equations (6) and (8). Similarly,

$$\tan\frac{\theta}{2} = \frac{\sin\tfrac{1}{2}\theta}{\cos\tfrac{1}{2}\theta}\frac{2\cos\tfrac{1}{2}\theta}{2\cos\tfrac{1}{2}\theta} = \frac{\sin\theta}{1 + \cos\theta}$$

We can also use the addition formulas to obtain the product and sum formulas. For example, straightforward computation yields

$$\sin(u + v) + \sin(u - v) = \sin u\cdot\cos v + \cos u\cdot\sin v$$
$$+ \sin u\cdot\cos v - \cos u\cdot\sin v$$
$$= 2\sin u\cdot\cos v$$

That is, certain products of trigonometric functions can be expressed as sums of trigonometric functions, a fact which proves useful in certain applications. The equation above is rewritten, along with others which are similarly derived, as follows:

$$\sin u\cdot\cos v = \frac{1}{2}[\sin(u + v) + \sin(u - v)]$$

$$\sin u\cdot\sin v = \frac{1}{2}[\cos(u - v) - \cos(u + v)] \qquad (9)$$

$$\cos u\cdot\cos v = \frac{1}{2}[\cos(u - v) + \cos(u + v)]$$

In other instances it is useful to express certain sums of trigonometric functions as products. We simply "reverse" Equations (9), substituting $u + v = x$ and $u - v = y$ so that $u = \tfrac{1}{2}(x + y)$ and $v = \tfrac{1}{2}(x - y)$. It is then an easy exercise to derive the following identities:

A6: Since $0 < t < \pi/2$, $\cos t = \sqrt{1 - \sin^2 t} = \sqrt{1 - 4/9} = \sqrt{5}/3$. Thus,

$$\sin 2t = 2 \cdot \frac{2}{3} \cdot \frac{\sqrt{5}}{3} = \frac{4\sqrt{5}}{9}$$

by Equation (6),

A7: $\cos 2t = \cos^2 t - \sin^2 t = \cos^2 t - (1 - \cos^2 t) = 2\cos^2 t - 1$. Thus,

$$\cos^2 t = \frac{1 + \cos 2t}{2}$$

A8: By Equation (8),

$$\sin^2 \frac{30°}{2} = \frac{1 - \cos 30°}{2}$$

So $\sin 15° = (1/2)\sqrt{2 - \sqrt{3}}$ and $\cos 15° = (1/2)\sqrt{2 + \sqrt{3}}$

$$\sin x + \sin y = 2 \sin\left(\frac{x + y}{2}\right) \cdot \cos\left(\frac{x - y}{2}\right)$$

$$\cos x + \cos y = 2 \cos\left(\frac{x + y}{2}\right) \cdot \cos\left(\frac{x - y}{2}\right)$$

$$\cos x - \cos y = -2 \sin\left(\frac{x + y}{2}\right) \cdot \sin\left(\frac{x - y}{2}\right) \tag{10}$$

$$\sin x - \sin y = 2 \sin\left(\frac{x - y}{2}\right) \cdot \cos\left(\frac{x + y}{2}\right)$$

PROBLEMS 4.6

1. Simplify each of the following:

(a) $\sin(t - \frac{3}{2}\pi)$

(b) $\sec(t + \frac{3}{2}\pi)$

(c) $\csc(\pi - t)$

(d) $\tan(\frac{1}{2}\pi - t)$

(e) $\dfrac{1 + \tan t}{1 + \cot t}$

(f) $\dfrac{\sec t + \tan t}{\sec t + \tan t - \cos t}$

2. Verify the following identities:

(a) $\tan x + \cot x = \sec x \cdot \csc x$

(b) $\tan x + \cot x = 2 \csc 2x$

(c) $\tan x \cdot \sin x + \cos x = \sec x$

(d) $\cos^4 x - \sin^4 x = \cos 2x$

(e) $\cot x = \dfrac{1 + \cos 2x}{\sin 2x} = \dfrac{\sin 2x}{1 - \cos 2x}$

(f) $\cot x - \tan x = 2 \cot 2x$

(g) $\dfrac{\tan x + 1}{\cot x + 1} = \tan x$

(h) $\csc x - \cot x = \tan \dfrac{x}{2}$

3. Use Equation (5) to derive an identity for sin $3t$ in terms of sin t and cos t.

4. Use Equation (5) to evaluate each of the following:

(a) $\cos \dfrac{\pi}{12}$ (b) $\cos \dfrac{5\pi}{12}$

(c) $\sin \dfrac{\pi}{12}$ (d) $\sin \dfrac{5\pi}{12}$

(e) $\tan \dfrac{\pi}{12}$

5. Use Equation (8) to compute the exact value of each of the following numbers. Compare your answers with the values in Table B.3.
(a) sin 22.5°   (b) $\cos \frac{7}{8}\pi$
✗(c) $\tan \frac{9}{8}\pi$

✗**6.** In Q-Question 5 we obtained the result that $\sin 15° = \dfrac{\sqrt{2}}{4}(\sqrt{3} - 1)$, but in Q-Question 8 we found that $\sin 15° = \frac{1}{2}\sqrt{2 - \sqrt{3}}$. Why are these answers different?

7. If sin $t = 2/3$ and $\pi/2 < t < \pi$, find (a) cos t, (b) sin $2t$, (c) cos $2t$, (d) tan $2t$, (e) sin $\frac{1}{2}t$, and (f) cos $\frac{1}{2}t$.

✓**8.** Find $P(2t)$ and $P(t/2)$ if (a) $P(t) = (-.6, .8)$, (b) $P(t) = (.6, -.8)$.

✗════════════════════════════════════

9. Verify each of the following identities:
(a) $\dfrac{\sin 2a + \sin 2b}{\cos 2a + \cos 2b} = \tan(a + b)$
(b) $\dfrac{\sin 2x + \sin 4x}{\cos 2x - \cos 4x} = \cot x$
+(c) $4 \sin^3 x = 3 \sin x - \sin 3x$
✗(d) $4 \sin x \cdot \sin 2x \cdot \sin 3x = \sin 2x + \sin 4x - \sin 6x$

**ANSWERS
4.6**

1. (a) cos t, (b) csc t, (c) csc t, (d) cot t, (e) tan t, (f) csc t

3. 3 sin t cos²t − sin³t or equivalent

4. (a) $\dfrac{\sqrt{2}}{4}(\sqrt{3} + 1)$ (b) $\dfrac{\sqrt{2}}{4}(\sqrt{3} - 1)$

(c) $\dfrac{\sqrt{2}}{4}(\sqrt{3} - 1)$ (d) $\dfrac{\sqrt{2}}{4}(\sqrt{3} + 1)$

(e) $2 - \sqrt{3}$

5. (a) $\dfrac{\sqrt{2 - \sqrt{2}}}{2}$ (b) $\dfrac{-\sqrt{2 + \sqrt{2}}}{2}$

(c) $\sqrt{\dfrac{2 - \sqrt{2}}{2 + \sqrt{2}}} = \sqrt{2} - 1$

6. They are equal.

7. (a) $-\sqrt{5}/3$, (b) $-4\sqrt{5}/9$, (c) $1/9$, (d) $-4\sqrt{5}$, (e) $\sqrt{\dfrac{3 + \sqrt{5}}{6}}$,

(f) $\sqrt{\dfrac{3 - \sqrt{5}}{6}}$

8. (a) $(-.28, -.96), (\sqrt{.2}, \sqrt{.8})$ (b) $(-.28, -.96), (-\sqrt{.8}, \overset{\frown}{\sqrt{.2}})$

Supplement 4.6

Problem S1.   Simplify $\csc(t + \pi)$.

Solution: Applying Equations (1) and (5), we obtain

$$\csc(t + \pi) = \frac{1}{\sin(t + \pi)} = \frac{1}{(\sin t)\cos \pi + (\cos t)\sin \pi}$$

Clearly $\cos \pi = -1$ and $\sin \pi = 0$. Hence, substituting, we obtain

$$\csc(t + \pi) = \frac{1}{-\sin t} = -\csc t$$

Thus, $\csc(t + \pi) = -\csc t$.

Problem S2.   Simplify $\dfrac{\sin t - \cos t}{\cos t} + 1$.

Solution: "Breaking up" the fraction yields

$$\frac{\sin t - \cos t}{\cos t} + 1 = \frac{\sin t}{\cos t} - \frac{\cos t}{\cos t} + 1 = \tan t - 1 + 1 = \tan t$$

Hence,

$$\frac{\sin t - \cos t}{\cos t} + 1 = \tan t$$

Problem S3.   If $\cos t = -2/3$ and $\pi/2 < t < \pi$, find $\sin 2t$.

Solution: First recall that $\sin 2t = 2 \sin t \cos t$ [Equation (6)]. Thus, if we can compute the number $\sin t$, we can use it to calculate $\sin 2t$. Since $\sin^2 t + \cos^2 t = 1$, $|\sin t| = \sqrt{1 - \cos^2 t}$. However, in our case, $\sin t = \sqrt{1 - \cos^2 t}$ because $\sin t > 0$ when $\pi/2 < t < \pi$. Thus, since $\cos t = -2/3$, $\sin t = \sqrt{1 - (-2/3)^2} = \sqrt{1 - 4/9} = \sqrt{5/9} = \sqrt{5}/3$. Using the identity $\sin 2t = 2 \sin t \cos t$, we obtain $\sin 2t = 2(\sqrt{5}/3)(-2/3) = -4\sqrt{5}/9$.

Alternate Solution: Since $\cos \theta = \dfrac{\text{side adj } \theta}{\text{hyp}}$, Figure 4.6.3 can be used to represent θ such that $\cos \theta = 2/3$. By the Pythagorean theorem, the third side has length $\sqrt{9-4} = \sqrt{5}$. Since $\sin \theta = \dfrac{\text{side opp } \theta}{\text{hyp}}$, $|\sin \theta| = \sqrt{5}/3$. (Recall that this method leads to an algebraic sign ambiguity.) Here $\sin t = \sqrt{5}/3$ since $\sin t > 0$ when $\pi/2 < t < \pi$. Thus, $\sin 2t$ then is found as in the first solution.

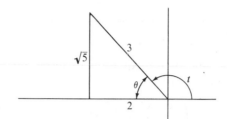

Figure 4.6.3

DRILL PROBLEMS

S4. Simplify $\cos(3\pi + t)$.

S5. Simplify $\dfrac{1}{\tan t} + \dfrac{\sin t}{\cos t - 1}$.

S6. If $\cos t = 3/4$ when $3\pi/2 < t < 2\pi$, find $\sin 2t$.

Answers

S4. $-\cos t$

S5. $-\csc t$

S6. $-\dfrac{3}{8}\sqrt{7}$

Problem S7. Compute the exact value of $\sin 22.5°$.

Solution: From Equation (8), we have that

$$\sin^2 \frac{\theta}{2} = \frac{1 - \cos \theta}{2} \quad \text{or} \quad \left|\sin \frac{\theta}{2}\right| = \sqrt{\frac{1 - \cos \theta}{2}}$$

Substituting $\theta = 45°$, we have

$$\sin \frac{45°}{2} = \sqrt{\frac{1 - \cos 45°}{2}} = \sqrt{\frac{1 - \sqrt{2}/2}{2}} = \sqrt{\frac{2 - \sqrt{2}}{4}} = \frac{1}{2}\sqrt{2 - \sqrt{2}}$$

Hence, $\sin 22.5° = \dfrac{1}{2}\sqrt{2 - \sqrt{2}}$. (Notice that the absolute value signs were discarded. Why?)

Problem S8.   Compute the exact value of $\cos(7\pi/12)$.

Solution: Observe that $7\pi/12 = 4\pi/12 + 3\pi/12 = \pi/3 + \pi/4$, so we apply Equation (5) with $u = \pi/3$ and $v = \pi/4$. Therefore,

$$\cos\frac{7\pi}{12} = \cos\left(\frac{\pi}{3} + \frac{\pi}{4}\right) = \cos\frac{\pi}{3}\cos\frac{\pi}{4} - \sin\frac{\pi}{3}\sin\frac{\pi}{4}$$

$$= \frac{1}{2}\frac{\sqrt{2}}{2} - \frac{\sqrt{3}}{2}\frac{\sqrt{2}}{2} = \frac{\sqrt{2}}{4}(1 - \sqrt{3})$$

Problem S9.   Verify the identity $(\tan^2 t + 1)\cos^2 t = 1$.

Solution: Changing one side of the equation to terms involving only sines and cosines often is helpful. For example,

$$(\tan^2 t + 1)\cos^2 t = \left[\left(\frac{\sin t}{\cos t}\right)^2 + 1\right]\cos^2 t = \left[\frac{\sin^2 t}{\cos^2 t} + \frac{\cos^2 t}{\cos^2 t}\right]\cos^2 t$$

$$= \frac{\sin^2 t + \cos^2 t}{\cos^2 t}\cdot\cos^2 t = \frac{1}{\cos^2 t}\cdot\cos^2 t = 1$$

Of course, you may have remembered that $\tan^2 t + 1 = \sec^2 t$, in which case the original equation becomes $\sec^2 t\cdot\cos^2 t = (1/\cos^2 t)\cdot\cos^2 t = 1$.

Problem S10. Verify the identity $\dfrac{1}{\csc t} + \cot t \cos t = \csc t$.

Solution: We proceed as in Problem S9. Thus, using the fact that $\sin^2 t + \cos^2 t = 1$, we obtain

$$\frac{1}{\csc t} + \cot t \cos t = \frac{1}{1/\sin t} + \frac{\cos t}{\sin t}\cdot\cos t$$

$$= \sin t + \frac{\cos^2 t}{\sin t} = \frac{\sin^2 t + \cos^2 t}{\sin t}$$

$$= \frac{1}{\sin t} = \csc t$$

Problem S11.   Verify the identity $\dfrac{\sec t}{\csc t(\sec t - 1)} - \dfrac{\csc t}{\sec t} = \csc t$.

Solution: We proceed as in Problem S9. Thus,

$$\frac{\sec t}{\csc t(\sec t - 1)} - \frac{\csc t}{\sec t} = \frac{1/\cos t}{(1/\sin t)[(1/\cos t) - 1]} - \frac{1/\sin t}{1/\cos t}$$

$$= \frac{1/\cos t}{(1/\sin t)[(1 - \cos t)/\cos t]} - \frac{\cos t}{\sin t}$$

$$= \frac{1}{(\cos t)}\frac{\sin t \cos t}{1 - \cos t} - \frac{\cos t}{\sin t}$$

$$= \frac{\sin t}{1 - \cos t} - \frac{\cos t}{\sin t}$$

$$= \frac{\sin^2 t - \cos t(1 - \cos t)}{\sin t(1 - \cos t)}$$

$$= \frac{\sin^2 t - \cos t + \cos^2 t}{\sin t(1 - \cos t)}$$

$$\overset{*}{=} \frac{1 - \cos t}{\sin t(1 - \cos t)}$$

$$= \frac{1}{\sin t} = \csc t$$

DRILL PROBLEMS

S12. Compute the exact value of $\sin(11\pi/12)$.

S13. Verify the identity $\cos x \left[\dfrac{\tan x}{\cos x \csc x} + 1 \right] = \sec x$.

Answers

S12. $\dfrac{\sqrt{2}}{4}(\sqrt{3}-1) = \dfrac{1}{2}\sqrt{2-\sqrt{3}}$

S13. $\cos x \left[\dfrac{\tan x}{\cos x \csc x} + 1 \right] = \cos x \left[\dfrac{\sin x/\cos x}{\cos x/\sin x} + 1 \right]$

$$= \cos x \left[\frac{\sin^2 x}{\cos^2 x} + \frac{\cos^2 x}{\cos^2 x} \right] = \left[\frac{\sin^2 x + \cos^2 x}{\cos x} \right]$$

$$= \sec x$$

SUPPLEMENTARY PROBLEMS

S14. Simplify $\sin\left(\dfrac{3\pi}{2} - t\right)$.

S15. Compute the exact value of $\cos(\pi/8)$.

S16. Verify the identity $\dfrac{1 + \cos 2t}{\sin 2t} = \cot t$.

Answers

S14. $-\cos t$

S15. $(1/2)\sqrt{2 + \sqrt{2}}$

S16. $\dfrac{1 + \cos 2t}{\sin 2t} = \dfrac{1 + \cos^2 t - \sin^2 t}{2 \sin t \cos t} = \dfrac{2\cos^2 t}{2 \sin t \cos t} = \cot t$

---

*In this step we used the fact that $\sin^2 t + \cos^2 t = 1$.

Achievement Test 4

You should be able to solve these problems without the use of Table B.3, except where indicated.

1. Find $P(-3\pi/4)$.

2. Evaluate csc $\frac{2}{3}\pi$.

3. Evaluate $\text{Tan}^{-1}(-1)$.

4. Using Table B.3, evaluate tan 125°.

5. Using Table B.3, evaluate tan$(-140°)$.

6. Evaluate $\sin(7\pi/2)$.

7. Evaluate $\text{Sin}^{-1}(-\sqrt{3}/2)$.

8. Find the degree measure of the angle with radian measure $(8\pi/5)$.

9. Sketch the graph of $y = \sin(x + \frac{1}{4}\pi)$.

10. Which of the following are true (list all)?
(a) cos 3 > sin 3
(b) cot$(-4) < 0$
(c) sin 10 > sin(-10)
(d) sin 4.3 < cos 4.3
(e) cot 10 > 0

11. Find $P(\frac{1}{2}\pi - t)$ if $P(t) = (a, b)$ and $0 \leq t \leq \frac{1}{2}\pi$.

12. Which of the following are true (list all)?
(a) $\cot x = \dfrac{1}{\tan x}$
(b) $\csc x = \dfrac{1}{\sec x}$
(c) $\sec(-x) = -\sec x$
(d) $\cot(t + \pi) = \cot \pi$
(e) $\csc(-t) = \csc t$
(f) cos 8 < cos 8°
(g) cot 2 < csc 2

370

13. Find the range of the function $f(x) = \sec x^2$.

14. Find the exact value of $\sin(7\pi/12)$.

15. Find the degree measure of an angle of 12 radians.

16. Find the exact value of $\cos t$ if $\sin t = .7$ and $0 \leq t \leq \pi/2$.

17. Prove or disprove the following identity: $\cot x - \tan x = 2 \cot 2x$.

18. Sketch the graph of $y = \pi + \cos x$.

19. State the range of f if $f(x) = \text{Tan}^{-1}x$.

20. Sketch the graph of $y = \csc x$.

21. Evaluate $\sec\left(\text{Sin}^{-1}\dfrac{1}{3}\right)$.

22. Sketch the graph of $y = \sin\left(\dfrac{1}{2}x + \pi\right)$.

23. Simplify $\tan x \cdot \sin x - \sec x$.

24. Use Table B.3 to find b in the triangle shown in Figure 4.A.1.

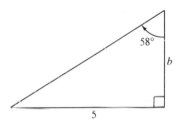

Figure 4.A.1

25. Use Table B.3 to find the length of side a in Figure 4.A.2.

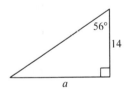

Figure 4.A.2

26. A balloonist hovering over point P sees point Q which is known to be 500 meters from P. If the angle from P to the balloon to Q measures 58°, how high is the balloonist? Use Table B.3.

27. Use Table B.3 to find the measure of angle β in Figure 4.A.3.

Figure 4.A.3

28. Sketch the graph of $y = \text{Cos}^{-1}(x - 1)$.

29. Find the exact value of $\sin\left(\text{Tan}^{-1}\left(\frac{\sqrt{2}}{2}\right)\right)$.

30. Sketch the graph of $\{(x, y) : x = \cos t, y = \sin t, 0 \le t \le \pi\}$.

Answers to Achievement Test 4

1. $\left(\dfrac{-\sqrt{2}}{2}, \dfrac{-\sqrt{2}}{2}\right)$

2. $\dfrac{2\sqrt{3}}{3}$

3. $-\dfrac{\pi}{4}$

4. -1.4281

5. $.8391$

6. -1

7. $-\dfrac{\pi}{3}$

8. $288°$

9. $y = \sin(x + \frac{1}{4}\pi)$. See Figure 4.A.4.

10. b, d, e

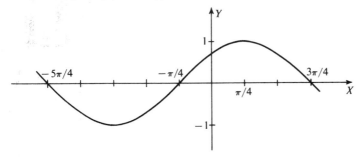

Figure 4.A.4

11. (b, a)

12. a, f, g

13. $[-\infty, -1] \cup [1, \infty]$

14. $\dfrac{\sqrt{2}}{4}(\sqrt{3} + 1) = \dfrac{1}{2}\sqrt{2 + \sqrt{3}}$

15. $\dfrac{12 \cdot 180}{\pi} = \dfrac{2160}{\pi}$

16. $\dfrac{\sqrt{51}}{10}$

17. $\cot x - \tan x = \dfrac{\cos x}{\sin x} - \dfrac{\sin x}{\cos x}$

$= \left(\dfrac{\cos^2 x - \sin^2 x}{\sin x \cdot \cos x}\right)\dfrac{2}{2}$

$= \dfrac{2 \cos 2x}{\sin 2x} = 2 \cot 2x$

18. See Figure 4.A.5.

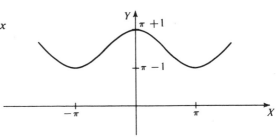

Figure 4.A.5

19. $(-\pi/2, \pi/2)$

20. See Figure 4.A.6.

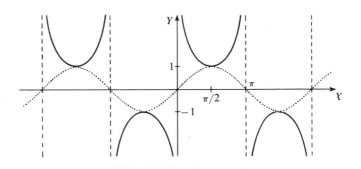

Figure 4.A.6

21. $3\sqrt{2}/4$

22. See Figure 4.A.7.

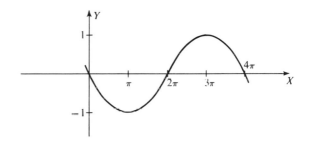

Figure 4.A.7

23. $-\cos x$

24. 3.12

25. about 21

26. 312

27. 24°

28. See Figure 4.A.8.

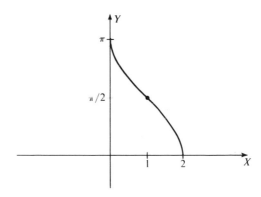

Figure 4.A.8

29. $\sqrt{3}/3$

30. See Figure 4.A.9.

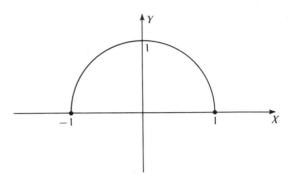

Figure 4.A.9

5 TRIGONOMETRY AND COMPLEX NUMBERS

5.0 INTRODUCTION

In this chapter we continue the study of trigonometry and introduce complex numbers, which are essential in many important applications in advanced mathematics, science, and engineering. Through the use of polar coordinates, we shall exploit trigonometry to aid in our study of the properties of complex numbers.

5.1 THE GRAPH OF
$$y = A \sin(ax + b)$$

In this section we shall use the mappings of Chapter 2 and some identities from Section 4.6 to help us graph certain equations involving trigonometric functions. Recall that the graph of $y = \sin x$ oscillates between the lines $y = 1$ and $y = -1$ (Figure 5.1.1). If we plot the graph of $y = 2 \sin x$, we see that it oscillates between 2 and -2 (Figure 5.1.2). Of course, we can obtain this graph by applying the distortion $YD/2$ to the graph $y = \sin x$. The graph of $y = -2 \sin x$ is the reflection YR of the graph of $y = 2 \sin x$ (see Figure 5.1.2). In general, if $A > 0$, we can obtain the graph of $y = A \sin x$ from the graph of $y = \sin x$ by applying the distortion YD/A; if $A < 0$, we apply the distortion $YD/|A|$ and follow with the reflection YR. In either case, the graph of

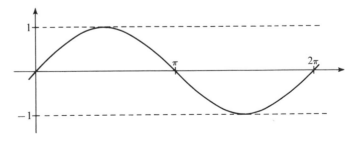

Figure 5.1.1

378

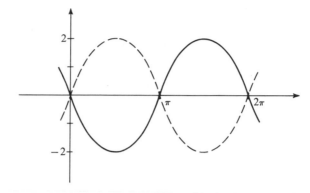

Figure 5.1.2

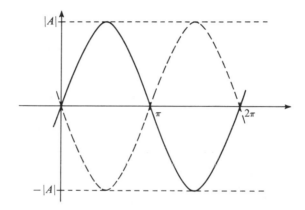

Figure 5.1.3

$y = A \sin x$ oscillates between the lines $y = |A|$ and $y = -|A|$ (Figure 5.1.3). In this case, the number $|A|$ is called the *amplitude* of this periodic function.

Next, let us sketch the graph of $y = \sin 3x$. Apply the distortion $XD/\frac{1}{3}$ to the graph of $y = \sin x$ to obtain the graph of $y = \sin 3x$. (Recall that the distortion $XD/\frac{1}{a}$ replaces x with ax everywhere in the original equation.) These graphs are shown in Figure 5.1.4. Notice that $f(x) = \sin 3x$ is a periodic function with period $2\pi/3$.

Now let us suppose that $a > 0$ throughout this section. The distortion $XD/\frac{1}{a}$ will map the graph of $y = \sin x$ onto the graph of $y = \sin ax$. In particular, the *cycle* of the graph of $y = \sin x$ between $x = 0$ and $x = 2\pi$ is mapped onto the region between 0 and $2\pi/a$ (see Figure 5.1.5). This suggests that the period of $f(x) = \sin ax$ is $2\pi/a$, and, indeed, it is easy to show that $f(x + 2\pi/a) = f(x)$.

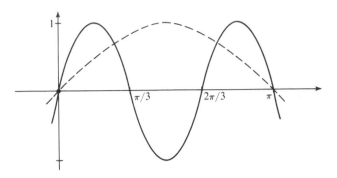

Figure 5.1.4

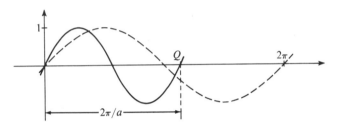

Figure 5.1.5

Q1: Find the amplitude and the period of $f(x) = 3 \sin 2x$, and sketch its graph.

**Example
5.1.1**

Sketch the graph of $y = 6 \sin x \cdot \cos x$.

Solution

Although this problem may appear rather formidable because of the product $\sin x \cdot \cos x$, we simply need recall the identity $\sin 2x = 2 \sin x \cdot \cos x$. Thus, $y = 6 \sin x \cdot \cos x = 3 \sin 2x$, and this is just the problem of Q1. In particular, the distortion $YD/3$ and the distortion $XD/\frac{1}{2}$ applied to the graph of $y = \sin x$ (in either order) will yield the graph of $y = 3 \sin 2x$ shown in Figure 5.1.6.

We can sketch the graph of $y = \sin(2x - \frac{1}{2}\pi)$ by applying an appropriate pair of mappings to the graph of $y = \sin x$. For example, $y = \sin x \xrightarrow{XD/\frac{1}{2}} y = \sin 2x$, as shown in Figure 5.1.7. Now to obtain the graph of $y = \sin(2x - \frac{1}{2}\pi)$ from the graph of $y = \sin 2x$, we first write $y = \sin(2x - \frac{1}{2}\pi) = \sin 2(x - \frac{1}{4}\pi)$. Notice that replacing x with $x - \frac{1}{4}\pi$ in $y = \sin 2x$ yields $y = \sin 2(x - \frac{1}{4}\pi) = \sin(2x - \frac{1}{2}\pi)$. Thus, we obtain the graph of $y = \sin(2x - \frac{1}{2}\pi)$ from the graph of $y = \sin 2x$ by applying the translation $XT/\frac{1}{4}\pi$, as shown in Figure 5.1.8.

2-0575

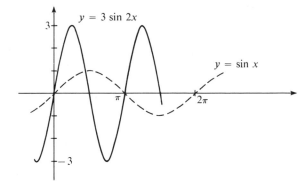

Figure 5.1.6

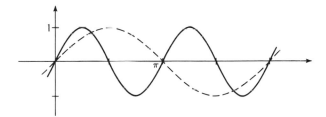

Figure 5.1.7

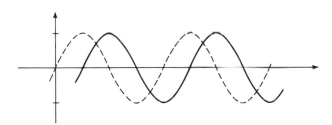

Figure 5.1.8

In general, if $a > 0$, we can obtain the graph of $y = \sin(ax + b)$ by applying the following sequence of mappings to the graph of $y = \sin x$:

$$y = \sin x \xrightarrow{XD/\frac{1}{a}} y = \sin ax \xrightarrow{XT/-\frac{b}{a}} y = \sin a\left(x + \frac{b}{a}\right)$$
$$= \sin(ax + b)$$

Notice that by writing $ax + b$ as $a\left(x + \dfrac{b}{a}\right)$, we see that the translation $XT/-\dfrac{b}{a}$ is needed when the mappings are performed

> A1: Amplitude is 3 and period is π. See Figure 5.1.6.

in this order. The number $-b/a$ is called the *phase shift*. Indeed, if A and a are both positive, we now have seen that the graph of $y = A \sin(ax + b)$ resembles the graph of $y = \sin x$ except that it possibly has been stretched (or squeezed) vertically and horizontally and shifted horizontally. These changes affect the amplitude, period, and phase shift of the sine curve. Figures 5.1.9 and 5.1.10 summarize the discussion above.

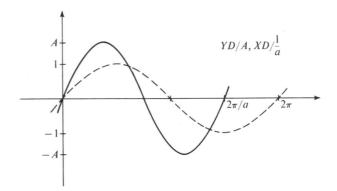

Figure 5.1.9

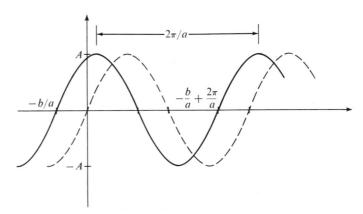

Figure 5.1.10

Example 5.1.2

Sketch the graph of $y = 2 \cos(3x + \tfrac{1}{2}\pi)$.

Solution

We proceed in a manner similar to that used in the discussion above. First observe that

$$y = \cos x \xrightarrow{XD/\frac{1}{3}} y = \cos 3x \xrightarrow{YD/2} y = 2 \cos 3x$$

Hence, we can obtain the graph of $y = 2 \cos 3x$ by horizontally compressing the graph of $y = \cos x$ by a factor of $1/3$ and then vertically stretching this graph by a factor of 2, obtaining the dashed graph of Figure 5.1.11. Now, since $2 \cos(3x + \frac{1}{2}\pi) = 2 \cos 3(x + \frac{1}{6}\pi)$, we apply the translation $XT/-\frac{1}{6}\pi$ to obtain the desired graph. That is,

$$y = 2 \cos 3x \xrightarrow{XT/-\frac{1}{6}\pi} y = 2 \cos 3\left(x + \frac{\pi}{6}\right) = 2 \cos\left(3x + \frac{\pi}{2}\right)$$

So shift the dashed graph in Figure 5.1.11 $\pi/6$ units to the left to obtain the solution.

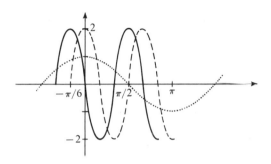

Figure 5.1.11

Now that we have proved the general results concerning the graph of $y = A \sin(ax + b)$, notice that we can check our work on any specific problem by checking the endpoints of one cycle. The graph of $y = \sin t$ traces one cycle for $0 \le t \le 2\pi$ (Figure 5.1.12). Thus, the graph of $y = A \sin(ax + b)$ will trace *one cycle* if $0 \le ax + b \le 2\pi$, that is, if $-b \le ax \le 2\pi - b$, or if

phase shift $\Longrightarrow$ $-\dfrac{b}{a} \le x \le -\dfrac{b}{a} + \dfrac{2\pi}{a}$

assuming $a > 0$. Figure 5.1.13 shows the left endpoint of the cycle at $-b/a$ (the phase shift) and the right endpoint at $-(b/a) + (2\pi/a)$, that is, phase shift + period.

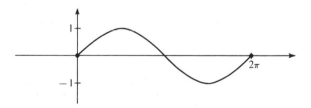

Figure 5.1.12

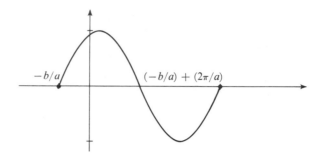

Figure 5.1.13

Example 5.1.3

Sketch the graph of $y = 3 \sin(2x + \pi)$.

Solution

If we follow the procedure above, one cycle of the desired graph will be traced when $0 \leq 2x + \pi \leq 2\pi$. Solving for x, we obtain $-\pi \leq 2x \leq \pi$ or $-\pi/2 \leq x \leq \pi/2$, as shown in Figure 5.1.14. Using the periodicity of the sine function and also taking into account the amplitude, we obtain the graph of $y = 3 \sin(2x + \pi)$ in Figure 5.1.15.

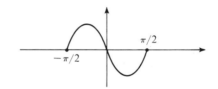

Figure 5.1.14

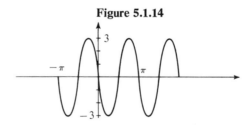

Figure 5.1.15

Q2: Sketch the graph of $y = 2 \cos(4x - \pi)$. Find the amplitude, period, and phase shift.

It is common to refer to any of the graphs displayed in this section as *sine waves* or *sine curves*. We shall use the term *sinusoidal* to describe any graph that can be obtained from the graph of $y = \sin x$ by any combination of *X*- and *Y*-translations and distortions. Thus, the graphs of $y = 2 \sin 3x$, $y = 4 \sin(\pi x - 1)$, and $y = \cos x = \sin(x + \frac{1}{2}\pi)$ are all sinusoidal curves. Likewise, a *sinusoidal function* is any function whose graph is a sinusoidal curve. In many applications, the input variable to a sinusoidal function is time. For example, if we measure time in seconds, then the period of the function $f(t) = 4 \sin(120\pi t - \beta)$ is $2\pi/120\pi = \frac{1}{60}$ seconds/cycle. The reciprocal of the period is called the *frequency;* in this example the frequency is 60 cycles/seconds. Indeed, the voltage available at an ordinary electrical outlet is 60-cycle.

One interesting mathematical fact regarding sinusoidal functions is the following: The sum of any two sinusoidal functions of the same period (or frequency) is also a sinusoidal function with that same period (frequency). Notice that this is independent of the amplitudes or phase shifts involved. In symbols, we are saying that if $f(x) = B \sin(ax + \beta) + C \sin(ax + \gamma)$, then there exist numbers A and α such that $f(x) = A \sin(ax + \alpha)$. The proof of this statement is rather awkward, and the details are left to the problems at the end of this section. We shall, however, illustrate the techniques in the following example.

Example
5.1.4

Find A and α such that $A \sin(x + \alpha) = \sqrt{3} \sin x + \cos x$.

Solution

Using an addition formula, Equation (5) in Section 4.6, we have

$$A \sin(x + \alpha) = A \sin x \cdot \cos \alpha + A \sin \alpha \cdot \cos x$$

We then see that the original equation will be satisfied if

$$A \cos \alpha = \sqrt{3} \quad \text{and} \quad A \sin \alpha = 1 \tag{1}$$

To find a solution to these two equations in two unknowns, we proceed as follows: (1) Dividing the two equations yields

$$\frac{A \sin \alpha}{A \cos \alpha} = \frac{1}{\sqrt{3}} \quad \text{or} \quad \tan \alpha = \frac{1}{\sqrt{3}}$$

(2) Squaring each equation and then summing yields

$$A^2 \sin^2 \alpha + A^2 \cos^2 \alpha = 1^2 + \sqrt{3}^2 \quad \text{or} \quad A^2 = 4$$

Thus, one solution to Equations (1) is $A = 2$ and $\alpha = \pi/6$. [Verify by substituting into Equation (1).] Also notice, for example, that $A = 2$, $\alpha = 7\pi/6$ does not satisfy Equations (1). Thus,

$$2 \sin\left(x + \frac{\pi}{6}\right) = \sqrt{3} \sin x + \cos x$$

Figure 5.1.16 portrays this fact graphically.

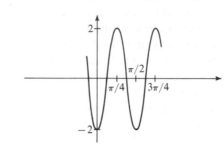

A2: 2, $\pi/2$, $\pi/4$. The graph is shown below.

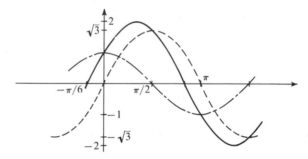

Figure 5.1.16

Example 5.1.5   Find a solution to the equation $2 = \sqrt{3} \sin x + \cos x$.

Solution   Using the results of Example 5.1.4, we have

$$2 = \sqrt{3} \sin x + \cos x = 2 \sin\left(x + \frac{\pi}{6}\right) \quad \text{or} \quad \sin\left(x + \frac{\pi}{6}\right) = 1$$

We obtain one solution by setting $x + \frac{1}{6}\pi = \pi/2$, or $x = \pi/3$.

PROBLEMS 5.1

1. Find the amplitude, period, frequency, and phase shift for

(a) $f(x) = \dfrac{\sin \pi x}{3}$ (b) $g(x) = 4 \sin(2x + 3)$

(c) $h(x) = 5 \sin\left(\dfrac{x}{3} - \dfrac{\pi}{3}\right)$

2. Sketch the graphs of each of the following equations:
(a) $y = 2 \sin(\pi x) - 1$ (b) $y = 2 \sin \pi(x - 1)$
(c) $y = \dfrac{\cos(3x - \pi)}{2}$ (d) $y = \cos 3\dfrac{(x - \pi)}{2}$

(e) $y = 1 - \sin\dfrac{\pi x}{2}$

(f) $y = 1 - x \sin\dfrac{\pi}{2}$

3. Sketch the graphs of each of the following equations:

(a) $y = 2\cos(\pi x - 1)$

(b) $y = 2\tan(2x - \tfrac{1}{4}\pi)$

(c) $y = \cos(2x + \tfrac{1}{3}\pi)$

(d) $y = 3\sin(2x - 3)$

(e) $y = 3\sin(3 - 2x)$

(f) $y = \sqrt{1 - \cos x}$ (Hint: Use a half-angle identity.)

---

4. Find real numbers A and α such that $A \sin(x + \alpha)$ equals

(a) $\sin x - \sqrt{3}\cos x$

(b) $\sin x + \cos x$

(c) $3\sin x - 4\cos x$

(d) $2\sin x + 3\cos x$

5. In this problem we shall show that it is always possible to find real numbers A and α such that

$$A \sin(ax + \alpha) = B \sin(ax + \beta) + C \sin(ax + \gamma)$$

(a) Use addition formula (4) to show that the equation above can be rewritten as

$$(A\cos\alpha)\sin ax + (A\sin\alpha)\cos ax = (B\cos\beta + C\cos\gamma)\sin ax$$
$$+ (B\sin\beta + C\sin\gamma)\cos ax$$

(b) Clearly, the equation above is satisfied if

$$A\cos\alpha = B\cos\beta + C\cos\gamma \quad \text{and} \quad A\sin\alpha = B\sin\beta + C\sin\gamma$$

Now show that these equations are satisfied if

$$\tan\alpha = \frac{B\sin\beta + C\sin\gamma}{B\cos\beta + C\cos\gamma}$$

$$\text{and} \quad A^2 = (B\cos\beta + C\cos\gamma)^2 + (B\sin\beta + C\sin\gamma)^2$$

(Hint: See Example 5.1.4.)

(c) Show that these last two equations are satisfied if

$$A = [B^2 + C^2 + 2BC\cos(\beta - \gamma)]^{1/2}$$

and, since $A \geq 0$,

$$\alpha = \begin{cases} \text{Tan}^{-1}\dfrac{N}{D} & \text{if } D > 0 \\[2mm] \pi + \text{Tan}^{-1}\dfrac{N}{D} & \text{if } D < 0 \\[2mm] \dfrac{\pi}{2} & \text{if } D = 0 \quad \text{and} \quad N > 0 \\[2mm] -\dfrac{\pi}{2} & \text{if } D = 0 \quad \text{and} \quad N < 0 \end{cases}$$

where $N = B\sin\beta + C\sin\gamma$ and $D = B\cos\beta + C\cos\gamma$.

(d) Finally, find real numbers A and α so that

$$A\sin(2x + \alpha) = 3\sin\left(2x + \frac{\pi}{2}\right) + 2\sin\left(2x + \frac{\pi}{6}\right)$$

ANSWERS
5.1

1. The answers are given in the following table:

	Amplitude	Period	Frequency	Phase Shift
(a)	1/3	2	1/2	0
(b)	4	π	$1/\pi$	$-3/2$
(c)	5	6π	$1/6\pi$	π

2. (a) See Figure 5.1.17.

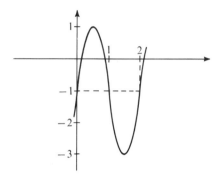

Figure 5.1.17

(c) See Figure 5.1.18.

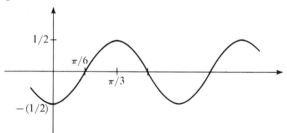

Figure 5.1.18

(e) See Figure 5.1.19.

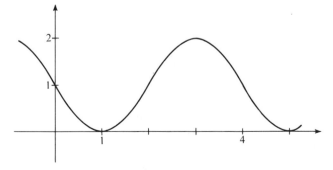

Figure 5.1.19

3. (a) See Figure 5.1.20.

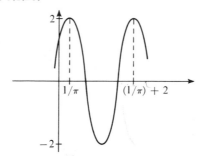

Figure 5.1.20

(c) See Figure 5.1.21.

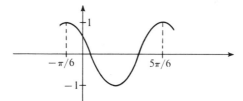

Figure 5.1.21

(e) See Figure 5.1.22.

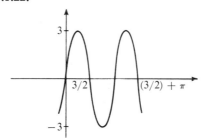

Figure 5.1.22

4. (a) $A = 2$, $\alpha = -\pi/3$ (b) $A = \sqrt{2}$, $\alpha = \pi/4$

   (c) $A = 5$, $\alpha = \text{Tan}^{-1}(-4/3)$ (d) $\sqrt{13}$, $\alpha = \text{Tan}^{-1}(3/2)$

Supplement 5.1

Problem S1. Sketch the graph of $y = \sin 4x$.

Solution: Refer to the discussion preceding Figure 5.1.12 and 5.1.13. The graph of $y = \sin 4x$ is sinusoidal (related to the sine function) and traces a complete cycle when $0 \leq 4x \leq 2\pi$. We must solve this inequality for x to see precisely what end-points are involved in the cycle. Thus, if $0 \leq x \leq \pi/2$, $y = \sin 4x$ traces one complete cycle (see Figure 5.1.23). Notice that the period is $(\pi/2) - 0 = \pi/2$, the length of one cycle. Also notice that we can obtain the graph from the sine graph by using the following distortion:

$$y = \sin x \xrightarrow{XD/\frac{1}{4}} y = \sin 4x$$

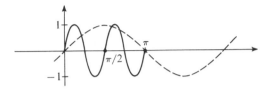

Figure 5.1.23

Problem S2.   Sketch the graph of $y = \sin\left(\dfrac{x}{2}\right)$.

Solution: We proceed in a manner similar to that used in Problem S1. So $y = \sin\left(\dfrac{x}{2}\right)$ traces one complete cycle when $0 \leq x/2 \leq 2\pi$ or when $0 \leq x \leq 4\pi$. Thus, the period is $4\pi - 0 = 4\pi$. Also notice that we can obtain the graph from the sine graph by applying the following mapping:

$$y = \sin x \xrightarrow{XD/2} y = \sin \frac{x}{2}$$

The graph is shown in Figure 5.1.24.

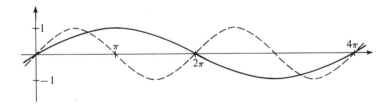

Figure 5.1.24

Problem S3.   Sketch the graph of $y = 2 \sin(x - \pi)$.

Solution: Here the amplitude is 2, so the graph of $y = 2 \sin(x - \pi)$ oscillates between the lines $y = -2$ and $y = 2$. Also, $y = 2 \sin(x - \pi)$ traces one complete cycle when $0 \leq x - \pi \leq 2\pi$ or when $\pi \leq x \leq 3\pi$. Thus, the phase shift is π and the period is $3\pi - \pi = 2\pi$. We can also obtain the graph by applying the following mappings to the graph of $y = \sin x$:

$$y = \sin x \xrightarrow{YD/2} y = 2 \sin x \xrightarrow{XT/\pi} y = 2 \sin(x - \pi)$$

The graph is shown in Figure 5.1.25. Notice that the distortion factor, 2, is the amplitude, and that the translation factor, π, is the phase shift.

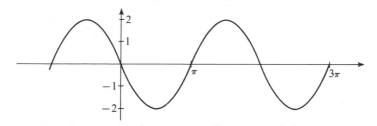

Figure 5.1.25

Problem S4. Sketch the graph of $y = \cos(\pi x/2)$.

Solution: First note that the remarks concerning the graph of $y = \sin(ax + b)$ are equally applicable to the cosine function. Thus, $y = \cos(\pi x/2)$ traces one complete cycle when $0 \leq (\pi x/2) \leq 2\pi$ or $0 \leq x \leq 4$. Notice that the period is $4 - 0 = 4$, the length of one cycle. We can also obtain the graph from the cosine graph using the following mapping:

$$y = \cos x \xrightarrow{XD/\frac{2}{\pi}} y = \cos \frac{x}{2/\pi} = \cos \frac{\pi}{2}x$$

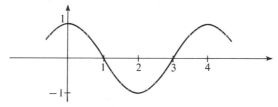

Figure 5.1.26

Problem S5. Sketch the graph of $y = \cos\left(\dfrac{x}{2} - \dfrac{\pi}{2}\right)$.

Solution: See Problem S4. Thus, $y = \cos\left(\dfrac{x}{2} - \dfrac{\pi}{2}\right)$ traces one complete cycle when

$$0 \leq \frac{x}{2} - \frac{\pi}{2} \leq 2\pi \quad \text{or} \quad \frac{\pi}{2} \leq \frac{x}{2} \leq \frac{5\pi}{2} \quad \text{or} \quad \pi \leq x \leq 5\pi$$

Therefore, the phase shift is π and the period is $5\pi - \pi = 4\pi$. Observe that a distortion by a factor equal to the period (4π) divided by 2π and a translation by a factor equal to the phase shift will also produce the graph of $y = \cos\left(\dfrac{x}{2} - \dfrac{\pi}{2}\right)$.

$$y = \cos x \xrightarrow{XD/2} y = \cos \frac{x}{2} \xrightarrow{XT/\pi} y = \cos \frac{1}{2}(x - \pi) = \cos\left(\frac{x}{2} - \frac{\pi}{2}\right)$$

The graph is shown in Figure 5.1.27.

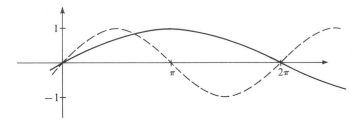

Figure 5.1.27

Problem S6.   Sketch the graph of $y = \frac{1}{2}\cos(\pi x + 2)$.

Solution: We will sketch $y = \frac{1}{2}\cos(\pi x + 2)$ in a manner similar to that used in Problem S5. Thus, $y = \frac{1}{2}\cos(\pi x + 2)$ traces one complete cycle when

$$0 \le \pi x + 2 \le 2\pi \quad \text{or} \quad -2 \le \pi x \le 2\pi - 2 \quad \text{or} \quad -\frac{2}{\pi} \le x \le 2 - \frac{2}{\pi}$$

Therefore, the phase shift is $-2/\pi$ and the period is $\left(2 - \frac{2}{\pi}\right) - \left(\frac{-2}{\pi}\right) = 2.$

Notice that the distortion $XD/\dfrac{1}{\pi}$ (resulting in a period of 2), the translation $XT/-\dfrac{2}{\pi}$ (the phase shift), and the distortion $YD/\frac{1}{2}$ (the amplitude) can be used to obtain the desired graph.

$$y = \cos x \xrightarrow{XD/\frac{1}{\pi}} y = \cos \pi x \xrightarrow{XT/-\frac{2}{\pi}} y = \cos \pi\left(x + \frac{2}{\pi}\right) = \cos(\pi x + 2)$$

$$\xrightarrow{YD/\frac{1}{2}} y = \frac{1}{2}\cos(\pi x + 2)$$

The graph is shown in Figure 5.1.28.

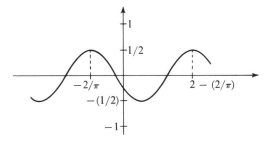

Figure 5.1.28

Problem S7.   Sketch the graph of $y = -\tan(x - \frac{1}{2}\pi)$.

Solution: First, remember that $y = \tan x$ traces one complete cycle when $-\frac{1}{2}\pi \le x \le \frac{1}{2}\pi$ (period π, not 2π). Thus, $y = -\tan(x - \frac{1}{2}\pi)$ traces one complete cycle when

$$-\frac{\pi}{2} \le x - \frac{\pi}{2} \le \frac{\pi}{2} \quad \text{or} \quad 0 \le x \le \pi$$

Therefore, the phase shift is $\pi/2$ and the period is $\pi - 0 = \pi$. Now, we can obtain the graph of $y = -\tan(x - \frac{1}{2}\pi)$ by applying the reflection YR, a reflection in the Y-direction. (See Figure 5.1.29.) Observe that we also could have obtained the graph by first translating by a factor of $\frac{1}{2}\pi$ and then reflecting in the Y-direction. That is,

$$y = \tan x \xrightarrow{XT/\frac{1}{2}\pi} y = \tan\left(x - \frac{\pi}{2}\right) \xrightarrow{YR} y = -\tan\left(x - \frac{\pi}{2}\right)$$

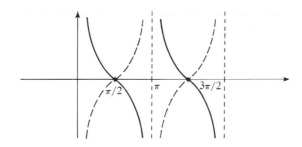

Figure 5.1.29

SUPPLEMENTARY PROBLEMS

S8. Sketch the graph of $y = 1 - \sin\frac{1}{2}\pi x$.

S9. Sketch the graph of $y = 3\cos(3x + 2)$.

S10. Sketch the graph of $y = \dfrac{\cos(2x - \pi)}{2}$.

S11. Sketch the graph of $y = -\tan(x + 1)$.

Answers

S8. See Figure 5.1.30.

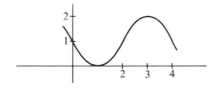

Figure 5.1.30

S9. The graph is shown in Figure 5.1.31.

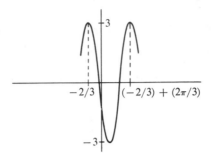

Figure 5.1.31

S10. See Figure 5.1.32.

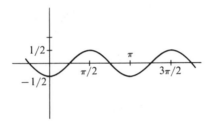

Figure 5.1.32

S11. The graph is shown in Figure 5.1.33.

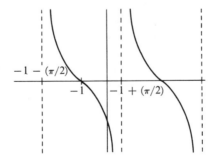

Figure 5.1.33

5.2 TRIGONOMETRIC EQUATIONS

In this section we present several examples to illustrate how the many properties and identities which we have encountered thus far in trigonometry are used to solve trigonometric equations. Although no one technique will work for every problem, some patterns will emerge to provide you with additional insight into problem solving. Needless to say, there is no substitute for experience.

Example 5.2.1

Find the solution set for $\sin t = 1$.

Solution

Since $\sin t$ is the second coordinate of the trigonometric point, $P(t)$ must be the point $(0, 1)$ on the unit circle. Thus, $t = \pi/2$ is one solution, and, using the periodic property of the trigonometric point function, we obtain the solution set:

$$\left\{ t = \frac{\pi}{2} + 2k\pi : k \in Z \right\}$$

See Figure 5.2.1.

Example 5.2.2

Find the solution set for $\sin 2x = 1$.

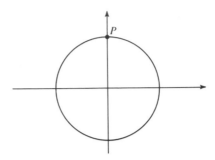

Figure 5.2.1

Solution   Using the results of Example 4.4.1 with $2x = t$, we have $2x = (\pi/2) + 2k\pi$ where $k \in Z$. Thus, the solution set is $\{\frac{1}{4}\pi + (k\pi) : k \in Z\}$. Notice that there is no reason to use the double-angle formula $\sin 2x = 2 \sin x \cdot \cos x$.

Q1: Find the solution set to $\sin 3x = 1$.

Example 5.2.3   Find the solution set for $2 \sin t = 1$.

Solution   The equation can be written as $\sin t = 1/2$. Clearly, there are two points U and V on the unit circle whose second coordinate equals $1/2$ (see Figure 5.2.2). From our previous work we know that $P(\pi/6)$ coincides with V and $P(5\pi/6)$ coincides with U. Again using the periodic property, we obtain the solution set:

$$\left\{\frac{\pi}{6} + 2k\pi : k \in Z\right\} \cup \left\{\frac{5\pi}{6} + 2k\pi : k \in Z\right\}$$

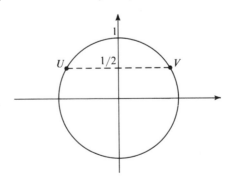

Figure 5.2.2

Q2: Find the solution set for $2 \sin 3t = 1$.

Example 5.2.4

Find the solution set to $3 \sin t = 1$.

Solution

Proceeding as in Example 5.2.3, we locate points U and V on the unit circle whose second coordinates equal $1/3$. (See Figure 5.2.3.) Without tables, we can say only that V coincides with $P(t)$ when $t = \operatorname{Sin}^{-1}\frac{1}{3}$ (about .34 using tables). Likewise, U coincides with $P(t)$ when $t = \pi - \operatorname{Sin}^{-1}\frac{1}{3} \approx 3.14 - .34 = 2.8$, since the shaded arcs have equal lengths. Thus, the solution set is

$$\left\{ \operatorname{Sin}^{-1}\frac{1}{3} + 2k\pi : k \in Z \right\} \cup \left\{ \pi - \operatorname{Sin}^{-1}\frac{1}{3} + 2k\pi : k \in Z \right\}$$

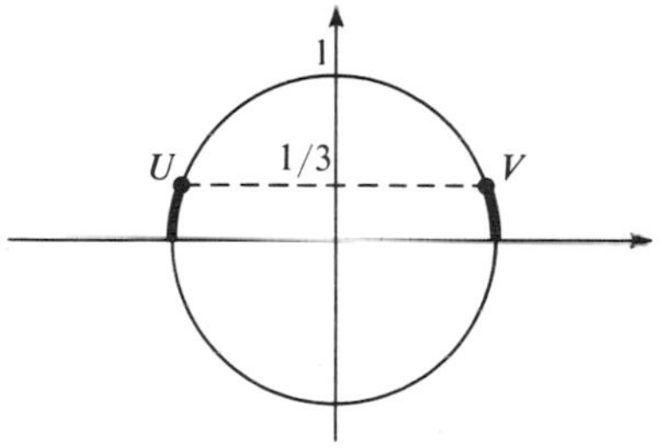

Figure 5.2.3

Q3: Find the solution set for $3 \sin t = 2$.

Example 5.2.5

Find the solution set for $3 \cos 2x = 1$.

Solution

Writing the equation as $\cos 2x = 1/3$, we see that we need to find those points on the unit circle whose first coordinates equal $1/3$. If $t = 2x = \operatorname{Cos}^{-1}\frac{1}{3} + 2k\pi$ where $k \in Z$, then $P(t)$ will coincide with U. By symmetry, if $t = 2x = -\operatorname{Cos}^{-1}\frac{1}{3} + 2k\pi$ where $k \in Z$, then $P(t)$ will coincide with V. Thus, the solution set is

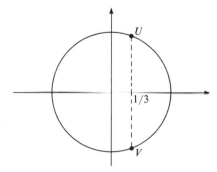

Figure 5.2.4

A1: $\left\{\frac{\pi}{6} + \frac{2k\pi}{3} : k \in Z\right\}$

A2: $\left\{\frac{\pi}{18} + \frac{2k\pi}{3} : k \in Z\right\} \cup \left\{\frac{5\pi}{18} + \frac{2k\pi}{3} : k \in Z\right\}$

A3: $\left\{\text{Sin}^{-1}\frac{2}{3} + 2k\pi : k \in Z\right\} \cup \left\{\pi - \text{Sin}^{-1}\frac{2}{3} + 2k\pi : k \in Z\right\}$

$\left\{\frac{1}{2}\text{Cos}^{-1}\frac{1}{3} + k\pi : k \in Z\right\} \cup \left\{-\frac{1}{2}\text{Cos}^{-1}\frac{1}{3} + k\pi : k \in Z\right\}$

Q4: Find the solution set for $\cos(2x - 3) = 1$.

Example 5.2.6

Find the solution set for $2\sin^2 t + 3\sin t - 2 = 0$.

Solution

We can factor this equation (which is quadratic in sin t) as $(2\sin t - 1)(\sin t + 2) = 0$. Thus, $2\sin t = 1$ or $\sin t = -2$. The solution set for the equation $2\sin t = 1$ was obtained in Example 5.2.3. The solution set for $\sin t = -2$ is empty since $|\sin t| \le 1$ for all $t \in R$. The solution set for the original equation is the union of these sets, namely,

$$\left\{\frac{\pi}{6} + 2k\pi : k \in Z\right\} \cup \left\{\frac{5\pi}{6} + 2k\pi : k \in Z\right\}$$

Q5: Find the solution set to the equation $3\sin t = 2\cos^2 t$ by first substituting $\cos^2 t = 1 - \sin^2 t$.

Example 5.2.7

Find the solution set for $\sin 2t = \sin t$.

Solution

Since the equation involves trigonometric function values of two different "angles" (t and $2t$), we shall first convert to function values of the same "angle." Thus, the equation becomes $2\sin t \cdot \cos t = \sin t$. Clearly, $\sin t = 0$ satisfies the equation. Furthermore, if $\sin t \ne 0$, we can divide the equation by it, obtaining $2\cos t = 1$. In symbols,

$$\{t : \sin 2t = \sin t\} = \{t : 2 \sin t \cdot \cos t = \sin t\}$$
$$= \{t : \sin t = 0\} \cup \{t : \cos t = \tfrac{1}{2}\}$$
$$= \{k\pi : k \in Z\} \cup \{\pm \tfrac{1}{3}\pi + 2k\pi : k \in Z\}$$

Example
5.2.8

Find the solution set for $3 \sin x = x$.

Solution

Unfortunately, we cannot obtain an explicit expression for every number x which satisfies the equation above. We can, however, estimate the solutions using the following graphical technique. We simply sketch the graphs of $y = 3 \sin x$ and $y = x$ on the same coordinate axes. The X-coordinates of the points of intersection of these graphs clearly satisfy the equation $3 \sin x = x$. In Figure 5.2.5 we see that the solution set is $\{0, a, -a\}$ where a is about $2\tfrac{1}{4}$. Using Table B.2, we find (by trial and error) that $a = 2.28$ is a good approximation since

$$3 \sin(2.28) \approx 3 \sin(.86) = 3(.7578) = 2.2734 \approx 2.28$$

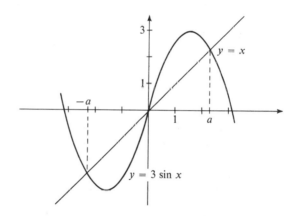

Figure 5.2.5

If you study calculus, you will encounter an iterative technique for approximating solutions to such equations. To use this method, however, you must have a reasonably close first approximation. We can often use the graphical technique of Example 5.2.8 to help us make such a first approximation.

PROBLEMS
5.2

1. Find the solution set for each of the following:
(a) $\cos(\text{Cos}^{-1}x) = x$ (b) $\text{Cos}^{-1}(\cos x) = x$
(c) $\sin(\sin x) = 0$ (d) $\cos(\cos t) = 1$
(e) $2 \cos 3t = 1$ ✗(f) $\sin t = \sin(t - \pi)$
(g) $3 \text{ Sin}^{-1}2x = \pi$ (h) $\sin(\log_b x) = 0$

A4: $\left\{\dfrac{3}{2} + k\pi : k \in Z\right\}$

A5: The solution is the same as that for Example 5.2.6.

(i) $\sin 2x = \cos 2x$

(k) $\sin^2 x - 4 \sin x + 3 = 0$

(m) $\tan^2 t = \sec t + 1$

(j) $\sin 2t < \cos 2t$

(l) $\cos^2 x = 4(1 - \sin x)$

(n) $\sin t + \cos t = 1$

2. Find the solution set for each of the following:

(a) $\cos 2t = \cos t$

χ(c) $2 \sin 2t - \tan t = 0$

(b) $\sin 2t - 2 \tan t = 0$

(d) $\cos t = 1 + \cos 2t$

3. Sketch the graph of each of the following:

(a) $y = \sin(\text{Sin}^{-1}x)$

(c) $y = \frac{1}{2}\pi - \text{Cos}^{-1}x$

(b) $y = \text{Sin}^{-1}(\sin x)$

(d) $y = 2\,\text{Cos}^{-1}|x|$

χ **4.** Use the technique of Example 5.2.8 to estimate the solution set for each of the following:

χ (a) $\tan x = x$

(c) $x \sin x = 1$

(e) $4 \sin x + x = 2$

(b) $3 \sin 2x - x^2 = 0$

(d) $\text{Cos}^{-1}x = x + 1$

(f) $2^{-x} = \log_2 x$

ANSWERS
5.2

1. (a) $[-1, 1]$, (b) $[0, \pi]$, (c) $\{k\pi\}$, (d) $\{\frac{1}{2}\pi + k\pi\}$, (e) $\{\pm(\pi/9) + (2k\pi/3)\}$ (f) $\{k\pi\}$, (g) $\{\sqrt{3}/4\}$, (h) $\{b^{k\pi}\}$, (i) $(\pi/8) + k(\pi/2)\}$, (j) $\{t : -(3\pi/8) + k\pi < t < (\pi/8) + k\pi\}$, (k) $\{(\pi/2) + 2k\pi\}$, (l) $\{(\pi/2) + 2k\pi\}$, (m) $\{(2k + 1)\pi\} \cup \{\pm(\pi/3) + 2k\pi\}$, (n) $\{2k\pi\} \cup \{(\pi/2) + 2k\pi\}$

2. (a) $\{2k\pi\} \cup \{(\pm 2\pi/3) + 2k\pi\}$, (b) $\{k\pi\}$, (c) $\{k\pi/3\}$, (d) $\{(\pm\pi/3) + 2k\pi\} \cup \{(\pi/2) + k\pi\}$

3. (a) See Figure 5.2.6.   (b) See Figure 5.2.7.

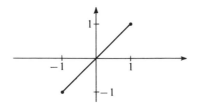

Figure 5.2.6

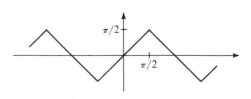

Figure 5.2.7

(c) See Figure 5.2.8.   (d) See Figure 5.2.9.

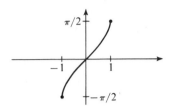

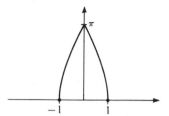

Figure 5.2.8   **Figure 5.2.9**

4. (a) $\{0, \pm 4.49, \text{about} \pm[(\pi/2) + k\pi] \text{ for } k \geq 2\}$
(b) $\{0, 1.28\}$
(c) $\{\pm 1.1, \pm 2.8, \text{about} \pm k\pi \text{ for } k \geq 2\}$
(d) $\{.28\}$
(e) $\{.41, 3.54, 5.31\}$
(f) $\{1.32\}$

Supplement 5.2

Problem S1. Find the solution set for $\cos t = -1$.

Solution: Recall that $\cos t$ is the *first* coordinate of $P(t)$. The only point on the unit circle with first coordinate -1 is $(-1, 0)$. Thus, $t = \pi$ is one solution since $P(\pi) = (-1, 0)$ (see Figure 5.2.10). However, notice that $P(\pi + 2\pi)$ and $P(\pi + 4\pi)$ also generate solutions since $P(\pi + 2\pi)$ and $P(\pi + 4\pi)$ coincide with $P(\pi)$. In general, by the periodic property of the trigonometric point function, $P(\pi) = P(\pi + k \cdot 2\pi)$ when k assumes all integer values. Therefore, $\cos(\pi + k \cdot 2\pi) = -1$ for every $k \in Z$, and the desired solution set is $\{\pi + 2k\pi : k \in Z\}$.

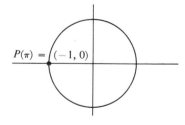

Figure 5.2.10

Problem S2.   Find the solution set for $2 \cos 2t = 1$.

Solution: First, divide both sides of the equation by 2 to obtain $\cos 2t = 1/2$. One way to proceed is to let $u = 2t$ and then solve $\cos u = 1/2$. Again recall that $\cos u$ is the first coordinate of $P(u)$. Observe that there are *two* points on the unit circle whose first coordinate equals $1/2$, namely, $P(\pi/3)$ and $P(-\pi/3)$. Thus, $u = \pi/3$ and $u = -\pi/3$ are solutions to $\cos u = 1/2$. By the periodic property of the trigonometric point function, $P(\tfrac{1}{3}\pi + k \cdot 2\pi) = P(\tfrac{1}{3}\pi)$ and $P(-\tfrac{1}{3}\pi + k \cdot 2\pi) = P(-\tfrac{1}{3}\pi)$ for every $k \in Z$. Thus, $u = \tfrac{1}{3}\pi + 2k\pi$ and $u = -\tfrac{1}{3}\pi + 2k\pi$ are also solutions for any integer k. Since $u = 2t$, $t = \tfrac{1}{2}u = \tfrac{1}{2}\cdot\tfrac{1}{3}\pi + \tfrac{1}{2}\cdot 2k\pi$ and $t = \tfrac{1}{2}(-\tfrac{1}{3})\pi + \tfrac{1}{2}\cdot 2k\pi$ are solutions. Therefore, the solution set is

$$\left\{\frac{\pi}{6} + k\pi : k \in Z\right\} \cup \left\{-\frac{\pi}{6} + k\pi : k \in Z\right\}$$

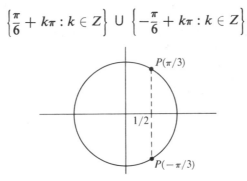

Figure 5.2.11

Problem S3.   Find the solution set for $4 \cos t = 1$.

Solution: The equation above is equivalent to $\cos t = \tfrac{1}{4}$. One exact solution to this equation is $t = \text{Cos}^{-1}\tfrac{1}{4}$. From Table B.3, $t = \text{Cos}^{-1}\tfrac{1}{4} \approx 1.31$ (locate $\tfrac{1}{4} = .25$ down the cosine column and read across the row). Recall again that $\cos t$ is the first coordinate of $P(t)$. Observe that there are two points on the unit circle (Figure 5.2.12) with first coordinate $\tfrac{1}{4}$, namely, $P(\text{Cos}^{-1}\tfrac{1}{4})$ and $P(-\text{Cos}^{-1}\tfrac{1}{4})$. Finally, accounting for the infinite number of solutions generated by periodic property of the trigonometric point function, we obtain the desired solution set:

$$\left\{\text{Cos}^{-1}\frac{1}{4} + 2k\pi : k \in Z\right\} \cup \left\{-\text{Cos}^{-1}\frac{1}{4} + 2k\pi : k \in Z\right\}$$

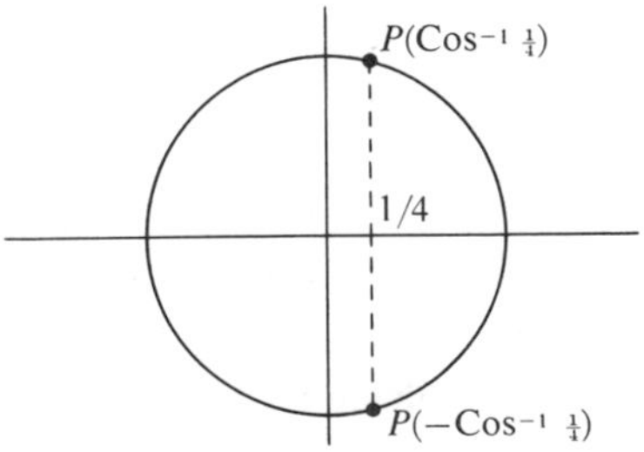

Figure 5.2.12

Problem S4.   Find the solution set for cos(sin *t*) = 1.

Solution: In a problem such as this it is often helpful to make a substitution. For example, let $u = \sin t$. It follows that cos(sin *t*) = cos *u* = 1. One solution to cos *u* = 1 is *u* = 0 since cos *u* is the first coordinate of $P(u)$ and $P(0) = (1, 0)$. See Figure 5.2.13. However, remember that $u = \sin t$ and so $-1 \leq u \leq 1$. Thus, *u* = 0 is the only possible solution. (The other solutions of cos *u* = 1, namely, $u = 0 + 2k\pi$, $k \in Z$, lie outside the interval $-1 \leq u \leq 1$.)

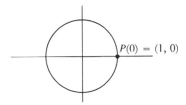

P(0) = (1, 0)

Figure 5.2.13

Since *u* = 0 and $u = \sin t$, the original problem is equivalent to finding the solution set to sin *t* = 0. Recall that sin *t* is the second coordinate of $P(t)$. Clearly, there are two points on the unit circle with second coordinate zero, namely, $P(0)$ and $P(\pi)$ (see Figure 5.2.14). Hence, *t* = 0 and $t = \pi$ are solutions to sin *t* = 0. By the periodic property of the trigonometric point function, $P(0 + k \cdot 2\pi) = P(0)$ and $P(\pi + k \cdot 2\pi) = P(\pi)$. Thus, the desired solution set is

$$\{2k\pi : k \in Z\} \cup \{\pi + 2k\pi : k \in Z\} = \{k\pi : k \subset Z\}$$

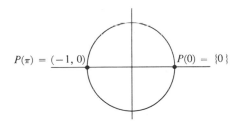

P(π) = (−1, 0)   P(0) = {0}

Figure 5.2.14

Problem S5.   Find the solution set for (a) Sin⁻¹(sin *x*) = *x* and (b) sin(Sin⁻¹*x*) = *x*.

Solution: (a) Let $t = \sin x$. Then Sin⁻¹(sin *x*) = *x* is equivalent to Sin⁻¹*t* = *x*. It follows that Sin *x* = *t* by Definition 4.5.1. Next, note that we let $t = \sin x$ in the beginning of the problem. Thus, $t = \text{Sin } x = \sin x$ and Sin *x* = sin *x* precisely when $-\pi/2 \leq x \leq \pi/2$ (see Definition 4.5.1). Thus, the desired solution set is $[-\pi/2, \pi/2]$, in interval notation.

(b) We would proceed in a manner similar to that used in Part (a). However, we will illustrate another approach in solving this problem. Recall that $f(f^{-1}(x)) = x$ is true for every *x* in the domain of f^{-1} (see Section 3.1). Thus, Sin(Sin⁻¹*x*) = *x* for $x \in [-1, 1]$, the domain of Sin⁻¹. Therefore, the solution set for sin(Sin⁻¹*x*) = *x* is also $[-1, 1]$ since if Sin *t* exists, then Sin *t* = sin *t*.

Problem S6.   Find the solution set for $\sin x < \cos(x - \pi)$.

Solution: First note by Equation (5) of Section 4.6 that $\cos(x - \pi) = \cos x \cos \pi +$ $\sin x \sin \pi = (\cos x)(-1) + (\sin x)(0) = -\cos x$. So the inequality becomes $\sin x <$ $-\cos x$. Next, graph $y = \sin x$ and $y = -\cos x$. Observe in the shaded regions of Figure 5.2.15 that the sine graph is *below* the graph of $y = -\cos x$. In these regions, $\sin x < -\cos x$. We can find the precise interval represented by the shaded regions by solving the equation $\sin x = -\cos x$. Observe that $\sin \frac{3}{4}\pi = -\cos \frac{3}{4}\pi$ and $\sin \frac{7}{4}\pi = -\cos \frac{7}{4}\pi$, so $(\frac{3}{4}\pi, \frac{7}{4}\pi)$ is the solution in the interval $[0, 2\pi]$. By the periodic property of the trigonometric point function, any number in the interval $(\frac{3}{4}\pi + 2k\pi, \frac{7}{4}\pi + 2k\pi)$ for any integer k will also be a solution of $\sin x < -\cos x$. The solution set is the union of the intervals $(\frac{3}{4}\pi + 2k\pi, \frac{7}{4}\pi + 2k\pi)$ for $k \in Z$.

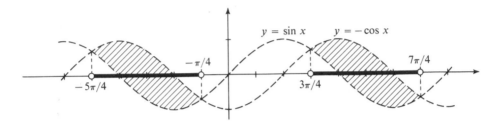

Figure 5.2.15

Problem S7.   Find the solution set for $\sin^2 t - 2 \sin t = 0$.

Solution: This equation represents a quadratic equation in $\sin t$. That is, if $u = \sin t$, then the above equation is equivalent to $u^2 - 2u = 0$. Thus factoring yields $\sin^2 t - 2 \sin t = (\sin t)(\sin t - 2) = 0$. Hence, $\sin t = 0$ or $\sin t - 2 = 0$ (or $\sin t = 2$). Clearly, $\sin t = 2$ has no real solutions since $-1 \le \sin t \le 1$ for any $t \in R$. Also, $\sin t = 0$ was solved in Problem S4. Hence, the desired solution set is

$$\{2k\pi : k \in Z\} \cup \{\pi + 2k\pi : k \in Z\} = \{k\pi : k \in Z\}$$

Problem S8.   Find the solution set for $2 \cos^2 x + \cos x - 1 = 0$.

Solution: Again we first try to factor the quadratic equation in $\cos x$. Factoring yields $2 \cos^2 x + \cos x - 1 = (2 \cos x - 1)(\cos x + 1) = 0$. Thus, $2 \cos x - 1 = 0$ or $\cos x + 1 = 0$. Therefore, $\cos x = 1/2$ or $\cos x = -1$. The solution to $\cos x = -1$ was found in Problem S1 and the solution to $\cos x = 1/2$ was discussed in Problem S2. Therefore, the desired solution set is

$$\{\pi + 2k\pi : k \in Z\} \cup \left\{\frac{\pi}{3} + 2k\pi : k \in Z\right\} \cup \left\{-\frac{\pi}{3} + 2k\pi : k \in Z\right\}$$

Problem S9.   Find the solution set for $\cos 2x + \cos x = 0$.

Solution: In a problem of this type you should first try to express all terms as functions of the same "angle." For example, in this case $\cos 2x = \cos^2 x - \sin^2 x$.

Thus, substituting, we obtain $\cos 2x + \cos x = \cos^2 x - \sin^2 x + \cos x = 0$. The new equation looks more complicated. However, if it can be changed to a quadratic equation involving only one trigonometric function, then its solution can be readily determined. For example, since $\sin^2 x = 1 - \cos^2 x$, substitution yields $\cos^2 x - \sin^2 x + \cos x = \cos^2 x - (1 - \cos^2 x) + \cos x = 0$. Observe that this equation involves only the cosine function. Simplifying, we obtain

$$\cos 2x + \cos x = \cos^2 x - (1 - \cos^2 x) + \cos x = 2\cos^2 x + \cos x - 1 = 0$$

This equation was solved in Problem S8.

Problem S10. Find the solution set for $2 \cot t = \csc t - 2 \sin t$.

Solution: In a problem of this type you should try to write an equivalent equation involving only one trigonometric function. Thus, using basic identities for $\cot t$ and $\csc t$, we obtain

$$\frac{\cos t}{\sin t} = \frac{1}{\sin t} - 2 \sin t \quad \text{or} \quad \cos t = 1 - 2 \sin^2 t$$

providing $\sin t \neq 0$.

Next recall that $\sin^2 t + \cos^2 t = 1$, so $\sin^2 t = 1 - \cos^2 t$. By substitution, we have $\cos t = 1 - 2 \sin^2 t = 1 - 2(1 - \cos^2 t) = 1 - 2 + 2\cos^2 t$. Simplifying yields $2\cos^2 t - \cos t - 1 = 0$, a quadratic equation in $\cos x$. Factoring, we obtain $2\cos^2 t - \cos t - 1 = (2\cos t + 1)(\cos t - 1) = 0$. Thus, $2\cos t + 1 = 0$ or $\cos t - 1 = 0$. This means that $\cos t = -1/2$ or $\cos t = 1$. Remember, earlier we assumed that $\sin t \neq 0$. Notice that if $\cos t = 1$, then $\sin t = 0$ (why?). Thus, $\cos t = 1$ will not yield valid solutions. Recall that $\cos t$ is the first coordinate of $P(t)$. Then $t = 2\pi/3$ or $t = 4\pi/3$ (see Figure 5.2.16) are solutions to $\cos t = -1/2$ since both $P(\frac{2}{3}\pi)$ and $P(\frac{4}{3}\pi)$ have first coordinate $-1/2$. By the periodic property of the trigonometric point function, it follows that the desired solution set to $\cot t = \csc t - 2 \sin t$ is

$$\left\{ \frac{2}{3}\pi + 2k\pi : k \in Z \right\} \cup \left\{ \frac{4}{3}\pi + 2k\pi : k \in Z \right\}$$

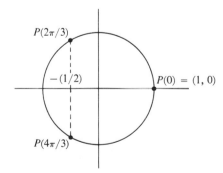

Figure 5.2.16

SUPPLEMENTARY PROBLEMS

S11. Find the solution set for $2 \sin 2t = \sqrt{3}$.

S12. Find the solution set for $1 = \csc x - \cot x$.

S13. Find the solution set for $2 \sin^2 t + 3 \sin t + 1 = 0$.

S14. Find the solution set for $(\cos t)(\sin 3t) = \cos t$.

S15. Find the solution set for $(\sin x)(\tan x) + \cos x = 1$.

Answers

S11. $\{\frac{1}{6}\pi + \pi k : k \in Z\} \cup \{\frac{1}{3}\pi + k\pi : k \in Z\}$

S12. $\{\frac{1}{2}\pi + 2k\pi : k \in Z\}$

S13. $\{\frac{7}{6}\pi + 2k\pi : k \in Z\} \cup \{\frac{11}{6}\pi + 2k\pi : k \in Z\} \cup \{\frac{3}{2}\pi + 2k\pi : k \in Z\}$

S14. $\{\frac{1}{2}\pi + k\pi : k \in Z\} \cup \{\frac{1}{6}\pi + \frac{2}{3}k\pi : k \in Z\}$

S15. $\{2k\pi : k \in Z\}$

5.3 THE LAWS OF SINES AND COSINES

In previous sections we have observed the relationship between the trigonometric functions and ratios of lengths of sides of a right triangle. For example, in Figure 5.3.1, $\sin \alpha = a/c$. In this section we shall derive two sets of formulas that apply to any plane triangle (including right triangles).

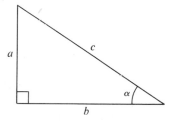

Figure 5.3.1

We shall denote each side of a triangle by its length a, b, or c. The measure of the angle opposite each side is denoted by the corresponding Greek letter α, β, or γ (see Figure 5.3.2). To derive the first result, we drop a perpendicular from the vertex opposite c to side c or its extension. Figure 5.3.3 shows two possibilities. If the length of this perpendicular is h, we can use the two right triangles formed to write $h = b \sin \alpha$ and $h = a \sin \beta$ (Figure 5.3.4). Equating these two expressions for h and dividing by ab yields $(\sin \alpha)/a = (\sin \beta)/b$. Of course, our choice of side c as the base of the triangle was arbitrary. Thus, the ratio of the sine of an angle of a triangle to the length of the side opposite it is constant for any given plane triangle. This result is called the *Law of Sines:*

$$\frac{\sin \alpha}{a} = \frac{\sin \beta}{b} = \frac{\sin \gamma}{c} \qquad \text{(in Figure 5.3.2)} \qquad (1)$$

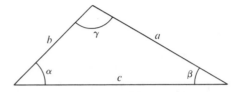

Figure 5.3.2

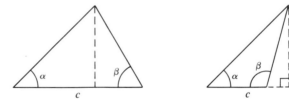

Figure 5.3.3

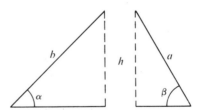

Figure 5.3.4

Given sufficient information, we can use the Law of Sines to help solve triangles or to show that such a triangle could not exist.

Example 5.3.1

If $\alpha = 50°$, $\beta = 60°$, and $a = 4$, find γ, b, and c.

Solution

Since $\alpha + \beta + \gamma = 180°$, then $\gamma = 70°$. Using the Law of Sines, we obtain

$$\frac{\sin 50°}{4} = \frac{\sin 60°}{b} = \frac{\sin 70°}{c}$$

Thus,

$$b = \frac{4 \sin 60°}{\sin 50°} = \frac{4(.866)}{.766} = 4.52$$

and

$$c = \frac{4 \sin 70°}{\sin 50°} = \frac{4(.9397)}{.766} = 4.91$$

Example 5.3.2

Find β, γ, and c if $\alpha = 50°$, $a = 4$, and $b = 5$.

Solution

Since we know α, a, and b, the Law of Sines yields

$$\sin \beta = \frac{b \sin \alpha}{a} = \frac{5(.766)}{4} = .958$$

Thus $\beta \approx 73°$ or $\beta \approx 107°$ since $\sin 73° = \sin(180 - 73)°$. If $\beta = 73°$, then

$$\gamma = (180 - 50 - 73)° = 57°$$

and

$$c = \frac{a \sin \gamma}{\sin \alpha} = \frac{4(.8387)}{.766} = 4.38$$

If $\beta = 107°$, then $\gamma = 23°$ and $c = 4(.3907)/.766 = 2.04$. Thus, there exist *two* triangles which satisfy the given conditions as shown in Figure 5.3.5.

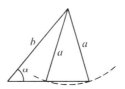

Figure 5.3.5

Example 5.3.3

Find β, γ, and c if $a = 3$, $b = 4$, and $\alpha = 60°$.

Solution

Since we know a, b, and α, the Law of Sines yields

$$\sin \beta = \frac{b \sin \alpha}{a} = \frac{4(.866)}{3} = 1.155$$

But this is clearly impossible since $|\sin \beta| \leq 1$. Therefore, no such triangle exists (Figure 5.3.6).

Notice that to use the Law of Sines we must know both the length of one side and the measure of the angle opposite it, in addition to either one other side length or angle. However, if we

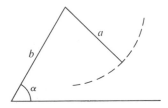

Figure 5.3.6

are given two sides and the included angle (α, b, and c in Figure 5.3.7), the Law of Sines will not yield the remaining parts of the uniquely determined triangle. To solve this problem we shall now derive the *Law of Cosines* which is basically a generalization of the Pythagorean theorem.

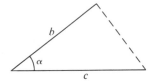

Figure 5.3.7

We first place the triangle in a Cartesian coordinate system with angle α in standard position (Figure 5.3.8). The coordinates of the vertex of angle β are clearly $(c, 0)$. The coordinates of the vertex of angle γ are $(b \cos \alpha, b \sin \alpha)$, since this point lies on a circle of radius b. We can now use these coordinates to find the distance between these two points. This distance also must equal a. Thus,

$$a = \sqrt{(b \cos \alpha - c)^2 + (b \sin \alpha - 0)^2}$$

So
$$a^2 = b^2\cos^2\alpha - 2bc \cdot \cos \alpha + c^2 + b^2\sin^2\alpha$$
$$= b^2(\sin^2\alpha + \cos^2\alpha) + c^2 - 2bc \cdot \cos \alpha$$

That is,

$$a^2 = b^2 + c^2 - 2bc \cos \alpha \qquad (2)$$

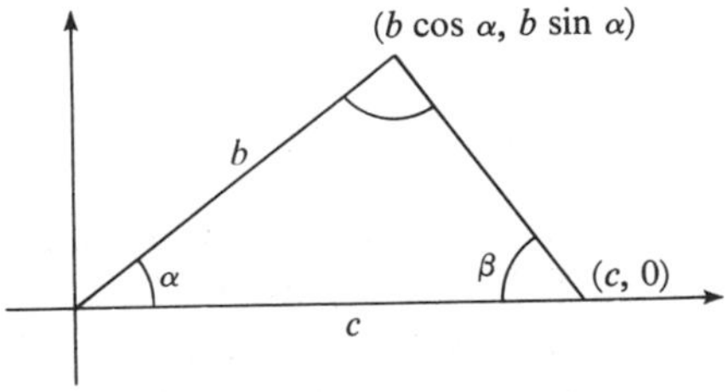

Figure 5.3.8

Notice that if $\alpha = \pi/2$, then cos $\alpha = 0$ and Equation (2) reduces to the Pythagorean theorem. Furthermore, by relabeling the sides and angles, we could write the Law of Cosines in the equivalent forms

$$b^2 = a^2 + c^2 - 2ac \cos \beta \quad \text{or} \quad c^2 = a^2 + b^2 - 2ab \cos \gamma$$

Example 5.3.4

Find a, β, and γ if $\alpha = 60°$, $b = 3$, and $c = 4$.

Solution

Since cos α = cos $60°$ = $1/2$, the Law of Cosines yields $a^2 = 3^2 + 4^2 - 2 \cdot 3 \cdot 4 \cdot (1/2) = 9 + 16 - 12 = 13$. Thus, $a = \sqrt{13}$. Now that we know a and α, the Law of Sines implies

$$\sin \beta = \frac{b \sin \alpha}{a} = \frac{3(.866)}{\sqrt{13}} = .72$$

Thus β = Sin^{-1}.72 or about $46°$ and $\gamma = (180 - 60 - 46)° = 74°$ is one solution. Notice that sin β = .72 is also true for $\beta = 134°$. But if $\beta = 134°$, then $\alpha + \beta = 194° > 180°$ and no such triangle would exist. This result is to be expected since two sides and the included angle uniquely determine a triangle.

Example 5.3.5

Find the smallest angle of the triangle whose sides have lengths 4, 5, and 6.

Solution

Since the smallest angle will be opposite the smallest side, Figure 5.3.9 and the Law of Cosines yields $4^2 = 5^2 + 6^2 - 2 \cdot 5 \cdot 6 \cdot \cos \alpha$. Thus,

$$\cos \alpha = \frac{25 + 36 - 16}{60} = \frac{3}{4}$$

so α = Cos$^{-1}\frac{3}{4}$ = .72 or about $41°$.

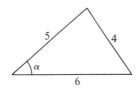

Figure 5.3.9

PROBLEMS 5.3

1. Given the following conditions, solve the triangle in Figure 5.3.10.
   (a) $\alpha = 79°$, $\beta = 33°$, and $a = 7$
   (b) $\alpha = 85°$, $a = 6$, and $b = 4$
   (c) $\alpha = 79°$, $b = 7$, and $c = 9$

(d) $\alpha = 34°$, $\beta = 74°$, and $c = 5$
(e) $a = 5$, $b = 7$, and $c = 6$
(f) $a = 5$, $b = 8$, and $\beta = 110°$

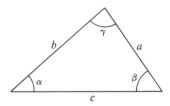

Figure 5.3.10

2. Given the following conditions, solve the triangle in Figure 5.3.11. Be careful to find all solutions or show that none exist. Illustrate your results with a sketch.
(a) $a = 6$, $b = 3$, and $\beta = 30°$
(b) $a = 8$, $b = 3$, and $\beta = 30°$
(c) $a = 5$, $b = 3$, and $\beta = 30°$
(d) $b = 13$, $c = 15$, and $\beta = 110°$
(e) $a = 4$, $b = 9$, and $c = 5$

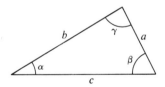

Figure 5.3.11

3. Find the radian measure of the largest angle of the triangle with sides of length 4, 5, and 6.

4. The sides of a parallelogram are of lengths 18 feet and 26 feet, and one angle is 39°. Find the length of the longest diagonal.

5. If $a = 2$ and $\beta = 45°$ (see Figure 5.3.12), determine the values of b for which α has (a) two values, (b) one value, and (c) no values.

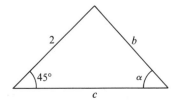

Figure 5.3.12

6. Two observers, 600 feet apart, are on opposite sides of a flagpole. The angles of elevation from the observers to the top of the pole are 19° and 21°. How high is the flagpole?

7. A blimp is sighted simultaneously by two observers, A, at the top of a 650′ tower, and B, at the base of the tower. Find the distance of the blimp from observer A if the angle of elevation (from the horizontal) viewed by A is 32° and the angle of elevation viewed by B is 56°. How high is the blimp?

ANSWERS
5.3

1. (a) $\gamma = 68°$, $b = 3.88$, $c = 6.61$
(b) $\beta = 42°$, $\gamma = 53°$, $c = 4.81$
(c) $a = 10.3$, $\beta = 42°$, $\gamma = 59°$
(d) $\gamma = 72°$, $a = 2.94$, $b = 5.05$
(e) $\alpha = 44.4°$, $\beta = 78.4°$, $\gamma = 57.2°$
(f) $\alpha = 36°$, $\gamma = 34°$, $c = 4.8$

2. (a) $\alpha = 90°$, $\gamma = 60°$, $c = 5.2$
(b) No solution
(c) $\alpha = 56°$, $\gamma = 94°$, $c = 5.99$ or $\alpha = 124°$, $\gamma = 26°$, $c = 2.63$
(d) No solution
(e) No solution

3. 1.45

4. 40.3

5. (a) $\sqrt{2} < b < 2$, (b) $b = \sqrt{2}$ or $b \geq 2$, (c) $b < \sqrt{2}$

6. 109 feet

7. 894 feet, 1124 feet

Supplement 5.3

Problem S1. Solve the triangle in Figure 5.3.13 if $\alpha = 40°$, $\beta = 75°$, and $b = 2$.

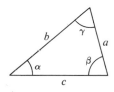

Figure 5.3.13

Solution: In this problem you are asked to find angle γ, side a, and side c. First, since $\alpha + \beta + \gamma = 180°$, in this triangle, $\gamma = 180° - 40° - 75° = 65°$. Next, we use the Law of Sines [Equation (1)] to find the lengths of sides a and c:

$$\frac{\sin 75°}{2} = \frac{\sin 40°}{a} \quad \text{and} \quad \frac{\sin 75°}{2} = \frac{\sin 65°}{c}$$

Hence,

$$a = \frac{2 \sin 40°}{\sin 75°} \quad \text{and} \quad c = \frac{2 \sin 65°}{\sin 75°}$$

Finally, use Table B.3 to obtain

$$a = \frac{2 \sin 40°}{\sin 75°} = \frac{2(.6428)}{.9659} = 1.33$$

and

$$c = \frac{2 \sin 65°}{\sin 75°} = \frac{2(.9063)}{.9659} = 1.88$$

In summary, $\gamma = 65°$, $a = 1.33$, and $c = 1.88$.

Problem S2.   Solve the triangle in Figure 5.3.14 if $\beta = 53°$, $b = 2.5$, and $c = 6$.

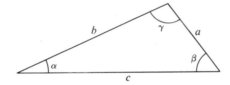

Figure 5.3.14

Solution: In this case we are given two sides and an opposite angle (β). Therefore, a direct application of the Law of Sines [Equation (1)] yields

$$\frac{\sin 53°}{2.5} = \frac{\sin \gamma}{6}$$

Notice that we cannot yet use the Law of Sines to find the length of side a (why not?).

Next we solve the equation above for $\sin \gamma$ and use Table B.3 to obtain

$$\sin \gamma = \frac{6 \sin 53°}{2.5} = \frac{6(.7986)}{2.5} = 1.9166$$

It should be obvious that such a γ does not exist because $-1 \leq \sin \theta \leq 1$ for any angle θ. Therefore, no such triangle exists in this case.

Problem S3.   Solve the triangle in Figure 5.3.15 if $\gamma = 37°$, $b = 2$, and $c = 1.8$.

Figure 5.3.15

Solution: This problem is similar to Problem S2. Therefore, direct application of the Law of Sines yields

$$\frac{\sin 37°}{1.8} = \frac{\sin \beta}{2}$$

Thus,

$$\sin \beta = \frac{2 \sin 37°}{1.8} = \frac{2(.6018)}{1.8} = .6687$$

One such β is $\text{Sin}^{-1}.67$. Now obtain a value for β (degrees) by using Table B.3. (You should do this now.) You should obtain $\beta \approx 42°$ since $\sin 42° = .6691$. (Henceforth, we will drop the use of $\approx$ and simply use $=$ with the understanding that the equality represents an approximation.)

The next step is easily overlooked. Observe that $\sin \beta = .6687$ has two solutions in the interval $0° \le \beta \le 180°$ because $\sin(180° - \beta) = \sin \beta$ (Figure 5.3.16). So $\beta = 42°$ or $\beta = 180° - 42° = 138°$. Therefore, (case 1) if $\beta = 42°$, then $\alpha = 180° - 42° - 37° = 101°$ and (case 2) if $\beta = 138°$, then $\alpha = 180° - 138° - 37° = 5°$.

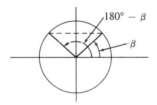

Figure 5.3.16

The two cases are illustrated in Figure 5.3.17. The smaller triangle represents case 2.

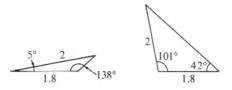

Figure 5.3.17

Finally, we shall find the length of a in *both* cases. Again, by direct application of the Law of Sines, we obtain (case 1) $\dfrac{\sin 37°}{1.8} = \dfrac{\sin 101°}{a}$ and (case 2) $\dfrac{\sin 37°}{1.8} = \dfrac{\sin 5°}{a}$. Solving each equation for a yields (case 1)

$$a = \frac{1.8 \sin 101°}{\sin 37°} = \frac{1.8 \sin(180° - 101°)}{\sin 37°}$$

$$= \frac{1.8 \sin 79°}{\sin 37°} = \frac{1.8(.9816)}{.6018} = 2.94$$

and (case 2)

$$a = \frac{1.8 \sin 5°}{\sin 37°} = \frac{1.8(.0872)}{.6018} = .26$$

To summarize, two solutions exist: $\beta = 42°$, $\alpha = 101°$, $a = 2.94$; and $\beta = 138°$, $\alpha = 5°$, $a = .26$.

Problem S4.   Solve the triangle in Figure 5.3.18 if $a = 2$, $b = 5$, and $c = 4$.

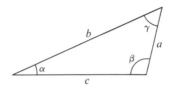

Figure 5.3.18

Solution: In this problem we are given only the lengths of the three sides. Thus, we can directly apply the Law of Cosines [Equation (2)] to find the angles. First we find angle α. Substitution of the given values yields $2^2 = 5^2 + 4^2 - 2 \cdot 5 \cdot 4 \cos \alpha$. After solving for $\cos \alpha$, we obtain

$$\cos \alpha = \frac{5^2 + 4^2 - 2^2}{2 \cdot 5 \cdot 4} = \frac{25 + 16 - 4}{2 \cdot 5 \cdot 4} = \frac{37}{40} = .9250$$

Using Table B.3, we have $\alpha = \text{Cos}^{-1}.9250$ or $22°$. Next we must find β and γ. There are two ways to proceed. Since two sides (a and b) and an opposite angle (α) are known, we could use the Law of Sines to find β; or we could again use the Law of Cosines. We illustrate both approaches.

Law of Cosines:

$$5^2 = 2^2 + 4^2 - 2 \cdot 2 \cdot 4 \cdot \cos \beta \quad \text{or} \quad \cos \beta = \frac{4 + 16 - 25}{16} = -.3125$$

Notice that this means $\beta > 90°$ since $\cos \beta < 0$. In Table B.3 we see that $\cos 72° = .309$. Thus, $\beta = 180° - 72° = 108°$ (see Figure 5.3.19).

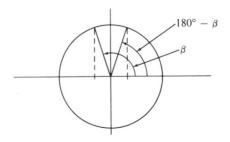

Figure 5.3.19

Law of Sines:

$$\frac{\sin 22°}{2} = \frac{\sin \beta}{5} \quad \text{or} \quad \sin \beta = \frac{5 \sin 22°}{2} = \frac{5(.3746)}{2} = .9365$$

Using Table B.3, we see that $\sin 70° = .9367$. Thus, $\beta = 70°$ and $\beta = 110°$ are possibilities. However, you should verify that $\beta = 70°$ is not valid. A sketch will help.

The discrepancy between $\beta = 108°$ and $\beta = 110°$ can be attributed to the lack of interpolating in the step "$\cos \beta = .9250$ implies $\beta = 22°$." The actual result is quite close to $108°$, the result obtained by using the Law of Cosines.

Finally, since $\gamma = 180° - \alpha - \beta$, $\gamma = 180° - 22° - 108° = 50°$. Thus, the solution is $\alpha = 22°$, $\beta = 108°$, and $\gamma = 50°$.

Problem S5. Solve the triangle in Figure 5.3.20 if $a = 3$, $b = 2$, and $c = 5$.

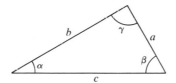

Figure 5.3.20

Solution: This problem is the same type as Problem S4. Therefore, the Law of Cosines [Equation (2)] yields $3^2 = 2^2 + 5^2 - 2 \cdot 2 \cdot 5 \cdot \cos \alpha$. Hence, $\cos \alpha = (4 + 25 - 9)/20 = 20/20 = 1$. This means that $\alpha = 0°$ and the desired triangle is degenerate (Figure 5.3.21), that is, it has no interior.

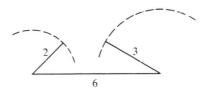

Figure 5.3.21

Observe that if we were given a value of c greater than 5, then no solution would exist. For example, if $c = 6$ (and $a = 3$ and $b = 2$), then $3^2 = 2^2 + 6^2 - 2 \cdot 2 \cdot 6 \cos \alpha$ or $\cos \alpha = (4 + 36 - 9)/24 = 31/24$. Thus, when $c = 6$, $\cos \alpha > 1$. Hence, there is no solution because $-1 \leq \cos \theta \leq 1$ for any angle θ. Figure 5.3.21 illustrates this case.

Problem S6.   Solve the triangle in Figure 5.3.22 if $\beta = 36°$, $a = 2$, and $c = 5$.

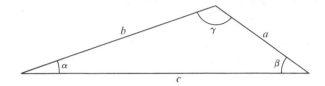

Figure 5.3.22

Solution: In this problem we are given lengths of two sides and the measure of the "included" angle. The Law of Sines cannot be applied directly in this case (why not?). However, the side opposite the included angle, b, can be determined by direct application of the Law of Cosines. Thus, $b^2 = 2^2 + 5^2 - 2 \cdot 2 \cdot 5 \cdot \cos 36°$ or $b^2 = 29 - 20(.8090) = 12.82$ (Figure 5.3.23). Hence, $b = \sqrt{12.82} = 3.58$.

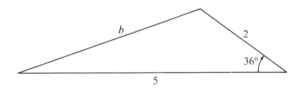

Figure 5.3.23

Now that we know a, b, and β, we can use the Law of Sines to find α. Thus, $\dfrac{\sin 36°}{3.58} = \dfrac{\sin \alpha}{2}$. Hence,

$$\sin \alpha = \frac{2 \sin 36°}{3.58} = \frac{2(.5878)}{3.58} = .3284$$

Thus, it appears that $\alpha = \text{Sin}^{-1}(.3284) = 19°$ or $\alpha = (180 - 19)° = 161°$. However, if $\alpha = 161°$, then $\alpha + \beta = 161° + 36° = 197° > 180°$ which is impossible (why?). Thus, $\alpha \neq 161°$; so $\alpha = 19°$. Now γ is easily computed since $\gamma = 180° - \alpha - \beta = 180° - 19° - 36° = 125°$. In summary, the solution is $b = 3.58$, $\alpha = 19°$, and $\gamma = 125°$.

It is also possible to use the Law of Cosines to find α after b is determined. For example,

$$2^2 = (3.58)^2 + 5^2 - 2(3.58) \cdot 5 \cos \alpha \quad \text{or} \quad \cos \alpha = \frac{12.82 + 25 - 4}{35.8} = .9447$$

Using Table B.3, we have $\alpha = \text{Cos}^{-1}.9447 = 19°$.

Problem S7.   If $a = 3$ and $\beta = 58°$, determine the values of b for which α has (a) two values, (b) one value, and (c) no values.

Solution: It is evident from Figure 5.3.24 that if side b is perpendicular to side c, then $\alpha = 90°$ and there is exactly one solution. In this case it should be clear that $\sin 58° = b/3$. Hence, $b = 3 \sin 58° = 3(.8481) = 2.54$. Also observe (see Figure 5.3.25) that if $b \geq 3$ (the length of a), then there is exactly one value of α determined. Thus, when $b = 2.54$ or $b \geq 3$, α has exactly one value.

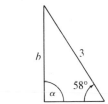

Figure 5.3.24

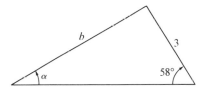

Figure 5.3.25

Next notice that if $b < 2.54$, then there is no solution because no triangle is determined as shown in Figure 5.3.26. Also, if $3 > b > 2.54$, then there are two possible angles (α_1 and α_2) that determine valid triangles as illustrated in Figure 5.3.27. Observe that $\alpha_1 + \alpha_2 = 180°$ and that α_1 is obtuse ($\alpha_1 > 90°$) and α_2 is acute ($\alpha_2 < 90°$). In summary, answer (a) is $3 > b > 2.54$, answer (b) is $b = 2.54$ or $b \geq 3$, and answer (c) is $b < 2.54$.

Figure 5.3.26

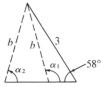

Figure 5.3.27

Problem S8.   Two observers, 400 feet apart, are on opposite sides of a tree. If the angles of elevation from the observers to the top of the tree are 15° and 20°, then how tall is the tree?

Solution: In Figure 5.3.28, x represents the height of the tree. Observe that if we could find the length of side a, then $\sin 20° = x/a$ will determine x (why?). Now note that $\gamma = 180° - 20° - 15° = 145°$. So, using the Law of Sines, we obtain

$$\frac{\sin 145°}{400} = \frac{\sin 15°}{a} \quad \text{or} \quad a = \frac{400 \sin 15°}{\sin 145°} = \frac{400(.2588)}{.5736} = 180.47$$

Thus, $\sin 20° = x/180.47$ or $x = 180.47(.3420) = 61.72$, so the tree is about 62 feet tall.

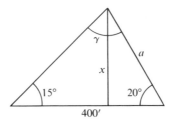

Figure 5.3.28

SUPPLEMENTARY PROBLEMS

S9. Solve the triangle in Figure 5.3.29.
(a) $\alpha = 65°$, $b = 6$, $c = 7$
(b) $a = 5$, $\alpha = 35°$, $\beta = 70°$
(c) $a = 4$, $b = 10$, $c = 9$

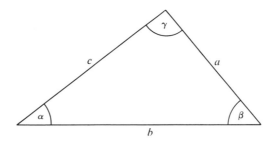

Figure 5.3.29

S10. If $a = 5$ and $\beta = 33°$, determine the values of b for which α has (a) two values, (b) one value, and (c) no values.

S11. In a parallelogram, two adjacent sides meet at an angle of 35° and are 3 feet and 8 feet in length. What is the length of the shortest diagonal of the parallelogram?

Answers

S9. (a) $a = 7.03$, $\beta = 51°$, $\gamma = 64°$
(b) $b = 8.17$, $c = 8.42$, $\gamma = 75°$
(c) $\alpha = 24°$, $\beta = 92°$, $\gamma = 64°$
S10. (a) $5 > b > 2.72$
(b) $b = 2.72$ or $b \geq 5$
(c) $b < 2.72$
S11. 5.8 feet

5.4 ALGEBRA OF COMPLEX NUMBERS

We have been working with R, the set of real numbers. In Chapter 1 we mentioned certain axioms and properties satisfied by R, for example, if $x \in R$, then $x^2 \geq 0$. Thus, we could conclude that there are no real number solutions for the equation $x^2 + 1 = 0$.

In the remainder of this chapter we shall study C, the set of complex numbers. The set C is "larger" than R in that it "contains" R, as we shall see subsequently. We shall discover solutions to the equation $x^2 + 1 = 0$ which are members of C, and we shall also discover some interesting properties about the roots of complex numbers.

We build on the properties of R by considering a complex number z to be an ordered pair of real numbers, together with the operations of addition and multiplication. We represent the complex number z by $a + bi$ or $a + ib$ interchangeably. The first member of the pair, a, is called the *real part* of z; b is called the *imaginary part* of z. Thus, the real part of $3 - 5i$ is 3 and the imaginary part of $3 - 5i$ is -5. This latter convention is perhaps an unfortunate choice of words as there is nothing "imaginary" about b, for b is indeed, by itself, a real number but is called the imaginary part of the complex number $a + bi$.

Q1: Find the real and imaginary parts of $z = 3 - 4i$.

422

Two complex numbers are *equal* if both of their real parts are equal and if both imaginary parts are equal. That is, if $x + iy = 3 - 4i$, then $x = 3$ and $y = -4$. Thus, one equality involving complex numbers is equivalent to two equalities involving real numbers.

Q2: Find a and b if $(a + b) + (a - b)i = 5 + 1i$

The set of complex numbers is closed under addition. That is, the sum of two complex numbers is a complex number. We add the real parts to obtain the real part of the sum. Likewise, the imaginary part of the sum is the sum of the imaginary parts. Thus,

$$(3 - 4i) + (2 + 5i) = (3 + 2) + (-4 + 5)i$$
$$= 5 + 1i = 5 + i.$$

By convention, we write $a + 0i = a$, $0 + bi = bi$, $1i = i$, and $a + (-b)i = a - bi$.

Multiplication is slightly more complicated. The real part of the product $(a + bi)(c + di)$ is $ac - bd$, and the imaginary part is $ad + bc$. These formulas need not be remembered if we make the following observations. First notice that $i^2 = -1$ using Q3.

Q3: Use the definition of multiplication to compute the product $i \cdot i = i^2 = (0 + 1i)(0 + 1i)$.

Next, we multiply $a + bi$ by $c + di$, as we would any two binomials, and replace i^2 by -1. That is,

$$(a + bi)(c + di) = ac + adi + bci + bdi^2$$
$$= ac + (ad + bc)i + bd(-1)$$
$$= (ac - bd) + i(ad + bc)$$

which yields the desired result.

We indicate repeated multiplications by exponents. That is, $z^n = \underbrace{z \cdot z \cdot \ldots \cdot z}_{n}$ for $n \in N$.

Q4: Compute (a) $(2 - 3i)(2 + 3i)$, (b) $2 - 3i(2 + 3i)$.

Let us summarize the discussion above with Definition 5.4.1.

A1: The real part is 3, and the imaginary part is -4.

A2: $a = 3$ and $b = 2$

A3: $(0 \cdot 0 - 1 \cdot 1) + (0 \cdot 1 + 1 \cdot 0)i = -1 + 0i = -1$

A4: (a) 13, (b) $11 - 6i$

Definition 5.4.1

The set of complex numbers, denoted by C, is $\{a + bi : a, b \in R\}$ on which equality, addition, and multiplication are defined as follows: If $u = a + bi \in C$ and $v = c + di \in C$, then

$$u = v \quad \text{if} \quad a = c \quad \text{and} \quad b = d \tag{1}$$
$$u + v = (a + c) + (b + d)i \tag{2}$$
$$u \cdot v = (ac - bd) + (ad + bc)i \tag{3}$$

The complex numbers C serve as an extension of the real numbers in several ways. First, C satisfies the field axioms of Appendix A. That is, closure, commutativity, associativity, etc., hold for C under addition and multiplication. (We leave the proof of this statement as an exercise.) Next, complex numbers of the form $a + 0i$ behave like real numbers. That is, $(a + 0i) + (b + 0i) = (a + b) + 0i$ just as $(a) + (b) = (a + b)$. Likewise, $(a + 0i)(b + 0i) = ab + 0i$ just as $(a) \cdot (b) = ab$. Thus, we shall not distinguish between the real number a and the complex number $a + 0i$, just as we take the integer k and the rational number $\frac{k}{1}$ to be the same. Formally, we say the real numbers are embedded in the complex numbers, as are the integers in the rationals.

Finally, the complex numbers admit solutions to certain equations which have no real number solutions, as we see in Example 5.4.1.

Example 5.4.1

Find all complex numbers z such that $z^2 + 1 = 0$.

Solution

Here, 0 means $0 + 0i$. Let $z = x + yi$ where $x, y \in R$. Then $(x + yi)^2 + 1 = x^2 - y^2 + 2xyi + 1 = 0 + 0i$. Equating real and imaginary parts, we obtain the two equations $x^2 - y^2 + 1 = 0$ and $2xy = 0$. Remember, x and y are real numbers. From the second of these equations, either $x = 0$ or $y = 0$. If $y = 0$, the first equation then becomes $x^2 + 1 = 0$, which has no *real* number solution. If $x = 0$, the first equation becomes $-y^2 + 1 = 0$ with (real number) solution set $\{1, -1\}$. Thus, the complex number solution set to the original equation is $\{0 + 1i, 0 - 1i\}$ or simply $\{i, -i\}$.

Q5: Find all complex numbers z such that $z^2 = 2i$.

In order to define division for complex numbers, we first define the conjugate of a complex number.

Definition
5.4.2

If $z = a + bi \in C$, then the *conjugate* of z, denoted $\bar{z}$, is the complex number $\bar{z} = \overline{a + bi} = a - bi$.

Thus, if $z = 2 - 3i$, then $\bar{z} = 2 + 3i$.

Q6: Find (a) $\overline{(3 + 4i)} + \overline{(2 - 5i)}$, (b) $\overline{(3 + 4i) + (2 - 5i)}$.

Notice what happens if we multiply any complex number $v = c + di$ by its conjugate. We obtain $v\bar{v} = (c + di)(c - di) = c^2 + cdi - cdi - d^2i^2 = c^2 + d^2$, a real number! Furthermore, $v\bar{v} = c^2 + d^2 = 0$ only if (both real numbers) $c = d = 0$. That is, $v\bar{v} = 0$ only if $v = 0$. We can use this fact to generate a single complex number for the quotient u/v where $u = a + bi$ and $v = c + di \neq 0$. We simply multiply both numerator and denominator of u/v by $\bar{v}$:

$$\frac{u}{v} = \frac{u \cdot \bar{v}}{v \cdot \bar{v}} = \frac{(a + bi)(c - di)}{(c + di)(c - di)} = \frac{(ac + bd) + (bc - ad)i}{c^2 + d^2}$$

which is now of the form $x + yi$ where

$$x = \frac{ac + bd}{c^2 + d^2} \quad \text{and} \quad y = \frac{bc - ad}{c^2 + d^2}$$

For example,

$$\frac{(-5 + 10i)}{3 + 4i} = \frac{(-5 + 10i)(3 - 4i)}{(3 + 4i)(3 - 4i)} = \frac{25 + 50i}{9 + 16} = 1 + 2i$$

Q7: Simplify $\dfrac{2 + 3i}{3 - 2i}$.

We conclude this section with some additional properties of the complex conjugate. We leave the proofs for the problems.

Theorem
5.4.1

If $u, v \in C$, then
(a) $\overline{u + v} = \bar{u} + \bar{v}$,
(b) $\overline{uv} = \bar{u} \cdot \bar{v}$,
(c) $\bar{u} = u$ if, and only if, u is real,
(d) $\bar{\bar{u}} = u$.

A5: Proceeding as in Example 5.4.1 yields $x^2 - y^2 = 0$ and $2xy = 2$. Since $x^2 - y^2 = 0$, either $x = y$ or $x = -y$. If $x = y$, the equation $2xy = 2$ has solutions $x = y = 1$ and $x = y = -1$. If $x = -y$, the equation $2xy = 2$ has no solutions. Thus, the solution set is $\{1 + i, -1 - i\}$.

A6: (a) $3 - 4i + 2 + 5i = 5 + i$ (b) $\overline{5 - i} = 5 + i$

A7: i

PROBLEMS 5.4

1. Write each of the following in the form $a + bi$:
(a) $2 - 3i - (4 + 5i)$
(b) $(2 - 3i)(4 + 5i)$
(c) $\dfrac{2 - 3i}{4 - 5i}$
(d) $\dfrac{2 - 3i}{i}$
(e) $2 - 3i(4 + 5i)$
(f) $\dfrac{2 - 3i}{4 - 5i} + \dfrac{1 - i}{4 + 5i}$

2. Find the complex number solution set in each of the following equations:
(a) $(2 - i)z + (3 + 2i) = 0$
(b) $z + 2\bar{z} = 3 + 2i$
(c) $2z + \bar{z} = iz + 4$
(d) $z^2 + \bar{z}^2 = 0$
(e) $z^2 + 4 = 0$
(f) $z^2 + 4i = 0$

3. Prove Theorem 5.4.1.

4. If $z = a + bi \in C$, let $Re(z) = a$ and $Im(z) = b$. Prove or disprove each of the following for $z_1, z_2 \in C$:
(a) $Re(z_1 + z_2) = Re(z_1) + Re(z_2)$
(b) $Im(z_1 + z_2) = Im(z_1) + Im(z_2)$
(c) $Re(\bar{z}) = Re(z)$
(d) $Re(z_1 z_2) = Re(z_1)Re(z_2)$
(e) $Re(z) = \dfrac{1}{2}(z + \bar{z})$
(f) $Im(z) = \left(\dfrac{\bar{z} - z}{2}\right)i$

5. For each $z \in C$, let $f(z) = \bar{z}$ define a function f.
(a) What is the domain of f?
(b) What is the range of f?
(c) Does $\dfrac{f(u)}{f(v)} = f\left(\dfrac{u}{v}\right)$?
(d) What is $f(f(z))$?
(e) Is f a one-to-one function?
(f) Does f^{-1} exist? If so, find a formula for $f^{-1}(z)$.

6. Show that the set of complex numbers C satisfies the field axioms of appendix A with the word *real* replaced by *complex* everywhere. Although it is not easy to prove, the complex numbers are not "ordered" — that is, if $u, v \in C$, we cannot assign a meaning to $u < v$ consistent with the order axioms.

ANSWERS
5.4

1. (a) $-2 - 8i$, (b) $23 - 2i$, (c) $\dfrac{23}{41} - \dfrac{2}{41}i$, (d) $-3 - 2i$, (e) $17 - 12i$,

(f) $\dfrac{22}{41} - \dfrac{11}{41}i$

2. (a) $\left\{ -\dfrac{4}{5} - \dfrac{7}{5}i \right\}$, (b) $\{1 - 2i\}$, (c) $\{1 + i\}$, (d) $\{x + iy : |x| = |y|\}$,

(e) $\{\pm 2i\}$, (f) $\{\pm(\sqrt{2} - \sqrt{2}i)\}$

4. Only (d) is not true.

5. (a) C, (b) C, (c) Yes, (d) z, (e) Yes, (f) Yes, $f^{-1}(z) = \bar{z}$.

Supplement 5.4

Problem S1.　(a) Simplify $3 - 2i(3 + 2i)$, and (b) solve $|x| + yi = 3 + i \log_{10}100$.

Solution:　(a) $3 - 2i(3 + 2i) = 3 - 6i - 4i^2 = 3 - 6i - 4(-1) = 7 - 6i$ since $i^2 = -1$. Notice that $3 - 2i(3 + 2i) \ne (3 - 2i)(3 + 2i)$.

(b) Two complex numbers are equal precisely when both of their real parts are equal and both of their imaginary parts are equal (see Definition 5.4.1). Thus, if $|x| + yi = 3 + i \log_{10}100$, then $|x| = 3$ and $y = \log_{10}100$ or $x = \pm 3$ and $y = 2$. Therefore, the solutions are $x = 3$ and $y = 2$ or $x = -3$ and $y = 2$.

Problem S2.　Simplify (a) $(3 - 2i)(4 + 5i)$ and (b) $\dfrac{3 - 4i}{2 + 5i}$.

Solution:　(a) The product can be easily determined by usual "polynomial" multiplication. That is,

$$(3 - 2i)(4 + 5i) = 3 \cdot 4 + 3 \cdot 5i - 2 \cdot 4i - 2 \cdot 5 \cdot i^2$$
$$= 12 + 15i - 8i - 10i^2$$

Now since $i^2 = -1$, $(3 - 2i)(4 + 5i) = 12 + 7i - 10(-1) = 22 + 7i$.

(b) To simplify this quotient, we use the fact that the product of the complex number $a + bi$ and its conjugate $a - bi$ is a *real* number. That is, $(a + bi)(a - bi) = a^2 - abi + abi - b^2i^2 = a^2 + b^2$. Therefore, to divide two complex numbers, first multiply the numerator and the denominator by the conjugate of the denominator, and then simplify. Thus, in this case, we multiply top and bottom by the conjugate of $2 + 5i$, namely, $2 - 5i$. Hence,

$$\left(\dfrac{3 - 4i}{2 + 5i}\right)\left(\dfrac{2 - 5i}{2 - 5i}\right) = \dfrac{6 - 15i - 8i + 20i^2}{2^2 + 5^2} = \dfrac{-14 - 23i}{29}$$

Finally,

$$\dfrac{3 - 4i}{2 + 5i} = -\dfrac{14}{29} - \dfrac{23}{29}i$$

DRILL PROBLEMS

S3. Simplify $(2 + 3i) - 2(-4 + i)$.

S4. Simplify (a) $\sqrt{12}i(2\sqrt{3} - 5i\sqrt{3})$ and (b) $\dfrac{2 - 3i}{1 + 4i}$

Answers

S3. $10 + i$

S4. (a) $30 + 12i$, (b) $-\dfrac{10}{17} - \dfrac{11}{17}i$

Problem S5. Find the complex number solution set for $(3 + 2i)z + 2i - 3 = 0$.

Solution: We solve for z. Thus,

$$(3 + 2i)z = 3 - 2i \quad \text{or} \quad z = \frac{3 - 2i}{3 + 2i}$$

Next we express z in the form $a + bi$ [see Problem S2 (b)]:

$$z = \frac{(3 - 2i)(3 - 2i)}{(3 + 2i)(3 - 2i)} = \frac{9 - 6i - 6i + 4i^2}{9 + 4} = \frac{5 - 12i}{13} = \frac{5}{13} - \frac{12}{13}i$$

Thus, the solution set is $\left\{ \dfrac{5}{13} - \dfrac{12}{13}i \right\}$.

Problem S6. Find the complex number solution set for $z + 3\bar{z} = i + 4$.

Solution: Let $z = a + bi$. Remember that $\bar{z}$ is the conjugate of z, so $\bar{z} = a - bi$. Then $z + 3\bar{z} = i + 4$ becomes $(a + bi) + 3(a - bi) = i + 4$. Simplifying, we obtain

$$a + bi + 3a - 3bi = i + 4 \quad \text{or} \quad 4a - 2bi = 4 + i$$

We equate both real parts and both imaginary parts (Definition 5.4.1) to obtain $4a = 4$ and $-2b = 1$. Clearly, $a = 1$ and $b = -\frac{1}{2}$. Thus, the solution set is $\{1 - \frac{1}{2}i\}$.

Problem S7. Find the complex number solution set for $z + 3i\bar{z} = 11 + i$.

Solution: Let $z = a + bi$. Thus, $z + 3i\bar{z} = a + bi + 3i(a - bi) = a + bi + 3ai - 3bi^2 = a + 3b + (3a + b)i = 11 + i$. Next, equating real and imaginary parts yields $a + 3b = 11$ and $3a + b = 1$. Using the first equation, we obtain $a = 11 - 3b$. Then, substituting for a in the second equation, we have $3(11 - 3b) + b = 1$. Thus, $33 - 9b + b = 1$ or $-8b = -32$ or $b = 4$, so $a = 11 - 3\cdot4 = -1$. Therefore, the solution set is $\{-1 + 4i\}$.

Problem S8. Find the complex number solution set for $z^2 - iz = 0$.

Solution: We could let $z = a + bi$ and proceed in a manner similar to that used in Problem S7. However, in this case, factoring is much easier. Clearly, $z^2 - iz = z(z - i)$. We will verify in the next section that if u and v are complex numbers, then $u \cdot v = 0$ implies $u = 0$ or $v = 0$. (It is understood that $0 = 0 + 0i$.) Thus, $z^2 - iz = z(z - i) = 0$ implies $z = 0$ or $z - i = 0$. Hence, the solution set is $\{0, i\}$.

Problem S9. Find the complex number solution set for $z^2 + 5i = 0$.

Solution: Let $z = a + bi$. Then $z^2 + 5i = (a + bi)^2 + 5i = a^2 + 2abi + b^2i^2 + 5i = a^2 - b^2 + (2ab + 5)i = 0 = 0 + 0i$. Equating real and imaginary parts, we obtain $a^2 - b^2 = 0$ and $2ab + 5 = 0$. The first equation implies that $a^2 = b^2$ or $a = \pm b$. Substituting $a = b$ in the second equation yields $2b^2 + 5 = 0$. Clearly, this equation has no real solution (why?). If $a = -b$, the second equation yields $2(-b)b + 5 = 0$ or $-2b^2 = -5$ or $b = \pm\sqrt{5/2}$. Since $a = -b$,

$$z = \sqrt{\frac{5}{2}} - \sqrt{\frac{5}{2}}i \quad \text{and} \quad z = -\sqrt{\frac{5}{2}} + \sqrt{\frac{5}{2}}i$$

are valid solutions. The solution set is

$$\left\{ \pm\sqrt{\frac{5}{2}} \mp \sqrt{\frac{5}{2}}i \right\}$$

DRILL PROBLEMS
S10. (a) Find the complex number solution set for $\bar{z} + 4 = 2z + 3(i + 3)$. (b) Find the complex number solution set for $z - 2i\bar{z} = 1 - 5i$. (c) Find the complex number solution set for $z^2 + 2z + 3 = 0$.

Answers

S10. (a) $\{-5 - i\}$, (b) $\{3 + i\}$, (c) $\{-1 \pm \sqrt{2}i\}$

Problem S11. Let $Re(z)$ denote the real part of the complex number z. That is, if $z = a + bi$, $Re(z) = a$. Similarly, $Im(z) = b$. Prove or disprove each of the following for $z_1, z_2 \in C$:
(a) $Re(z_1 + z_2) = Re(z_1) + Re(z_2)$ (b) $Re(z_1 z_2) = Re(z_1) Re(z_2)$
(c) $Im(z) = Im(\bar{z})$

Solution: (a) Let $z_1 = a_1 + b_1 i$ and $z_2 = a_2 + b_2 i$. It follows that $Re(z_1 + z_2) = Re[(a_1 + b_1 i) + (a_2 + b_2 i)] = Re[(a_1 + a_2) + (b_1 + b_2)i] = a_1 + a_2$. Also it is clear that $Re(z_1) = a_1$ and $Re(z_2) = a_2$, so $a_1 + a_2 = Re(z_1) + Re(z_2)$. Thus, we have proved that $Re(z_1 + z_2) = Re(z_1) + Re(z_2)$.

Note: A specific example would illustrate the assertion but *not prove* it. If you have no idea regarding the validity of an assertion (conjecture), then you should try several arbitrary examples. If they prove valid, then you should look for a general proof, possibly based on the examples that you have worked.

(b) This equation is not valid for arbitrary complex numbers. For example, if $z_1 = i = z_2$, then $Re(z_1 z_2) = Re(-1) \neq 0 = Re(z_1) \cdot Re(z_2)$. Therefore, we have exhibited a *counter example* (a specific example disproving the assertion) to the claim that $Re(z_1 z_2) = Re(z_1) Re(z_2)$ for $z_1, z_2 \in C$.

(c) This assertion is false. For example, if $z = 1 + i$, then $\bar{z} = 1 - i$ and $Im(z) = 1$ and $Im(\bar{z}) = -1$. Clearly, $Im(z) \neq Im(\bar{z})$ for $z = 1 + i$. Thus, we have exhibited a counter example.

SUPPLEMENTARY PROBLEMS

S12. Simplify

(a) $2 - 3i(3 + i)$

(b) $(2 - 3i)(3 + i)$

(c) $\dfrac{3 + 2i}{-1 - i}$

S13. Solve

(a) $z - 2\bar{z} + i = 2$

(b) $z + 2i\bar{z} = 2 - 2i$

(c) $z^2 + 3i = 0$

S14. Prove or disprove

(a) $Re(z^2) = \left(\dfrac{z + \bar{z}}{2}\right)^2$

(b) $[Re(z)]^2 = \left(\dfrac{z + \bar{z}}{2}\right)^2$

Answers

S12. (a) $5 - 9i$, (b) $9 - 7i$, (c) $-\dfrac{5}{2} + \dfrac{1}{2}i$

S13. (a) $\left\{-2 - \dfrac{1}{3}i\right\}$, (b) $\{-2 + 2i\}$, (c) $\left\{\pm\left(\sqrt{\dfrac{3}{2}} - \sqrt{\dfrac{3}{2}}i\right)\right\}$

S14. (a) False, (b) true

5.5 GEOMETRY OF COMPLEX NUMBERS

We have previously observed the one-to-one correspondence between real numbers and points on a number line, as well as that between ordered pairs of real numbers and points in a Cartesian plane. Since a complex number $a + bi$ can be identified by the ordered pair of real numbers (a, b), there is a natural one-to-one correspondence between complex numbers and points in a plane. That is, the point (a, b) is associated with the complex number $z = a + bi$, and conversely. In this setting we call $a + bi$ the *rectangular* or Cartesian form of z and refer to the plane as the *complex plane*. (See Figure 5.5.1.)

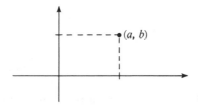

Figure 5.5.1

Q1: Indicate each of the following numbers as points in the complex plane:

$$u = 2 - i, \quad v = i, \quad w = \frac{2}{i}$$

It is easy to graphically portray the addition of two complex numbers in rectangular form using the parallelogram as shown in Figure 5.5.2. The heavy line indicates the sum $w = u + v$ where $u = 2 - i$ and $v = 1 + 2i$.

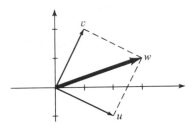

Figure 5.5.2

Q2: Graphically portray the addition of $2 - i$ and $-3 - i$.

Unfortunately, it is not so easy to picture the multiplication of two complex numbers in rectangular form. However, we shall now introduce a different description of the point (a, b) to help us. In Figure 5.5.3 we have drawn the line segment from $(0, 0)$ to (a, b). Let r denote its length. Thus,

$$r = \sqrt{a^2 + b^2} \tag{1}$$

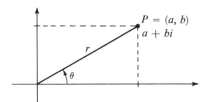

Figure 5.5.3

Definition 5.5.1

Let $P = (a, b)$ be a point in a Cartesian plane and $r = \sqrt{a^2 + b^2}$. If $r \neq 0$, let θ be an angle, in standard position, whose terminal side contains P. The numbers r and θ are called the *polar coordinates* of P.

Notice that the polar coordinates of P are not unique, for there are infinitely many numbers that "work" for θ, any two of which differ by an integral multiple of 2π. Also, if $r = 0$, then θ is undefined.

If $r \neq 0$, we can write $z = a + bi = r\left(\dfrac{a}{r} + \dfrac{b}{r}\right)i$. Since $\dfrac{a}{r} = \cos\theta$

and $\dfrac{b}{r} = \sin\theta$ in Figure 5.5.3, then

$$a = r\cos\theta \quad \text{and} \quad b = r\sin\theta \tag{2}$$

Thus,

$$z = a + bi = r(\cos\theta + i\sin\theta) \tag{3}$$

For example, let $z = \sqrt{3} + i$. Then

$$z = 2\left(\frac{\sqrt{3}}{2} + \frac{1}{2}i\right) = 2\left(\cos\frac{\pi}{6} + i\sin\frac{\pi}{6}\right)$$

since $\cos\frac{1}{6}\pi = \frac{1}{2}\sqrt{3}$ and $\sin\frac{1}{6}\pi = \frac{1}{2}$. (See Figure 5.5.4.) The expression $z = r(\cos\theta + i\sin\theta)$ is called the *trigonometric* or polar form of the complex number z, and is often written $z = r\operatorname{cis}\theta$.

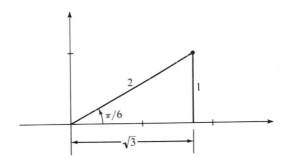

Figure 5.5.4

Example 5.5.1

Plot the number $3 - 3i$ in the complex plane and write $3 - 3i$ in trigonometric form.

Solution

Plot $3 - 3i = (3, -3)$, as shown in Figure 5.5.5. Clearly, $r = \sqrt{3^2 + 3^2} = 3\sqrt{2}$, and $-\pi/4$ is one possibility for θ. Thus we can write

$$3 - 3i = 3\sqrt{2}\left[\cos\left(-\frac{\pi}{4}\right) + i\sin\left(-\frac{\pi}{4}\right)\right] = 3\sqrt{2}\operatorname{cis}\left(-\frac{\pi}{4}\right).$$

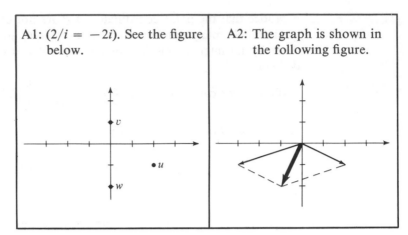

A1: $(2/i = -2i)$. See the figure below.

A2: The graph is shown in the following figure.

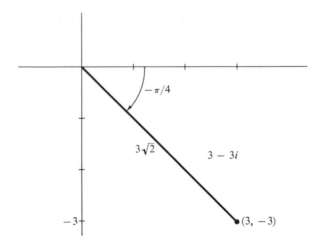

$-\pi/4$

$3\sqrt{2}$

$3 - 3i$

-3

$(3, -3)$

Figure 5.5.5

Q3: Plot each of the following numbers in the complex plane, and write each in the form $r(\cos \theta + i \sin \theta) = r \operatorname{cis} \theta$:

$$u = i, \quad v = 1 + i, \quad w = -1 - i, \quad z = \sqrt{3} - i$$

The number r is called the *modulus* or *absolute value* of z, written $|z| = r$. Notice that $|z|$ represents the distance between z and the origin in the complex plane, just as $|a|$ represents the distance between a and the origin on the number line. The measure of the angle θ (in degrees or radians) is called an *argument* of z and, as mentioned earlier, is not uniquely determined for a given number z.

Q4: Find the absolute value and argument of each of the following complex numbers:

(a) 3 cis 45°   (b) $2i$ (c) $2 - 2i$

Let us make two additional observations relating to the trigonometric form of a complex number. Let $u = r$ cis θ and $v = s$ cis φ. Then

$$\bar{u} = r \text{ cis } (-\theta) \tag{4}$$

$$u = v \quad \text{if, and only if,} \quad r = s \quad \text{and} \quad \theta = \varphi + k \cdot 2\pi, \, k \in Z \tag{5}$$

The second of these properties [Equation (5)] follows from the definition of trigonometric form. To prove the first [Equation (4)], simply observe that

$$\bar{u} = r(\cos \theta - i \sin \theta) = r(\cos(-\theta) + i \sin(-\theta)) = r \text{ cis}(-\theta)$$

The usefulness of the trigonometric form of complex numbers becomes obvious when we multiply two complex numbers and interpret the result geometrically. For example, let r cis θ and s cis φ be two complex numbers. Forming the product, we obtain

$$(r \text{ cis } \theta) \cdot (s \text{ cis } \varphi) = (r \cos \theta + i r \sin \theta)(s \cos \varphi + i s \sin \varphi)$$
$$= rs \cos \theta \cos \varphi - rs \sin \theta \sin \varphi$$
$$\quad + i(rs \cos \theta \sin \varphi + rs \sin \theta \cos \varphi)$$
$$= rs[(\cos \theta \cos \varphi - \sin \theta \sin \varphi)$$
$$\quad + i(\sin \theta \cos \varphi + \cos \theta \sin \varphi)]$$
$$= rs[(\cos(\theta + \varphi) + i \sin(\theta + \varphi)]$$
$$\text{[using Equations (5) in Section 4.6]}$$

Thus,

$$(r \text{ cis } \theta) \cdot (s \text{ cis } \varphi) = rs \text{ cis } (\theta + \varphi) \tag{6}$$

That is, we can multiply two complex numbers by multiplying their absolute values and adding their arguments (Figure 5.5.6). For example, (2 cis 60°)·(4 cis 30°) = 8 cis 90° = $8i$. Using rectangular form to check our result, we have

$$2\left(\frac{1}{2} + \frac{\sqrt{3}}{2}i\right) \cdot 4\left(\frac{\sqrt{3}}{2} + \frac{1}{2}i\right) = (1 + \sqrt{3}i)(2\sqrt{3} + 2i)$$
$$= 2\sqrt{3} + 2i + 2\sqrt{3} \cdot \sqrt{3}i + 2\sqrt{3}i^2$$
$$= 0 + (2 + 6)i = 8i$$

A3: For example,

$$u = 1\left(\cos\frac{\pi}{2} + i\sin\frac{\pi}{2}\right) = \operatorname{cis}\frac{\pi}{2}$$

$$v = \sqrt{2}\left(\cos\frac{\pi}{4} + i\sin\frac{\pi}{4}\right) = \sqrt{2}\operatorname{cis}\frac{\pi}{4}$$

$$w = \sqrt{2}\left(\cos\frac{5\pi}{4} + i\sin\frac{5\pi}{4}\right)$$

$$z = 2\operatorname{cis}\left(-\frac{\pi}{6}\right)$$

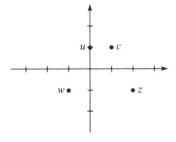

A4: (a) 3, 45°, (b) 2, $\pi/2$, (c) $2\sqrt{2}$, $-\pi/4$

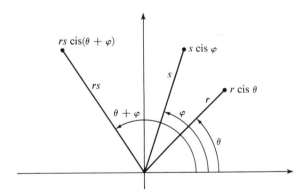

Figure 5.5.6

Q5: Find the trigonometric form of each product below:

(a) $(2\operatorname{cis}\frac{1}{4}\pi)\cdot 3\operatorname{cis}(-\frac{1}{6}\pi)$ (b) $(2 + 2i)(1 + \sqrt{3}i)$

There is a related rule for division of complex numbers in trigonometric form. Again, let $u = r\operatorname{cis}\theta$ and $v = s\operatorname{cis}\varphi \neq 0$. Then $\bar{v} = s\operatorname{cis}(-\varphi)$ and

$$\frac{u}{v} = \frac{u \cdot \bar{v}}{v \cdot \bar{v}} = \frac{r \text{ cis } \theta \cdot s \text{ cis}(-\varphi)}{s \text{ cis } \varphi \cdot s \text{ cis}(-\varphi)} = \frac{rs \text{ cis}(\theta - \varphi)}{s \cdot s \text{ cis}(\varphi - \varphi)} = \frac{r \text{ cis}(\theta - \varphi)}{s \cdot 1}$$

Thus,

$$\frac{r \text{ cis } \theta}{s \text{ cis } \varphi} = \frac{r}{s} \text{ cis}(\theta - \varphi) \qquad (s \neq 0) \qquad (7)$$

Q6: Find the trigonometric form of each quotient below:

$$\text{(a)} \quad \frac{12 \text{ cis } \frac{2}{3}\pi}{3 \text{ cis } \frac{1}{6}\pi} \qquad \text{(b)} \quad \frac{-2 + 2\sqrt{3}i}{1 + i}$$

Notice the result when Equation (6) is applied to repeated multiplications. If $z = r \text{ cis } \theta$, then

$$z^2 = z \cdot z = (r \text{ cis } \theta)(r \text{ cis } \theta)$$
$$= r \cdot r \text{ cis}(\theta + \theta) = r^2 \text{ cis } 2\theta$$

Thus,

$$z^3 = z^2 \cdot z = (r^2 \text{cis } 2\theta)(r \text{ cis } 0) = r^2 \cdot r \text{ cis}(2\theta + 0) = r^3 \text{cis } 3\theta$$

Similarly,

$$z^n = \underbrace{(r \text{ cis } \theta) \cdot \ldots \cdot (r \text{ cis } \theta)}_{n \text{ factors}} = \underbrace{r \cdot r \ldots \cdot r}_{n} \text{ cis } \underbrace{(\theta + \cdots + \theta)}_{n} = r^n \text{cis } n\theta$$

That is,

$$(r \text{ cis } \theta)^n = r^n \text{cis } n\theta \qquad (8)$$

If $r = 1$, the result is known as *De Moivre's theorem:*

$$(\cos \theta + i \sin \theta)^n = \cos n\theta + i \sin n\theta \qquad (9)$$

An important property of real numbers is that $ab = 0$ if, and only if, either $a = 0$ or $b = 0$ (or both). Thus, we know that the (real number) solution set to $(x - 2)(x + 3) = 0$ is $\{2, -3\}$. This important result is also true for complex numbers. That is, for $u, v \in C$,

$$u \cdot v = 0 \quad \text{if, and only if, } u = 0 \text{ or } v = 0 \qquad (10)$$

To verify this result, consider the trigonometric forms of u and v, say, $u = r \text{ cis } \theta$ and $v = s \text{ cis } \varphi$. Then $uv = rs \text{ cis}(\theta + \varphi) = 0$ if, and only if, $rs = 0$. But since $r, s \in R$, $rs = 0$ if, and only if $r = 0$ or $s = 0$, that is, if, and only if, $u = 0$ or $v = 0$. Equation (10) tells us, for example, that the solution set to $(z - 2)(z + i) = 0$ is $\{2, -i\}$.

A5: (a) $6 \text{ cis } \frac{1}{12}\pi$, (b) $(2\sqrt{2} \text{ cis } \frac{1}{4}\pi) \cdot (2 \text{ cis } \frac{1}{3}\pi) = 4\sqrt{2} \text{ cis } \frac{7}{12}\pi$

A6: (a) $4 \text{ cis } \frac{1}{2}\pi = 4i$, (b) $\dfrac{4 \text{ cis } \frac{2}{3}\pi}{\sqrt{2} \text{ cis } \frac{1}{4}\pi} = 2\sqrt{2} \text{ cis } \frac{5}{12}\pi$

Q7: Find the solution set to

  (a) $(z - 1 + i)(2z - 4i) = 0$ (b) $(z + i)(iz - 3 + i) = 0$

PROBLEMS
5.5

1. Write each of the following complex numbers in rectangular form and plot as points in the complex plane:

(a) $2 \text{ cis } \frac{1}{3}\pi$ (b) $4 \text{ cis}(-\frac{1}{2}\pi)$

(c) $\sqrt{2} \text{ cis } \frac{3}{4}\pi$ (d) $5 \text{ cis } \pi$

(e) $\frac{1}{2} \text{ cis } 50°$ (f) $3 \text{ cis } 1$

2. Plot each of the following complex numbers in the complex plane and write each in trigonometric form. You will need to use Table B.3 for Parts (e) and (f).

(a) $2 - 2i$ (b) $-\sqrt{3} + i$

(c) $-16i$ (d) $-(2 + \sqrt{12i})$

(e) $3 - 4i$ (f) $-3 + 4i$

3. Find $|z|$ for each of the following:

(a) $z = 2 - 5i$ (b) $z = (2 - 5i)(3 + 4i)$

(c) $z = \dfrac{2 - 5i}{3 + 4i}$

4. Show that each of the following is true:

(a) $|uv| = |u| \cdot |v|$ (b) $\left|\dfrac{u}{v}\right| = \dfrac{|u|}{|v|}$ (provided $v \neq 0$)

5. Simplify

(a) $(1 + i)^8$ (b) $(\sqrt{3} - i)^6$

(c) i^{1234} (d) $\dfrac{(1 + i)^8}{(\sqrt{3} - i)^6}$

(e) $\dfrac{(\sqrt{3} - i)^8}{(1 + i)^6}$

6. Find the solution set to each of the following:

(a) $(3z + i)(iz + 3) = 0$

(b) $\left(\dfrac{1}{2} + \dfrac{1}{2}i - z\right)(z + 2 - 3i) = 0$

(c) $z^4 - 16 = 0$
(d) $z^2 + (2 + 3i)z + 6i = 0$

7. Write each of the following in trigonometric form.
(a) $r(\cos \theta - i \sin \theta)$ (b) $r(\sin \theta - i \cos \theta)$

ANSWERS
5.5

1. (a) $1 + i\sqrt{3}$, (b) $-4i$, (c) $-1 + i$, (d) -5, (e) $.3214 + .383i$, (f) $1.62 + 2.52i$

2. (a) $2\sqrt{2} \text{ cis}(-\frac{1}{4}\pi)$, (b) $2 \text{ cis } \frac{5}{6}\pi$, (c) $16 \text{ cis } \frac{3}{2}\pi$, (d) $4 \text{ cis } \frac{4}{3}\pi$, (e) $5 \text{ cis}(-.93)$, (f) $5 \text{ cis } 2.21$

3. (a) $\sqrt{29}$, (b) $5\sqrt{29}$, (c) $\sqrt{29}/5$

5. (a) 16, (b) -64, (c) -1, (d) $-1/4$, (e) $32 \text{ cis}(-\frac{5}{6}\pi)$

6. (a) $\left\{ -\frac{1}{3}i, 3i \right\}$,   (b) $\left\{ \frac{1}{2} + \frac{1}{2}i, -2 + 3i \right\}$,   (c) $\{2, -2, 2i, -2i\}$,
(d) $\{-2, -3i\}$

7. (a) $r \text{ cis}(-\theta)$, (b) $r \text{ cis}(\theta - \frac{1}{2}\pi)$

Supplement 5.5

Problem S1.   Write each of the following complex numbers in rectangular form and plot each as a point in the complex plane:
(a) $3 \text{ cis } \frac{1}{2}\pi$ (b) $2 \text{ cis}(-\frac{2}{3}\pi)$

Solution: (a) First remember that $r \text{ cis } \theta$ is "shorthand" for $r(\cos \theta + i \sin \theta)$. Thus,

$$3 \text{ cis } \frac{\pi}{2} = 3\left(\cos \frac{\pi}{2} + i \sin \frac{\pi}{2}\right) = 3(0 + i \cdot 1) = 3i.$$

(See Figure 5.5.7.)

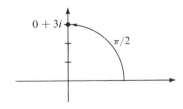

Figure 5.5.7

A7: (a) $\{1 - i, 2i\}$,   (b) $\{-i, -1 - 3i\}$

(b)

$$2 \operatorname{cis}\left(-\frac{2\pi}{3}\right) = 2\left[\cos\left(-\frac{2\pi}{3}\right) + i \sin\left(-\frac{2\pi}{3}\right)\right] = 2\left[-\frac{1}{2} + i\left(-\frac{\sqrt{3}}{2}\right)\right] = -1 - \sqrt{3}i$$

See Figures 5.5.8 and 5.5.9. Observe that we needed to know the coordinates of $P(-\frac{2}{3}\pi)$.

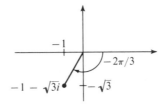

Figure 5.5.8

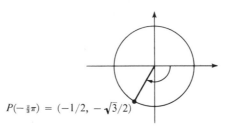

$P(-\frac{2}{3}\pi) = (-1/2, -\sqrt{3}/2)$

Figure 5.5.9

DRILL PROBLEMS

S2. Write each of the following complex numbers in rectangular form and plot as points in the complex plane:
(a) $4 \operatorname{cis} \frac{3}{4}\pi$ (b) $3 \operatorname{cis}(-150°)$

Answers

S2. (a) $(-2\sqrt{2}, 2\sqrt{2})$. See Figure 5.5.10.   (b) $(-3\sqrt{3}/2, -3/2)$. See Figure 5.5.11.

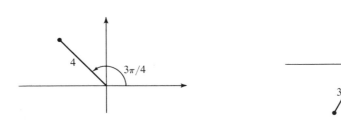

Figure 5.5.10   **Figure 5.5.11**

Problem S3. Plot each of the following complex numbers as points in the complex plane and write in trigonometric form: (a) i, (b) $-1 + i$, and (c) $1 - \sqrt{3}i$.

Solution: (a) Recall that to write a complex number $a + bi$ in trigonometric form, you must find r and θ so that $a + bi = r(\cos \theta + i \sin \theta) = r \operatorname{cis} \theta$. To do so, simply plot the point $P = (a, b)$ associated with the complex number $a + bi$ and then remember that r is the length of the segment OP and θ is an angle, in standard position, whose terminal side is the segment OP (see Figure 5.5.12). By periodicity,

$$a + bi = r \operatorname{cis} \theta = r \operatorname{cis}(\theta + 2k\pi), \; k \in Z$$

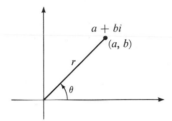

Figure 5.5.12

Thus, we plot $i = 0 + i = (0, 1)$ and observe that $r = 1$ and $\theta = \frac{1}{2}\pi$ (or 90°). See Figure 5.5.13. Hence, $i = 1 \cdot (\cos \frac{1}{2}\pi + i \sin \frac{1}{2}\pi) = \operatorname{cis} \frac{1}{2}\pi$. You should understand that this is not the only possible answer.

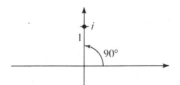

Figure 5.5.13

(b) Plot the point $-1 + i = (-1, 1)$, as shown in Figure 5.5.14. Clearly, $r = \sqrt{(-1)^2 + 1^2} = \sqrt{2}$ and $\theta = 135°$. Thus, $-1 + i = \sqrt{2} \operatorname{cis} 135°$.

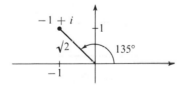

Figure 5.5.14

(c) Plot the point $1 - \sqrt{3}i = (1, -\sqrt{3})$, as shown in Figure 5.5.15. Clearly, $r = \sqrt{1^2 + (-\sqrt{3})^2} = 2$ and $\theta = -\pi/3$. Thus, $1 - \sqrt{3}i = 2 \operatorname{cis}(-\pi/3)$.

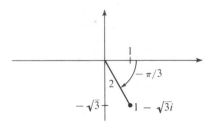

Figure 5.5.15

Problem S4.   Write the complex number $-2 + 3i$ in trigonometric form and plot it as a point in the complex plane.

Solution: Here again we want to find r and θ so that $-2 + 3i = r$ cis θ. Recall that $r = \sqrt{a^2 + b^2}$, $\cos \theta = a/r$ and $\sin \theta = b/r$ [Equations (1) and (2)]. In this case, $a = -2$ and $b = 3$. Computing, we obtain

$$r = \sqrt{(-2)^2 + 3^2} = \sqrt{13} = 3.605$$
$$\cos \theta = \frac{-2}{\sqrt{13}} = -.5548$$
$$\sin \theta = \frac{3}{\sqrt{13}} = .8322$$

(Remember, these calculations are all approximations.) $\sin \theta = .8322$ and Table B.3 imply that $\theta = 56°$ or $180 - 56 = 124°$. Clearly, $\theta = 124°$ since $\cos \theta < 0$. Therefore, $-2 + 3i = \sqrt{13}$ cis $124°$. The number is plotted in Figure 5.5.16.

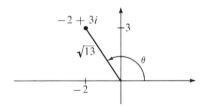

Figure 5.5.16

DRILL PROBLEMS

S5. Write the following complex numbers in trigonometric form and plot as points in the complex plane:
(a) $-\sqrt{3} + i$ (b) $-3 - 3i$

Answers

S5. (a) See Figure 5.5.17.   (b) See Figure 5.5.18.

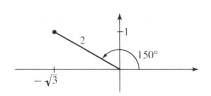

Figure 5.5.17

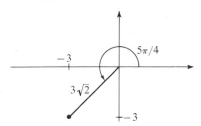

Figure 5.5.18

Problem S6. Find $|z|$ for each of the following:
(a) $z = -3 + 4i$ (b) $z = (-3 + 4i)(2 + 3i)$

Solution: (a) If $z = a + bi$, then $|z| = r$ where $r = \sqrt{a^2 + b^2}$. Hence, if $z = -3 + 4i$, then $a = -3$ and $b = 4$ and thus

$$|z| = r = \sqrt{(-3)^2 + 4^2} = \sqrt{25} = 5$$

(b) One way to proceed is to use the result that if $z = (r \text{ cis } \theta)(s \text{ cis } \varphi)$, then $z = rs \text{ cis}(\theta + \varphi)$ [see Equation (6)]. In particular, this means that we can find the absolute value of the product of two complex numbers by simply multiplying the individual absolute values. In other words, if $z = u \cdot v$ where $|u| = r$ and $|v| = s$, then $|z| = |uv| = r \cdot s = |u||v|$. This is essentially the proof of the result in Problem 4(a). Therefore,

$$|z| = |-3 + 4i| \cdot |2 + 3i| = \sqrt{(-3)^2 + 4^2} \cdot \sqrt{2^2 + 3^2} = 5\sqrt{13}$$

Problem S7. (a) Prove that, for any two complex numbers u and v, $\left|\dfrac{u}{v}\right| = \dfrac{|u|}{|v|}$, if $v \neq 0$. (b) Use Part (a) to find $|z|$ where $z = \dfrac{-3 + 4i}{2 + 3i}$.

Solution: (a) Let $u = r \text{ cis } \theta$ and $v = s \text{ cis } \varphi$ ($r \neq 0$). Then from Equation (7),

$$\frac{u}{v} = \frac{r \text{ cis } \theta}{s \text{ cis } \varphi} = \frac{r}{s} \text{ cis}(\theta - \varphi)$$

Now by the definition of absolute value of a complex number, $|u| = r$, $|v| = s$ and $\left|\dfrac{u}{v}\right| = \dfrac{r}{s}$ $\left(\text{since } \dfrac{u}{v} = \dfrac{r}{s} \text{ cis}(\theta - \varphi)\right)$. Thus, $\left|\dfrac{u}{v}\right| = \dfrac{r}{s} = \dfrac{|u|}{|v|}$, provided that $v \neq 0$.
(b) Using Part (a), we obtain

$$|z| = \left|\frac{-3 + 4i}{2 + 3i}\right| = \frac{|-3 + 4i|}{|2 + 3i|} = \frac{\sqrt{(-3)^2 + 4^2}}{\sqrt{2^2 + 3^2}} = \frac{5}{\sqrt{13}}$$

Problem S8. Find the solution set to each of the following:
(a) $(2z - 5i)(z - 4i + 1) = 0$ (b) $z^2 + (4 - 3i)z + 12i = 0$

Solution: (a) Here we use the result that $u \cdot v = 0$ implies $u = 0$ or $v = 0$ where u and v are complex numbers [see Equation (10)]. Thus, $(2z - 5i)(z - 4i + 1) = 0$ implies $2z - 5i = 0$ or $z - 4i + 1 = 0$. Hence, $z = (5/2)i$ or $z = -1 + 4i$. Therefore, the solution set is $\{(5/2)i, -1 + 4i\}$.

(b) In a problem of this type, first try to factor the expression and then apply Equation (10). Notice that $z^2 + (4 - 3i)z + 12i = (z - 3i)(z + 4) = 0$. So $z - 3i = 0$ or $z + 4 = 0$. Therefore, $z = 3i$ or $z = -4$ and the solution set is $\{-4, 3i\}$.

Problem S9. Simplify (a) $(-1 - i)^8$ and (b) i^{258}.

Solution: (a) To solve this problem, we could multiply $(-1 - i)$ by itself eight times — not a very pleasant task! However, we can apply the useful result that $(r \operatorname{cis} \theta)^n = r^n \operatorname{cis} n\theta$ [see Equation (8)]. Use of this result requires that $-1 - i$ be expressed in trigonometric form. Plotting $-1 - i$ (Figure 5.5.19), we see that $-1 - i = \sqrt{2} \operatorname{cis} \frac{5}{4}\pi$. Using Equation (8), we have

$$(-1 - i)^8 = \left(\sqrt{2} \operatorname{cis} \frac{5\pi}{4}\right)^8 = (\sqrt{2})^8 \operatorname{cis} \left(8 \cdot \frac{5\pi}{4}\right) = 2^4 \operatorname{cis} 10\pi$$
$$= 16(\cos 10\pi + i \sin 10\pi) = 16(1 + i \cdot 0) = 16$$

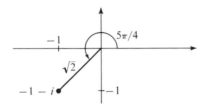

Figure 5.5.19

Note: The result $(-1 - i)^8 = 16$ means that $-1 - i$ is one solution to the equation $z^8 - 16 = 0$. You will see in the next section that there are seven other solutions to $z^8 - 16 = 0$.

(b) First write i in trigonometric form. Clearly, $i = \operatorname{cis} \frac{1}{2}\pi$. Thus,

$$i^{258} = \left(\operatorname{cis} \frac{\pi}{2}\right)^{258} = 1^{258} \operatorname{cis}\left(258 \cdot \frac{\pi}{2}\right) = \operatorname{cis} 129\pi = -1 + i \cdot 0 = -1$$

See Figure 5.5.20.

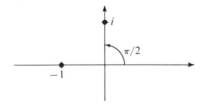

Figure 5.5.20

Problem S10. Simplify (a) $(\sqrt{3} + i)^6$ and (b) $\dfrac{(-1 - i)^8}{(\sqrt{3} + i)^6}$.

Solution: (a) Here again we will use Equation (8). First we must write $\sqrt{3} + i$ in trigonometric form. Then we want to find r and θ, so that $\sqrt{3} + i = r$ cis θ. Since $a = \sqrt{3}$ and $b = 1$, then $r = \sqrt{(\sqrt{3})^2 + 1^2} = 2$. Next, plot the point $\sqrt{3} + i = (\sqrt{3}, 1)$, as shown in Figure 5.5.21. Clearly, $\theta = \frac{1}{6}\pi$. Therefore, $\sqrt{3} + i = 2$ cis $\frac{1}{6}\pi$. So using Equation (8), we have

$$(\sqrt{3} + i)^6 = \left(2 \text{ cis } \frac{\pi}{6}\right)^6 = 2^6 \text{cis } \left(6 \cdot \frac{\pi}{6}\right) = 64 \text{ cis } \pi$$
$$= 64(\cos \pi + i \sin \pi) = 64(-1 + i \cdot 0) = -64$$

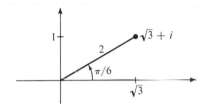

Figure 5.5.21

(b) $(-1 - i)^8 = 16$ [see Problem S9 (a)] and $(\sqrt{3} + i)^6 = -64$. [See Part (a) above] Thus,

$$\frac{(-1 - i)^8}{(\sqrt{3} + i)^6} = \frac{16}{-64} = -\frac{1}{4}$$

SUPPLEMENTARY PROBLEMS

S11. Write each of the following complex numbers in rectangular form and plot as points in the complex plane:
(a) 5 cis 12π (b) 4 cis $\frac{5}{8}\pi$

S12. Plot each of the following complex numbers as points in the complex plane and write in trigonometric form:
(a) $5 - 5i$ (b) $2 - 2i\sqrt{3}$

S13. Simplify the following (express the results in rectangular form):
(a) $\dfrac{8 \text{ cis } \frac{3}{2}\pi}{2 \text{ cis } \frac{2}{3}\pi}$ (b) $(6 \text{ cis } \frac{7}{6}\pi)(3 \text{ cis } \frac{2}{3}\pi)$

S14. Find $|z|$ for each of the following:
(a) $z = -3 - 5i$ (b) $z = (-3 - 5i)(2 - 6i)$

S15. Find the solution set in C for $z^2 + (5i - 3)z - 15i = 0$.

S16. Simplify
(a) $(1 - i)^8$ (b) i^{281}
(c) $(2 - 2i\sqrt{3})^{10}$

Answers

S11. (a) 5, (b) $-2\sqrt{3} + 2i$

S12. (a) $5\sqrt{2}$ cis$(-\tfrac{1}{4}\pi)$, (b) 4 cis$(-\tfrac{1}{3}\pi)$

S13. (a) $-2\sqrt{3} + 2i$, (b) $9\sqrt{3} - 9i$

S14. (a) $\sqrt{34}$, (b) $2\sqrt{340}$

S15. $\{3, -5i\}$

S16. (a) 16, (b) i, (c) 4^{10}cis $\dfrac{2\pi}{3}$

5.6 ROOTS OF COMPLEX NUMBERS

In the previous section, we discovered the ease of multiplying complex numbers which are expressed in trigonometric form. Recall that $(r \text{ cis } \theta) \cdot (s \text{ cis } \varphi) = rs \text{ cis}(\theta + \varphi)$. Let us now exploit this property and discover how to find all roots of any complex number. By *root*, we mean the following.

Definition 5.6.1

If $u^n = z$ where $n \in N$, then u is called an *nth root* of z.

Q1: Show that $1 + i$ is a 4th root of -4.

Let us begin with a specific example.

Example 5.6.1

Find all the 4th roots of -16, that is, find $\{u : u^4 = -16\}$.

Solution

We shall work with trigonometric form. Let $u = s \text{ cis } \varphi$, so $u^4 = s^4 \text{ cis } 4\varphi$ by Equation (8) in Section 5.5. Since $-16 = 16 \text{ cis } \pi$, the equation we wish to solve becomes

$$s^4 \text{cis } 4\varphi = 16 \text{ cis } \pi \qquad (1)$$

These two complex numbers are equal if, and only if, $s^4 = 16$ and $4\varphi = \pi + 2k\pi$ for some integer k, that is, if $s = 16^{1/4} = 2$

447

$$A1: (1 + i)^4 = (\sqrt{2}\ \text{cis}\ \tfrac{1}{4}\pi)^4 = (\sqrt{2})^4 \text{cis}\ \pi = 4\ \text{cis}\ \pi = -4$$

(remember, s is real and positive) and $\varphi = (\pi/4) + 2k\pi/4$, $k \in Z$. Notice that different integers k can yield different arguments φ and hence different solutions $u = s\ \text{cis}\ \varphi$. Let us index u and φ as u_k and φ_k and write some of the solutions:

$$u_0 = 2\ \text{cis}\ \varphi_0 = 2\ \text{cis}\ \frac{\pi}{4} = \sqrt{2} + i\sqrt{2} \qquad\qquad (k = 0)$$

$$u_1 = 2\ \text{cis}\ \varphi_1 = 2\ \text{cis}\left(\frac{\pi}{4} + \frac{\pi}{2}\right) = 2\ \text{cis}\ \frac{3\pi}{4} = -\sqrt{2} + i\sqrt{2} \qquad (k = 1)$$

$$u_2 = 2\ \text{cis}\ \varphi_2 = 2\ \text{cis}\left(\frac{\pi}{4} + 2\frac{\pi}{2}\right) = 2\ \text{cis}\ \frac{5\pi}{4} = -\sqrt{2} - i\sqrt{2} \quad (k = 2)$$

$$u_3 = 2\ \text{cis}\ \varphi_3 = 2\ \text{cis}\left(\frac{\pi}{4} + 3\frac{\pi}{2}\right) = 2\ \text{cis}\ \frac{7\pi}{4} = \sqrt{2} - i\sqrt{2} \qquad (k = 3)$$

$$u_4 = 2\ \text{cis}\ \varphi_4 = 2\ \text{cis}\left(\frac{\pi}{4} + 4\frac{\pi}{2}\right) = 2\ \text{cis}\ \frac{9\pi}{4} = \sqrt{2} + i\sqrt{2} \qquad (k = 4)$$

$$\cdot$$
$$\cdot$$
$$\cdot$$

Notice that in this example the roots u_k repeat in a cycle of 4, that is, $u_4 = u_0$, $u_5 = u_1$, etc. Thus, we have found four distinct 4th roots of -16. Furthermore, by examining the trigonometric form of these roots and plotting them in the complex plane (Figure 5.6.1), we observe that they all lie *evenly spaced along a circle* of radius 2.

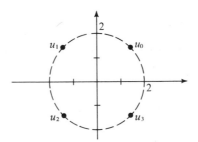

Figure 5.6.1

Example 5.6.1 illustrates the procedure of finding all nth roots of a complex number $z = r\ \text{cis}\ \theta \neq 0$. We let $u = s\ \text{cis}\ \varphi$ so that $u^n = s^n\ \text{cis}\ n\varphi$. Then, $u^n = z$ becomes $s^n\ \text{cis}\ n\varphi = r\ \text{cis}\ \theta$. This last equation is true if, and only if, both $s^n = r$ and $n\varphi = \theta + k \cdot 2\pi$ for $k \in Z$. That is, if, and only if, $s = r^{1/n}$ (since s and r are positive reals) and $\varphi_k = (\theta/n) + k \cdot 2\pi/n$ for $k \in Z$ (we have again indexed the arguments with k). Thus, when all nth roots $u_k = s\ \text{cis}\ \varphi_k$ are plotted in the complex plane, they lie on a circle of radius $s = r^{1/n}$. Furthermore, there are precisely n distinct points on this circle. To verify this last statement, notice that $\varphi_0, \varphi_1, \varphi_2, \ldots, \varphi_{n-1}$ yield n distinct points (since no two of these differ by an integer multiple of 2π), yet any other integer k yields a φ_k which differs by an integral multiple of 2π from one of these above (that is, $\varphi_n = \varphi_0 + 2\pi$, $\varphi_{n+1} = \varphi_1 + 2\pi$, $\varphi_{2n+1} = \varphi_1 + 2 \cdot 2\pi$, etc.). Example 5.6.1 illustrates this concept. For $n = 4$, $\varphi_4 = \varphi_0 + 2\pi$ and hence $u_4 = u_0$. In a similar fashion we can see $u_5 = u_1$, $u_6 = u_2$, etc. Finally, notice that one of the n nth roots of $z = r$ cis θ has argument $\varphi_0 = \theta/n$ and the remaining $n - 1$ roots are evenly spaced along the circle of radius $s = r^{1/n}$.

Example 5.6.2

Describe the location of all the cube roots of $8i$.

Solution

Since $8i = 8$ cis $\frac{1}{2}\pi$, the three cube roots are evenly spaced [i.e., $\frac{1}{3}(360°) = 120°$ apart] along a circle of radius $8^{1/3} = 2$, with one root having argument $\frac{1}{3} \cdot \frac{1}{2}\pi = \frac{1}{6}\pi$ or $30°$. The roots, shown in Figure 5.6.2, are 2 cis $30°$, 2 cis $150°$, and 2 cis $270°$, or $\sqrt{3} + i$, $-\sqrt{3} + i$, and $-2i$.

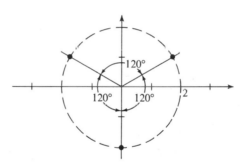

Figure 5.6.2

Q2: Locate the three cube roots of $-8i$.

In the preceding discussion we have essentially proved the following theorem.

Theorem 5.6.1

The complex number $z = r$ cis $\theta \neq 0$ has precisely n distinct nth roots, $u_k = s$ cis φ_k, where $s = r^{1/n}$ and $\varphi_k = (\theta/n) + k \cdot 2\pi/n$, $k = 0$, $1, \ldots, n - 1$.

Example 5.6.3

Find $\{u : u^5 = -1 + i\sqrt{3}\}$.

A2: Since $-8i = 8$ cis $270°$, the three roots are $8^{1/3}$cis$(270/3)° =$ 2 cis $90°$, 2 cis$(90 + 120)° = 2$ cis $210°$, and 2 cis $330°$. See the following figure.

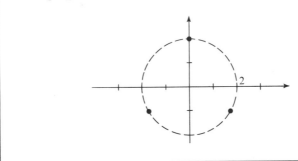

Solution

Using trigonometric form, we have $z = -1 + \sqrt{3}i = 2$ cis $120°$. Thus, the five 5th roots of z lie evenly spaced, or $(360/5)° = 72°$ apart, along a circle of radius $2^{1/5}$. One root has argument $(120/5)°$ or $24°$. Thus, we can write the roots in trigonometric form as

$$\{2^{1/5}\text{cis } 24°, 2^{1/5}\text{cis } 96°, 2^{1/5}\text{cis } 168°, 2^{1/5}\text{cis } 240°, 2^{1/5}\text{cis } 312°\}$$

or simply

$$\{2^{1/5}\text{cis}(24 + k \cdot 72)° : k = 0, 1, 2, 3, 4\}$$

PROBLEMS 5.6

1. Compute each of the following:

(a) $(1 - i)^8$ (b) $(1 + i)^8$
(c) $(-1 + i)^8$ (d) $(-1 - i)^8$
(e) $(\sqrt{2} \text{ cis } \frac{1}{2}\pi)^8$ (f) $(\sqrt{2} \text{ cis } 270°)^8$
(g) $(-\sqrt{2})^8$ (h) $(\sqrt{2})^8$

2. Find all cube roots of z for each of the following. Express your answers in rectangular form: (a) 1, (b) -8, and (c) $-64i$.

3. Find all cube roots of z for each of the following. Express your answers in trigonometric form:
(a) 6 cis $120°$ (b) $-1 + i$
(c) $2\sqrt{3} - 2i$

4. Find the solution set for each of the following. Express your answer in rectangular form:
(a) $u^6 = 1$ (b) $z^4 - 81 = 0$
(c) $z^8 = 16$ (d) $z^5 + 4z = 0$
(e) Compare Problem 4(c) with Problem 1.

5. Find the solution set to each of the following. Express your answer in trigonometric form.
(a) $u^5 = 10$ cis $100°$ (b) $z^5 - i = 0$
(c) $z^6 + 4z = 0$

ANSWERS
5.6

1. (a)–(h), 16

2. (a) $\left\{ 1, -\dfrac{1}{2} \pm \dfrac{\sqrt{3}}{2}i \right\}$, (b) $\{-2, 1 \pm \sqrt{3}i\}$, (c) $\{4i, \pm 2\sqrt{3} - 2i\}$

3. (a) $\{6^{1/3}\text{cis } 40°, 6^{1/3}\text{cis } 160°, 6^{1/3}\text{cis } 280°\}$
(b) $\{2^{1/6}\text{cis } 45°, 2^{1/6}\text{cis } 165°, 2^{1/6}\text{cis } 285°\}$
(c) $\{2^{2/3}\text{cis}(-10)°, 2^{2/3}\text{cis } 110°, 2^{2/3}\text{cis } 230°\}$

4. (a) $\left\{ \pm\left(\dfrac{1}{2} \pm \dfrac{\sqrt{3}}{2}\right)i, \pm 1 \right\}$

(b) $\{\pm 3, \pm 3i\}$
(c) $\{\pm(1 \pm i), \pm\sqrt{2}, \pm\sqrt{2}i\}$
(d) $\{0, \pm(1 \pm i)\}$

5. (a) $\{10^{1/5}\text{cis } 20°, 10^{1/5}\text{cis } 92°, 10^{1/5}\text{cis } 164°, 10^{1/5}\text{cis } 236°, 10^{1/5}\text{cis } 308°\}$
(b) $\{\text{cis } 18°, \text{cis } 90°, \text{cis } 162°, \text{cis } 234°, \text{cis } 306°\}$
(c) $\{0, 2^{2/5}\text{cis}(36 + k\cdot 72)° : k = 0, 1, 2, 3, 4\}$

Supplement 5.6

Problem S1. Find all the square roots of $2\sqrt{2} \quad 2\sqrt{2}i$.

Solution: In all problems of this type, you should work with the given complex number in trigonometric form. Thus, we want to find r and θ so that $2\sqrt{2} \quad 2\sqrt{2}i = r$ cis θ. Therefore,

$$r = \sqrt{(2\sqrt{2})^2 + (-2\sqrt{2})^2} = 4$$

Next, plot the point representing the complex number $2\sqrt{2} - 2\sqrt{2}i$. (See Figure 5.6.3.) Clearly, $\theta = 7\pi/4$ (or $-\pi/4$), so $2\sqrt{2} - 2\sqrt{2}i = 4 \text{ cis}(-\pi/4)$. Next, recognize that this problem is equivalent to finding all complex solutions to $z^2 = 2\sqrt{2} - 2\sqrt{2}i$. Suppose that $u = s \text{ cis } \varphi$ is such a solution. Then $u^2 = (s \text{ cis } \varphi)^2 = 2\sqrt{2} - 2\sqrt{2}i = 4 \text{ cis}(-\pi/4)$.

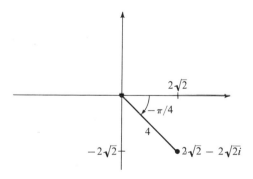

Figure 5.6.3

Now apply the result of Equation (8), Section 5.5, to obtain $(s \text{ cis } \varphi)^2 = s^2 \text{cis } 2\varphi = 4 \text{ cis}(-\pi/4)$. Remember that the two complex numbers are equal precisely when both of their absolute values are equal and both of their arguments are equal, *up to a multiple of* 2π. So $s^2 = 4$ and $2\varphi = (-\pi/4) + 2k\pi$, $k \in Z$. It is important to realize that $2\varphi = -\pi/4$ is *not* the only solution to cis $2\varphi = \text{cis}(-\pi/4)$. If necessary, you can refer to Equation (5), Section 5.5, for an explanation. Therefore, since $s^2 = 4$ and $s > 0$, $s = 2$. By dividing both sides of $2\varphi = (-\pi/4) + 2k\pi$, $k \in Z$, by 2 we obtain $\varphi = (-\pi/8) + k\pi$, $k \in Z$. Thus, the two square roots of $2\sqrt{2} - 2\sqrt{2}i$ are $u_0 = 2 \text{ cis}(-\pi/8)$ and $u_1 = 2 \text{ cis}(7\pi/8)$. These values are determined when $k = 0$ and $k = 1$. Notice that for any other $k \in Z$, 2 cis φ coincides with either $2 \text{ cis}(-\pi/8)$ or $2 \text{ cis}(7\pi/8)$.

We can also obtain these results directly by applying Theorem 5.6.1. Also observe that the two square roots of $2\sqrt{2} - 2\sqrt{2}i$ are *equally spaced* around the circle of radius 2 centered at the origin, as shown in Figure 5.6.4.

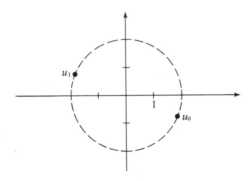

Figure 5.6.4

Problem S2.   Find all cube roots of 64.

Solution: First, write 64 in trigonometric form: $64 = 64 \text{ cis } 0$. Next, recognize that we want all complex solutions to the equation $z^3 = 64 = 64 \text{ cis } 0$. Let $u = r \text{ cis } \theta$ be a solution to the last equation. Then $u^3 = (r \text{ cis } \theta)^3 = 64 \text{ cis } 0$. Now apply Equation (8), Section 5.5, to obtain $u^3 = r^3 \text{cis } 3\theta = 64 \text{ cis } 0$. Next by Equation (5), Section 5.5, it follows that $r^3 = 64$ and $3\theta = 0 + 2k\pi$, $k \in Z$. So $r = \sqrt[3]{64} = 4$ and $\theta = (0/3) + (2k\pi/3) = (2k\pi/3)$, $k \in Z$. Thus, $u_k = 4 \text{ cis}(2k\pi/3)$, $k \in Z$ will generate all the solutions to $z^3 = 64$. Therefore, $u_1 = 4 \text{ cis}(2\pi/3)$, $u_2 = 4 \text{ cis}(4\pi/3)$, and $u_3 = 4 \text{ cis } 2\pi$ are the solutions to $z^3 = 64$, the three cube roots of 64. We find these solutions by evaluating $u_k = 4 \text{ cis}(2k\pi/3)$ for $k = 1$, 2, and 3. Note that $u_k = 4 \text{ cis}(2k\pi/3)$ for other values of $k \in Z$ which coincide with either u_1, u_2, or u_3.

Also observe that the solutions lie equally spaced along a circle of radius 4 centered at the origin (Figure 5.6.5). In rectangular form, the solutions are

$$u_1 = 4\left(\cos\frac{2\pi}{3} + i\sin\frac{2\pi}{3}\right) = 4\left(-\frac{1}{2} + \frac{\sqrt{3}}{2}i\right)$$

$$u_2 = 4\left(\cos\frac{4\pi}{3} + i\sin\frac{4\pi}{3}\right) = 4\left(-\frac{1}{2} - \frac{\sqrt{3}}{2}i\right)$$

$$u_3 = 4$$

Thus, the solution set in rectangular form is

$$\{-2 + 2\sqrt{3}i, -2 - 2\sqrt{3}i, 4\}$$

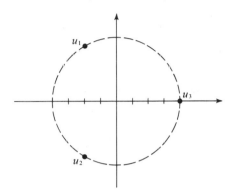

Figure 5.6.5

Problem S3. (a) Find all 4th roots of $-2 + 2i$. (b) Find the solution set in C for $z^4 + 2 - 2i = 0$.

Solution: (a) First express $-2 + 2i$ in trigonometric form. Thus, $-2 + 2i = r \text{ cis } \theta$ where $r = \sqrt{(-2)^2 + 2^2} = \sqrt{8} = 2\sqrt{2}$. Next, plot the complex number $-2 + 2i = (-2,2)$. Clearly, $\theta = 3\pi/4$ (Figure 5.6.6) and $-2 + 2i = 2\sqrt{2} \text{ cis}(3\pi/4)$.

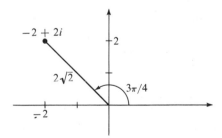

Figure 5.6.6

Now suppose that u is a 4th root of $-2 + 2i = 2\sqrt{2} \text{ cis}(3\pi/4)$. Then $u^4 = r - 2 + 2i = 2\sqrt{2} \text{ cis}(3\pi/4)$. Let $u = s \text{ cis } \varphi$, so $u^4 = (s \text{ cis } \varphi)^4 = 2\sqrt{2} \text{ cis}(3\pi/4)$. Applying Equation (8) Section 5, we obtain $u^4 = s^4 \text{cis } 4\varphi = 2\sqrt{2} \text{ cis}(3\pi/4)$. Thus by Equation (5), Section 5, $s^4 = 2\sqrt{2}$ or $s = (2\sqrt{2})^{1/4} = 2^{3/8}$ and $4\varphi = (3\pi/4) + 2k\pi$, $k \in Z$ or dividing by 4, $\varphi = (3\pi/16) + (\pi/2)k$, $k \in Z$. Therefore,

$$u_0 = 2^{3/8}\text{cis}\,\frac{3\pi}{16}$$

$$u_1 = 2^{3/8}\text{cis}\left(\frac{3\pi}{16} + \frac{\pi}{2}\cdot 1\right) = 2^{3/8}\text{cis}\,\frac{11\pi}{16}$$

$$u_2 = 2^{3/8}\text{cis}\left(\frac{3\pi}{16} + \frac{\pi}{2}\cdot 2\right) = 2^{3/8}\text{cis}\,\frac{19\pi}{16}$$

$$u_3 = 2^{3/8}\text{cis}\left(\frac{3\pi}{16} + \frac{\pi}{2}\cdot 3\right) = 2^{3/8}\text{cis}\,\frac{27\pi}{16}$$

Observe that $u_4 = 2^{3/8}\text{cis}\left(\frac{3\pi}{16} + \frac{\pi}{2}\cdot 4\right) = 2^{3/8}\text{cis}\left(\frac{3\pi}{16} + 2\pi\right) = u_0,\ u_5 = u_1$, etc. Thus, $u_k = 2^{3/8}\text{cis}\left(\frac{3\pi}{16} + \frac{\pi}{2}k\right)$, where $k = 0, 1, 2,$ and 3, generate the four 4th roots of $-2 + 2i$.

Again, notice that $u_0, u_1, u_2,$ and u_3 lie equally spaced along a circle of radius $2^{3/8}$ (Figure 5.6.7).

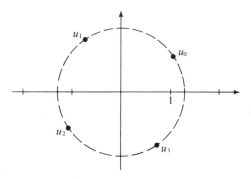

Figure 5.6.7

(b) The solution set to $z^4 + 2 - 2i = 0$ or $z^4 = -2 + 2i$ is precisely the four fourth roots $-2 + 2i$, because if u is a 4th root of $-2 + 2i$, then $u^4 = -2 + 2i$. So u is a solution to $z^4 = -2 + 2i$. Thus, the solution set is given in Part (a).

Problem S4.   Find the solution set in C for $z^6 = 2i$.

Solution: The solutions to the equation $z^6 = 2i$ are exactly the 6th roots of $2i$. In trigonometric form, $2i = 2\,\text{cis}(90°)$ (see Figure 5.6.8). In this example, we will apply Theorem 5.6.1 directly. (You should refer to it now.)

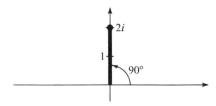

Figure 5.6.8

Thus, there are six 6th roots of $2i = 2$ cis 90°, and they are located 360°/6 = 60° apart on a circle of radius $2^{1/6}$ (Figure 5.6.9). Furthermore, one root is

$$u = 2^{1/6}\text{cis}\ \frac{90°}{6} = 2^{1/6}\text{cis}\ 15°$$

Hence, we can find the others by adding multiples of 60° to the argument of u. Therefore, the solution set is

$$\{2^{1/6}\text{cis}\ 15°,\ 2^{1/6}\text{cis}\ 75°,\ 2^{1/6}\text{cis}\ 135°,\ 2^{1/6}\text{cis}\ 195°,\ 2^{1/6}\text{cis}\ 255°,\ 2^{1/6}\text{cis}\ 315°\}$$

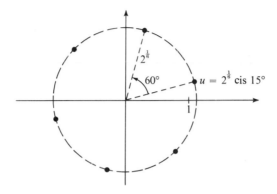

Figure 5.6.9

Problem S5.   Find z and the three cube roots of z if one cube root of z is $1 + \sqrt{3}i$.

Solution: First, express $1 + \sqrt{3}i$ in trigonometric form:

$$1 + \sqrt{3}i = r\ \text{cis}\ \theta \quad \text{where } r = \sqrt{1^2 + (\sqrt{3})^2} = \sqrt{1 + 3} = 2$$

Next, plot the complex number $1 + \sqrt{3}i = (1, \sqrt{3})$. Clearly, $\theta = 60°$, so $1 + \sqrt{3}i = 2$ cis 60° (Figure 5.6.10).

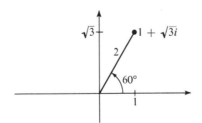

Figure 5.6.10

Let $u_1 = 2$ cis 60°. Thus, u_1 is one cube root of z. Recall that the other two cube roots of z lie *equally* spaced along a circle of radius 2. Thus, the three cube roots are 360°/3 = 120° apart (Figure 5.6.11). Now it is easy to find the other two cube roots. They are

$$2 \text{ cis}(60° + 120°) = 2 \text{ cis } 180° = -2$$
$$2 \text{ cis}(180° + 120°) = 2 \text{ cis } 300° = 1 - \sqrt{3}i$$

To find z, note that $z = u_1^3 = (2 \text{ cis } 60°)^3 = 2^3 \text{cis} (3 \cdot 60°) = 8 \text{ cis } 180° = -8.$

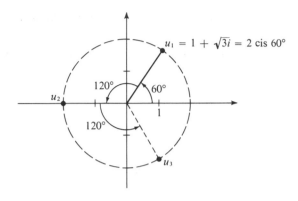

Figure 5.6.11

Problem S6. Given that $-2\sqrt{3} + 2i$ is one 5th root of z, find z, in rectangular form, and the other 5th roots of z.

Solution: First express $-2\sqrt{3} + 2i$ in trigonometric form:

$$-2\sqrt{3} + 2i = r \text{ cis } \theta \quad \text{where } r = \sqrt{(-2\sqrt{3})^2 + 2^2} = \sqrt{12 + 4} = 4$$

Plot the complex number $-2\sqrt{3} + 2i = (-2\sqrt{3}, 2)$. Clearly, $\theta = 150°$ (Figure 5.6.12). Thus, $-2\sqrt{3} + 2i = 4 \text{ cis } 150°$.

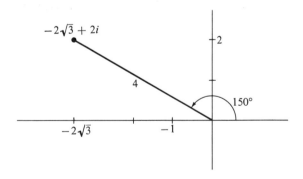

Figure 5.6.12

Now, recall that the other 5th roots of z lie $360°/5 = 72°$ apart along a circle of radius 4 (Figure 5.6.13). So to find the other fifth roots, simply add multiples of $72°$ to the argument of $u = 4 \text{ cis } 150°$.

Thus, the others are

$$4\operatorname{cis}(150° + 72°) = 4 \operatorname{cis} 222°$$
$$4\operatorname{cis}(222° + 72°) = 4 \operatorname{cis} 294°$$
$$4\operatorname{cis}(294° + 72°) = 4 \operatorname{cis} 366° = 4 \operatorname{cis} 6°$$
$$4\operatorname{cis}(6° + 72°) = 4 \operatorname{cis} 78°$$

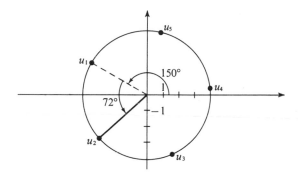

Figure 5.6.13

Finally, to find z, apply Equation (8), Section 5.5:
$$z = u^5 = (4 \operatorname{cis} 150°)^5 = 4^5\operatorname{cis}(5 \cdot 150°) = 4^5\operatorname{cis}(750°) = 4^5\operatorname{cis} 30°$$
Therefore, in rectangular form,

$$z = 4^5\cos 30° + i4^5 \sin 30° = 512\sqrt{3} + 512i$$

SUPPLEMENTARY PROBLEMS

S7. Find all fourth roots of $-81i$.
S8. Find the solution set in C for $z^4 - 3 + 3i = 0$.
S9. Given that $-4 + 4i$ is a cube root of z, then find z, in rectangular form, and the other two cube roots of z.

Answers

S7. $3 \operatorname{cis} \dfrac{3\pi}{8}$, $3 \operatorname{cis} \dfrac{7\pi}{8}$, $3 \operatorname{cis} \dfrac{11\pi}{8}$, and $3 \operatorname{cis} \dfrac{15\pi}{8}$

S8. $\left\{ (3\sqrt{2})^{1/4} \operatorname{cis}\left(-\dfrac{\pi}{16}\right), (3\sqrt{2})^{1/4} \operatorname{cis} \dfrac{7\pi}{16}, (3\sqrt{2})^{1/4} \operatorname{cis} \dfrac{15\pi}{16}, (3\sqrt{2})^{1/4} \operatorname{cis} \dfrac{23\pi}{16} \right\}$

S9. $\{4\sqrt{2} \operatorname{cis} 135°, 4\sqrt{2} \operatorname{cis} 255°, 4\sqrt{2} \operatorname{cis} 15°\}$, $z = (4\sqrt{2})^3 \operatorname{cis} 45° = 128 + 128i$

Achievement Test 5

1. Sketch the graph $y = 2 \cos(\pi x - 1)$.
2. Sketch the graph $y = \sin(\mathrm{Sin}^{-1}x)$.
3. Sketch the graph $y = \cos^2 x - \sin^2 x$.
4. Find the amplitude, period, and phase shift of $f(x) = 2 \cos(\pi x - 1)$.
5. Find the amplitude, period, and phase shift of $f(x) = 2 \sin x \cdot \cos x$.
6. Find the solution set for $2 \sin 3t = 1$.
7. Find the solution set for $2 \sin 3t < 1$.
8. Find the solution set for $\sin(\pi t - 2) = 1$.
9. Find the solution set for $\sin(\pi t - 1) = 2$.
10. Find the solution set for $\sin 2t = \sin t$.
11. Find the solution set for $\cos t = \sin(\pi - 3t)$.
12. Find c if $a = 20$, $b = 12$, and $\gamma = 30°$ in Figure 5.A.1.

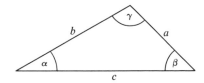

Figure 5.A.1

458

13. Find c if $a = 20$, $\beta = 75°$, and $\gamma = 60°$.

14. Find c if $a = \sqrt{7}$, $b = 2$, and $\alpha = 30°$.

15. Find c if $a = 3$, $b = 4$ and $\alpha = 60°$.

16. Simplify $2 - i(2 + i)$.

17. Simplify $\dfrac{3 - 5i}{i + 4}$.

18. Simplify $(3 + 5i)(2 - 4i)$.

19. Simplify $i^{123} - i^{456}$.

20. Simplify $(1 + i)^{10}$.

21. Simplify $(\sqrt{3} + i)^{20}$.

22. Write $2 - 2i$ in trigonometric form.

23. Write $\sqrt{12} - 2i$ in trigonometric form.

24. Write $(2 - 2i)^8$ in trigonometric form.

25. Write $(\sqrt{3} + i)(2 - 2i)$ in trigonometric form.

26. Which of the following are true for all complex numbers u and v:
(a) $|u + v| = |u| + |v|$ (b) $\overline{u} = \bar{u}$
(c) $|u\bar{u}| = u\bar{u}$ (d) $u\bar{v} = \bar{u}v$
(e) $u + \bar{u} = |u + \bar{u}|$

27. Find the complex number solution set for $z + 2z = 3 + 5i$.

28. Find the complex number set for $z - 2iz = 4 - 5i$.

29. Find the complex number solution set for $|z| + z = 1 + 3i$.

30. Find the complex number solution set for $|z| \cdot \bar{z} = 15 + 20i$.

Answers to Achievement Test 5

1. See Figure 5.A.2.

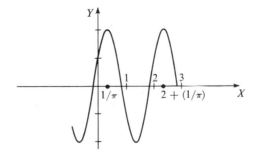

Figure 5.A.2

2. See Figure 5.A.3.

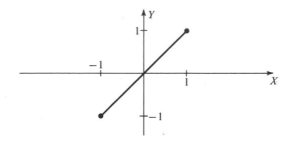

Figure 5.A.3

3. See Figure 5.A.4.   **4.** 2, 2, $1/\pi$

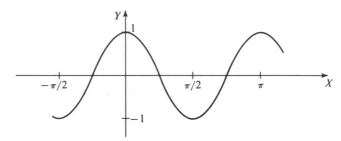

Figure 5.A.4

5. 1, π, 0

6. $\left\{\dfrac{\pi}{18} + \dfrac{2k\pi}{3}\right\} \cup \left\{\dfrac{5\pi}{18} + \dfrac{2k\pi}{3}\right\}$

7. $\left\{t : t \notin \left[\dfrac{\pi}{18} + \dfrac{2k\pi}{3}, \dfrac{5\pi}{18} + \dfrac{2k\pi}{3}\right]\right\}$, or we might write $\displaystyle\bigcup_{k \in Z} \left(\dfrac{5\pi}{18} + \dfrac{2k\pi}{3}, \dfrac{13\pi}{18} + \dfrac{2k\pi}{3}\right)$

8. $\left\{\dfrac{2}{\pi} + \dfrac{1}{2} + 2k\right\}$

9. $\varnothing$

10. $\{k\pi\} \cup \left\{\pm\dfrac{\pi}{3} + 2k\pi\right\}$

11. $\left\{\dfrac{\pi}{4} + k\pi\right\} \cup \left\{\dfrac{\pi}{8} + \dfrac{k\pi}{2}\right\}$

12. $\sqrt{544 - 240\sqrt{3}}$

13. $10\sqrt{6}$

14. If $\alpha = 60°$, $c = 3$. If $\alpha = 30°$, $c = \sqrt{3}(1 + \sqrt{2})$.

15. Impossible

16. $3 - 2i$

17. $\dfrac{7}{17} - \dfrac{23}{17}i$

18. $-14 - 22i$

19. $-1 - i$

20. $32i$

21. $-2^{19}(1 + \sqrt{3}i)$

22. $2\sqrt{2}$ cis $(-\pi/4)$

23. 4 cis $(-\pi/6)$

24. 2^{12}cis 0

25. $4\sqrt{2}$ cis$(-\pi/12)$

26. Only (c) is true.

27. $1 - 5i$

28. $2 - i$

29. $-4 + 3i$

30. $3 - 4i$

Appendix A

In this appendix we shall present the axioms used to describe the real number system and develop some familiar properties from them. Mathematicians describe the real number system as a *complete ordered field*. Let us see what each of these italicized words means. We begin with *field*.

A set S of objects (usually called numbers) together with two operations (usually called addition and multiplication, denoted $+$ and $\cdot$) is called a field if each of the following axioms is satisfied.

The Field Axioms

For any $x, y, z \in S$:

1A (closure): $x + y \in S$

1M (closure): $x \cdot y \in S$

2A (commutative): $x + y = y + x$

2M (commutative): $x \cdot y = y \cdot x$

3A (associative: $(x + y) + z = x + (y + z)$

3M (associative): $(x \cdot y) \cdot z = x \cdot (y \cdot z)$

4A (identity): There exists a number 0 (called the additive identity) such that $x + 0 = x$.

4M (identity): There exists a number 1 (called the multiplicative identity) such that $x \cdot 1 = x$.

5A (inverse): For any x, there exists a number $-x$ (called the additive inverse) such that $x + (-x) = 0$.

461

5M (inverse): For any $x \neq 0$, there exists a number x^{-1}
(called the multiplicative inverse) such that
$x \cdot x^{-1} = 1$.
6 (distributive): $x \cdot (y + z) = (x \cdot y) + (x \cdot z)$

Any set which satisfies the above list of axioms is called a *field*. Certain subsets of the real numbers do not satisfy all of the axioms. For example, N, the set of natural numbers, satisfies 1 through 3. However, N does not contain an additive identity (i.e., 0), and thus axiom 4A is not satisfied. The set $T = \{0, 1, 2, 3, \ldots\}$ overcomes this defect, but now the requirement of axiom 5A is not met since T does not contain an additive inverse for any element except 0.

Q1: Which of the field axioms are not satisfied by the set Z? By Q?

Next, notice that the field axioms explicitly involve the operations of addition and multiplication. We can define subtraction and division using the inverse axioms.

Definition A.1
The number $a - b$, called the difference of a and b, is defined by the equation $a - b = a + (-b)$.

Definition A.2
The number a/b $\left(\text{also written } \dfrac{a}{b}\right)$, called the quotient of a and b, is defined by the equation $a/b = a \cdot b^{-1}$.

Using only the field axioms, we can prove many elementary but important properties rather easily.

Theorem A.1
For any $a, b \in R$,
(1) $-(-a) = a$,
(2) $a \cdot 0 = 0$, and
(3) $a \cdot b = 0$
implies $a = 0$ or $b = 0$.

Proof of (1)

$(-a) + a = 0$	Identity axiom
$(-a) + (-(-a)) = 0$	Existence of inverse for $-a$, namely, $-(-a)$
$(-a) + a = (-a) + (-(-a))$	Combining above two equations
$(a + (-a)) + a = (a + (-a)) + (-(-a))$	Adding a to both sides and the associative axiom

$$0 + a = 0 + (-(-a)) \qquad \text{Inverse axiom}$$
$$a = -(-a) \qquad\qquad\qquad \text{Identity axiom}$$

Proof of (2)

$$a + 0 = a = a \cdot 1 = a \cdot (1 + 0) \qquad \text{Identity axioms}$$
$$a \cdot (1 + 0) = a \cdot 1 + a \cdot 0 = a + a \cdot 0 \qquad \text{Distributive and identity axioms}$$

Thus,

$$a + 0 = a + a \cdot 0 \qquad \text{From the above two equations}$$

so

$$0 = a \cdot 0 \qquad \text{Adding } -a \text{ to both sides and simplifying}$$

Proof of (3)

If $a = 0$ we have finished, so assume that $a \neq 0$.

Then a^{-1} exists. \qquad Inverse axiom

$$a^{-1}(a \cdot b) = a^{-1} \cdot 0 \qquad \text{Multiplying and using property of equality}$$

$$(a^{-1}a)b = a^{-1} \cdot 0 = 0 \qquad \text{Associative axiom and Part (2)}$$

$$1 \cdot b = b = 0 \qquad \text{Inverse and identity axioms}$$

As we have seen in the answer to Q1, the set of rational numbers is a field. Apparently we need more axioms to distinguish between the rational number system and the real number system. However, you should first realize that there are indeed numbers which are not rational. $\sqrt{2}$ is such a number.

To prove this statement, suppose that $\sqrt{2} \in Q$. Then $\sqrt{2} = a/b$ where a and b are integers and relatively prime. (This means that the fraction a/b is "reduced to lowest terms.")

Then $2 = a^2/b^2$ or $2b^2 = a^2$. This means that a^2 is an even number, since one of its factors is 2. If a^2 is even, it can be shown that a is even. Q2 demonstrates this fact. Thus, let $a = 2k$, where k is an integer. Then $a^2 = 4k^2$, and the above equation involving a^2 and b^2 becomes

$$2b^2 = 4k^2 \quad \text{or} \quad b^2 = 2k^2$$

This last statement says that b^2 is even, and, as before, b must be even. We now have that both a and b are even, but this contradicts the statement that a/b was "reduced." Thus, the assumption that $\sqrt{2} \in Q$ is false, and we conclude that $\sqrt{2} \notin Q$.

(The above indirect proof is an example of a proof by contradiction. Rather than use axioms or properties to arrive at the result directly, we assumed that the fact to be proven was not true and then showed that this assumption led to an impossible conclusion, thus contradicting our original assumption.)

A1: Z, 5M. Q, all are satisfied.

Q2: Prove that if $a \in Z$ and a^2 is even, then a is even.

Of course, there are other irrational numbers besides $\sqrt{2}$. In fact, there are "more" irrational numbers than rationals, although we shall neither prove that statement nor even try to explain it (as there are infinitely many of each). In the problems you are asked to show that any rational number has a repeating decimal expansion and, conversely, that any repeating decimal is a rational number. From this it follows that numbers like .10110111011110 . . . (where the number of 1's continually increases) are irrational.

To continue with our development of the real number system, we need to describe an ordered field. Recall that a field consists of a set, say, S, together with two operations which satisfy a certain collection of axioms. Such a field is said to be ordered if there exists a subset of S (let us call it P), with the following properties:

The Order Axioms

1. (trichotomy) For any $x \in S$, precisely one of the following three statements is true: $x \in P$, $-x \in P$, $x = 0$
2. (closure) If $x, y \in P$, then $x + y \in P$ and $x \cdot y \in P$.

If we are dealing with the real number system, P is the set of positive numbers. In Section 1.1 we introduced the concept of "less than" and the symbol $<$ by referring to a number scale. We can now finally define "less than" using the order axioms above.

Definition A.3

$a < b$ if, and only if, $b - a$ is a positive number.

Using this definition and the order axioms, we can prove the following important theorem stated in Section 1.4.

Theorem A.1

Suppose that $a < b$.
(1) If $b < c$, then $a < c$.
(2) For any number c, $a + c < b + c$.
(3) If c is a positive number, then $ca < cb$.
(4) If c is a negative number, then $ca > cb$.

Proof

We are given that $a < b$, so $b - a$ is a positive number.

(1) If $b < c$, then $c - b$ is a positive number. Since the sum of two positive numbers is a positive number, by Axiom 2, we can conclude that $(c - b) + (b - a) = c - a$ is a positive number.

Hence, $a < c$ by Definition A.3.

(2) $(b + c) - (a + c) = b - a$ is a positive number, so $a + c < b + c$.

(3) If c is a positive number, then $c(b - a) = cb - ca$ is a positive number. Thus, $ca < cb$.

(4) If c is a negative number, then $-c$ is a positive number. Since $b - a$ is also positive, $(-c)(b - a) = ca - cb$ is a positive number. It follows that $cb < ca$; that is, $ca > cb$.

In the problems you will show that both the real number system and the rational number system are examples of ordered fields. The axiom that distinguishes between these sets is the completeness property. Before we state this final axiom, we need to develop some terminology.

We say that a set of numbers is *bounded above* if there exists some real number which is greater than or equal to each number in the set. Such a number is called an *upper bound* for the set. For example, the set $A = \{x : x = 1/n : n \in N\} = \{1, 1/2, 1/3,\ldots\}$ is *bounded above* by 1, since $x \leq 1$ for every $x \in A$. Of course, there are other upper bounds for this set, $x \leq 2$ for every $x \in A$, $x \leq \sqrt{2}$ for any x in A, etc. However, the number 1 is significant for this set because it is the smallest of all the upper bounds for the set. (Any number smaller than 1 would not be an upper bound for this set.)

Definition A.4

A number b is an *upper bound* for a nonempty set of numbers A if $x \leq b$ for every $x \in A$. Furthermore, b is a *least upper bound* if $x \leq b$ for every $x \subset A$ (i.e., b is an upper bound for A) and $b \leq c$ where c is any other upper bound for A.

> Q3: Find least upper bounds for each of the following:
> $$A = (0, 4), \quad B = [0, 4]$$
> $$C = \left\{x : x = \frac{n}{n + 1}, n \in N\right\} = \left\{\frac{1}{2}, \frac{2}{3}, \frac{3}{4}, \ldots\right\}$$
> $$D = \{x : x = (-1)^n, n \in N\}$$
> $$E = \left\{x : x = \frac{n}{n^2 + 8}, n \in N\right\}$$

We are now ready to define completeness.

Definition A.5

A nonempty set S is said to be *complete* if every nonempty subset of S which is bounded above has a least upper bound in S.

The final axiom necessary to describe the real number system is the completeness axiom:

A2: Again we shall use a proof by contradiction. Assume that $a \in Z$ and a is *not* even. Then a is odd, which means we can write $a = 2k + 1$ where $k \in Z$ (i.e., k is some integer). Then $a^2 = (2k + 1)^2 = 4k^2 + 4k + 1 = 2(2k^2 + 2k) + 1$. But this last number is clearly odd, since $2k^2 + 2k$ is an integer and two times an integer plus one is an odd number. But, by hypothesis, a^2 is even, not odd. Thus, a cannot be odd and must, therefore, be even.

A3: The least upper bound for A and B is 4; the least upper bound for C and D is 1; and for E it is 3/17.

The Completeness Axiom

The set of real numbers is complete.

The completeness axiom has the geometrical interpretation that there are no "holes" on the real number line. That is to say, if but one point were removed, the resulting set would not be complete. The proof of this fact is illustrated in the following example.

Example A.1

Show that the set $S = \{x : x \in R,\ x \neq 3\}$ is not complete.

Solution

Consider the set $A = \{x : x < 3,\ x \in R\}$. Clearly, A is a nonempty subset of S, and A is bounded above in S. Now if S is complete, then A has a least upper bound *in* S by the completeness axiom. Call this least upper bound b. Now $b \neq 3$, since $3 \notin S$, so $b < 3$ or $b > 3$. First we show that if $b < 3$, b cannot be a bound, for we can always find a number larger than b but less than 3. For example, if $b < 3$, then $b + b < b + 3 < 6$ or $b < (b + 3)/2 < 3$. These last inequalities state that $(b + 3)/2 \in S$ (since $(b + 3)/2 < 3$), so b cannot be an upper bound for S (since $b < (b + 3)/2$). Thus, b cannot be less than 3 and still be a bound for S. Similarly, if $b > 3$, $b + b > 3 + b > 3 + 3$ or $b > (3 + b)/2 > 3$. This says that $(3 + b)/2$ is a bound for S which is less than b. This contradicts the fact that b is a least upper bound for S. We have shown that $b = 3$, $b < 3$, and $b > 3$ are all impossible. Therefore, S cannot be complete.

It is the completeness axiom which sets apart R from Q, for the rational numbers are not complete. We can show this, for example, by considering the set $T = \{x : x \in Q,\ x < \sqrt{2}\}$. Certainly T has many rational upper bounds. However, it can be shown that $\sqrt{2}$ is the least upper bound. But $\sqrt{2} \notin Q$, and hence Q is not complete.

Appendix B

x	0	1	2	3	4	5	6	7	8	9
1.0	.0000	.0043	.0086	.0128	.0170	.0212	.0253	.0294	.0334	.0374
1.1	.0414	.0453	.0492	.0531	.0569	.0607	.0645	.0682	.0719	.0755
1.2	.0792	.0828	.0864	.0899	.0934	.0969	.1004	.1038	.1072	.1106
1.3	.1139	.1173	.1206	.1239	.1271	.1303	.1335	.1367	.1399	.1430
1.4	.1461	.1492	.1523	.1553	.1584	.1614	.1644	.1673	.1703	.1732
1.5	.1761	.1790	.1818	.1847	.1875	.1903	.1931	.1959	.1987	.2014
1.6	.2041	.2068	.2095	.2122	.2148	.2175	.2201	.2227	.2253	.2279
1.7	.2304	.2330	.2355	.2380	.2405	.2430	.2455	.2480	.2504	.2529
1.8	.2553	.2577	.2601	.2625	.2648	.2672	.2695	.2718	.2742	.2765
1.9	.2788	.2810	.2833	.2856	.2878	.2900	.2923	.2945	.2967	.2989
2.0	.3010	.3032	.3054	.3075	.3096	.3118	.3139	.3160	.3181	.3201
2.1	.3222	.3243	.3263	.3284	.3304	.3324	.3345	.3365	.3385	.3404
2.2	.3424	.3444	.3464	.3483	.3502	.3522	.3541	.3560	.3579	.3598
2.3	.3617	.3636	.3655	.3674	.3692	.3711	.3729	.3747	.3766	.3784
2.4	.3802	.3820	.3838	.3856	.3874	.3892	.3909	.3927	.3945	.3962
2.5	.3979	.3997	.4014	.4031	.4048	.4065	.4082	.4099	.4116	.4133
2.6	.4150	.4166	.4183	.4200	.4216	.4232	.4249	.4265	.4281	.4298
2.7	.4314	.4330	.4346	.4362	.4378	.4393	.4409	.4425	.4440	.4456
2.8	.4472	.4487	.4502	.4518	.4533	.4548	.4564	.4579	.4594	.4609
2.9	.4624	.4639	.4654	.4669	.4683	.4698	.4713	.4728	.4742	.4757
3.0	.4771	.4786	.4800	.4814	.4829	.4843	.4857	.4871	.4886	.4900
3.1	.4914	.4928	.4942	.4955	.4969	.4983	.4997	.5011	.5024	.5038
3.2	.5051	.5065	.5079	.5092	.5105	.5119	.5132	.5145	.5159	.5172
3.3	.5185	.5198	.5211	.5224	.5237	.5250	.5263	.5276	.5289	.5302
3.4	.5315	.5328	.5340	.5353	.5366	.5378	.5391	.5403	.5416	.5428
3.5	.5441	.5453	.5465	.5478	.5490	.5502	.5514	.5527	.5539	.5551
3.6	.5563	.5575	.5587	.5599	.5611	.5623	.5635	.5647	.5658	.5670
3.7	.5682	.5694	.5705	.5717	.5729	.5740	.5752	.5763	.5775	.5786
3.8	.5798	.5809	.5821	.5832	.5843	.5855	.5866	.5877	.5888	.5899
3.9	.5911	.5922	.5933	.5944	.5955	.5966	.5977	.5988	.5999	.6010

Table B.1, Cont.

x	0	1	2	3	4	5	6	7	8	9
4.0	.6021	.6031	.6042	.6053	.6064	.6075	.6085	.6096	.6107	.6117
4.1	.6128	.6138	.6149	.6160	.6170	.6180	.6191	.6201	.6212	.6222
4.2	.6232	.6243	.6253	.6263	.6274	.6284	.6294	.6304	.6314	.6325
4.3	.6335	.6345	.6355	.6365	.6375	.6385	.6395	.6405	.6415	.6425
4.4	.6435	.6444	.6454	.6464	.6474	.6484	.6493	.6503	.6513	.6522
4.5	.6532	.6542	.6551	.6561	.6571	.6580	.6590	.6599	.6609	.6618
4.6	.6628	.6637	.6646	.6656	.6665	.6675	.6684	.6693	.6702	.6712
4.7	.6721	.6730	.6739	.6749	.6758	.6767	.6776	.6785	.6794	.6803
4.8	.6812	.6821	.6830	.6839	.6848	.6857	.6866	.6875	.6884	.6893
4.9	.6902	.6911	.6920	.6928	.6937	.6946	.6955	.6964	.6972	.6981
5.0	.6990	.6998	.7007	.7016	.7024	.7033	.7042	.7050	.7059	.7067
5.1	.7076	.7084	.7093	.7101	.7110	.7118	.7126	.7135	.7143	.7152
5.2	.7160	.7168	.7177	.7185	.7193	.7202	.7210	.7218	.7226	.7235
5.3	.7243	.7251	.7259	.7267	.7275	.7284	.7292	.7300	.7308	.7316
5.4	.7324	.7332	.7340	.7348	.7356	.7364	.7372	.7380	.7388	.7396
5.5	.7404	.7412	.7419	.7427	.7435	.7443	.7451	.7459	.7466	.7474
5.6	.7482	.7490	.7497	.7505	.7513	.7520	.7528	.7536	.7543	.7551
5.7	.7559	.7566	.7574	.7582	.7589	.7597	.7604	.7612	.7619	.7627
5.8	.7634	.7642	.7649	.7657	.7664	.7672	.7679	.7686	.7694	.7701
5.9	.7709	.7716	.7723	.7731	.7738	.7745	.7752	.7760	.7767	.7774
6.0	.7782	.7789	.7796	.7803	.7810	.7818	.7825	.7832	.7839	.7846
6.1	.7853	.7860	.7868	.7875	.7882	.7889	.7896	.7903	.7910	.7917
6.2	.7924	.7931	.7938	.7945	.7952	.7959	.7966	.7973	.7980	.7987
6.3	.7993	.8000	.8007	.8014	.8021	.8028	.8035	.8041	.8048	.8055
6.4	.8062	.8069	.8075	.8082	.8089	.8096	.8102	.8109	.8116	.8122
6.5	.8129	.8136	.8142	.8149	.8156	.8162	.8169	.8176	.8182	.8189
6.6	.8195	.8202	.8209	.8215	.8222	.8228	.8235	.8241	.8248	.8254
6.7	.8261	.8267	.8274	.8280	.8287	.8293	.8299	.8306	.8312	.8319
6.8	.8325	.8331	.8338	.8344	.8351	.8357	.8363	.8370	.8376	.8382
6.9	.8388	.8395	.8401	.8407	.8414	.8420	.8426	.8432	.8439	.8445

x	0	1	2	3	4	5	6	7	8	9
7.0	.8451	.8457	.8463	.8470	.8476	.8482	.8488	.8494	.8500	.8506
7.1	.8513	.8519	.8525	.8531	.8537	.8543	.8549	.8555	.8561	.8567
7.2	.8573	.8579	.8585	.8591	.8597	.8603	.8609	.8615	.8621	.8627
7.3	.8633	.8639	.8645	.8651	.8657	.8663	.8669	.8675	.8681	.8686
7.4	.8692	.8698	.8704	.8710	.8716	.8722	.8727	.8733	.8739	.8745
7.5	.8751	.8756	.8762	.8768	.8774	.8779	.8785	.8791	.8797	.8802
7.6	.8808	.8814	.8820	.8825	.8831	.8837	.8842	.8848	.8854	.8859
7.7	.8865	.8871	.8876	.8882	.8887	.8893	.8899	.8904	.8910	.8915
7.8	.8921	.8927	.8932	.8938	.8943	.8949	.8954	.8960	.8965	.8971
7.9	.8976	.8982	.8987	.8993	.8998	.9004	.9009	.9015	.9020	.9025
8.0	.9031	.9036	.9042	.9047	.9053	.9058	.9063	.9069	.9074	.9079
8.1	.9085	.9090	.9096	.9101	.9106	.9112	.9117	.9122	.9128	.9133
8.2	.9138	.9143	.9149	.9154	.9159	.9165	.9170	.9175	.9180	.9186
8.3	.9191	.9196	.9201	.9206	.9212	.9217	.9222	.9227	.9232	.9238
8.4	.9243	.9248	.9253	.9258	.9263	.9269	.9274	.9279	.9284	.9289
8.5	.9294	.9299	.9304	.9309	.9315	.9320	.9325	.9330	.9335	.9340
8.6	.9345	.9350	.9355	.9360	.9365	.9370	.9375	.9380	.9385	.9390
8.7	.9395	.9400	.9405	.9410	.9415	.9420	.9425	.9430	.9435	.9440
8.8	.9445	.9450	.9455	.9460	.9465	.9469	.9474	.9479	.9484	.9489
8.9	.9494	.9499	.9504	.9509	.9513	.9518	.9523	.9528	.9533	.9538
9.0	.9542	.9547	.9552	.9557	.9562	.9566	.9571	.9576	.9581	.9586
9.1	.9590	.9595	.9600	.9605	.9609	.9614	.9619	.9624	.9628	.9633
9.2	.9638	.9643	.9647	.9652	.9657	.9661	.9666	.9671	.9675	.9680
9.3	.9685	.9689	.9694	.9699	.9703	.9708	.9713	.9717	.9722	.9727
9.4	.9731	.9736	.9741	.9745	.9750	.9754	.9759	.9763	.9768	.9773
9.5	.9777	.9782	.9786	.9791	.9795	.9800	.9805	.9809	.9814	.9818
9.6	.9823	.9827	.9832	.9836	.9841	.9845	.9850	.9854	.9859	.9863
9.7	.9868	.9872	.9877	.9881	.9886	.9890	.9894	.9899	.9903	.9908
9.8	.9912	.9917	.9921	.9926	.9930	.9934	.9939	.9943	.9948	.9952
9.9	.9956	.9961	.9965	.9969	.9974	.9978	.9983	.9987	.9991	.9996

Table B.2, Cont.

$$\cos t, \sin t, \tan t : 0 \leq t \leq \pi/2$$

t	$\cos t$	$\sin t$	$\tan t$	t	$\cos t$	$\sin t$	$\tan t$
.00	1.0000	.0000	.0000	.45	.9004	.4350	.4831
.01	1.0000	.0100	.0100	.46	.8961	.4439	.4954
.02	.9998	.0200	.0200	.47	.8916	.4529	.5080
.03	.9996	.0300	.0300	.48	.8870	.4618	.5206
.04	.9992	.0400	.0400	.49	.8823	.4706	.5334
.05	.9988	.0500	.0500	.50	.8776	.4794	.5463
.06	.9982	.0600	.0601	.51	.8727	.4882	.5594
.07	.9976	.0699	.0701	.52	.8678	.4969	.5726
.08	.9968	.0799	.0802	.53	.8628	.5055	.5859
.09	.9960	.0899	.0902	.54	.8577	.5141	.5994
.10	.9950	.0998	.1003	.55	.8525	.5227	.6131
.11	.9940	.1098	.1104	.56	.8473	.5312	.6269
.12	.9928	.1197	.1206	.57	.8419	.5396	.6310
.13	.9916	.1296	.1307	.58	.8365	.5480	.6552
.14	.9902	.1395	.1409	.59	.8309	.5564	.6696
.15	.9888	.1494	.1511	.60	.8253	.5646	.6841
.16	.9872	.1593	.1614	.61	.8196	.5729	.6989
.17	.9856	.1692	.1717	.62	.8139	.5810	.7139
.18	.9838	.1790	.1820	.63	.8080	.5891	.7291
.19	.9820	.1889	.1923	.64	.8021	.5972	.7445
.20	.9801	.1987	.2027	.65	.7961	.6052	.7602
.21	.9780	.2085	.2131	.66	.7900	.6131	.7761
.22	.9759	.2182	.2236	.67	.7838	.6210	.7923
.23	.9737	.2280	.2341	.68	.7776	.6288	.8087
.24	.9713	.2377	.2447	.69	.7712	.6365	.8253
.25	.9689	.2474	.2553	.70	.7648	.6442	.8423
.26	.9664	.2571	.2660	.71	.7584	.6518	.8595
.27	.9638	.2667	.2768	.72	.7518	.6594	.8771
.28	.9611	.2764	.2876	.73	.7452	.6669	.8949
.29	.9582	.2860	.2984	.74	.7358	.6743	.9131
.30	.9553	.2955	.3094	.75	.7317	.6816	.9316
.31	.9523	.3051	.3203	.76	.7248	.6889	.9505
.32	.9492	.3146	.3314	.77	.7179	.6961	.9697
.33	.9460	.3240	.3425	.78	.7109	.7033	.9893
.34	.9428	.3335	.3537	.79	.7038	.7104	1.009
.35	.9394	.3429	.3650	.80	.6967	.7174	1.030
.36	.9359	.3523	.3764	.81	.6895	.7243	1.050
.37	.9323	.3616	.3879	.82	.6822	.7311	1.072
.38	.9287	.3709	.3994	.83	.6749	.7379	1.093
.39	.9249	.3802	.4111	.84	.6675	.7446	1.116
.40	.9211	.3894	.4228	.85	.6600	.7513	1.138
.41	.9171	.3986	.4346	.86	.6524	.7578	1.162
.42	.9131	.4078	.4466	.87	.6448	.7643	1.185
.43	.9090	.4169	.4586	.88	.6372	.7707	1.210
.44	.9048	.4259	.4708	.89	.7294	.7771	1.235

t	cos t	sin t	tan t	t	cos t	sin t	tan t
.90	.6216	.7833	1.260	1.35	.2190	.9757	4.455
.91	.6137	.7895	1.286	1.36	.2092	.9779	4.673
.92	.6058	.7956	1.313	1.37	.1994	.9799	4.913
.93	.5978	.8016	1.341	1.38	.1896	.9819	5.177
.94	.5898	.8076	.1369	1.39	.1798	.9837	5.471
.95	.5817	.8134	1.398	1.40	.1700	.9854	5.798
.96	.5735	.8192	1.428	1.41	.1601	.9871	6.165
.97	.5653	.8249	1.459	1.42	.1502	.9887	6.581
.98	.5570	.8305	1.491	1.43	.1403	.9901	7.055
.99	.5487	.8360	1.524	1.44	.1304	.9915	7.602
1.00	.5403	.8415	1.557	1.45	.1205	.9927	8.238
1.01	.5319	.8468	1.592	1.46	.1106	.9939	8.989
1.02	.5234	.8521	1.628	1.47	.1006	.9949	9.887
1.03	.5148	.8573	1.665	1.48	.0907	.9959	10.938
1.04	.5062	.8624	1.704	1.49	.0807	.9967	12.350
1.05	.4976	.8674	1.743	1.50	.0707	.9975	14.101
1.06	.4889	.8724	1.784	1.51	.0608	.9982	16.428
1.07	.4801	.8772	1.827	1.52	.0508	.9987	19.670
1.08	.4713	.8820	1.871	1.53	.0408	.9992	24.498
1.09	.4625	.8866	1.917	1.54	.0308	.9995	32.461
1.10	.4536	.8912	1.965	1.55	.0208	.9998	48.078
1.11	.4447	.8957	2.014	1.56	.0108	.9999	92.620
1.12	.4357	.9001	2.066	1.57	.0008	1.0000	1255.8
1.13	4267	,9044	2.120	1.58	−.0092	1.0000	−108.65
1.14	.4176	.9086	2.176	1.59	−.0192	.9998	−52.067
				1.60	−.0292	.9996	−34.233
1.15	.4085	.9128	2.234				
1.16	.3993	.9168	2.296				
1.17	.3902	.9208	2.360				
1.18	.3809	.9246	2.427				
1.19	.3717	.9284	2.498				
1.20	.3624	.9320	2.572				
1.21	.3530	.9356	2.650				
1.22	.3436	.9391	2.733				
1.23	.3342	.9425	2.820				
1.24	.3248	.9458	2.912				
1.25	.3153	.9490	3.010				
1.26	.3058	.9521	3.113				
1.27	.2963	.9551	3.224				
1.28	.2867	.9580	3.341				
1.29	.2771	.9608	3.467				
1.30	.2675	.9636	3.602				
1.31	.2579	.9662	3.747				
1.32	.2482	.9687	3.903				
1.33	.2385	.9711	4.072				
1.34	.2288	.9735	4.256				

Table B.3

$\cos \theta, \sin \theta, \tan \theta : 0° \leq \theta \leq 90°$

Radians	Degrees				Radians	Degrees			
t	θ	cos θ	sin θ	tan θ	t	θ	cos θ	sin θ	tan θ
.000	0	1.0000	.0000	.0000	.785	45	.7071	.7071	1.0000
.017	1	.9999	.0175	.0175	.803	46	.6947	.7193	1.0355
.035	2	.9994	.0349	.0349	.820	47	.6820	.7314	1.0724
.052	3	.9986	.0523	.0524	.838	48	.6691	.7431	1.1106
.070	4	.9976	.0698	.0699	.855	49	.6561	.7547	1.1504
.087	5	.9962	.0872	.0875	.873	50	.6428	.7660	1.1918
.105	6	.9945	.1045	.1051	.890	51	.6293	.7772	1.2349
.122	7	.9926	.1219	.1228	.908	52	.6157	.7880	1.2799
.140	8	.9903	.1392	.1405	.925	53	.6018	.7986	1.3270
.157	9	.9877	.1564	.1584	.942	54	.5878	.8090	1.3764
.175	10	.9848	.1737	.1763	.960	55	.5736	.8192	1.4281
.192	11	.9816	.1908	.1944	.977	56	.5591	.8290	1.4826
.209	12	.9782	.2079	.2126	.994	57	.5446	.8387	1.5399
.227	13	.9744	.2250	.2309	1.012	58	.5299	.8481	1.6003
.244	14	.9703	.2419	.2493	1.030	59	.5150	.8572	1.6643
.262	15	.9659	.2588	.2680	1.047	60	.5000	.8660	1.7321
.279	16	.9613	.2756	.2868	1.065	61	.4848	.8746	1.8040
.297	17	.9563	.2924	.3058	1.082	62	.4695	.8830	1.8807
.314	18	.9511	.3090	.3249	1.100	63	.4540	.8910	1.9626
.332	19	.9455	.3256	.3443	1.117	64	.4384	.8988	2.0503
.349	20	.9397	.3420	.3640	1.134	65	.4226	.9063	2.1445
.366	21	.9336	.3584	.3839	1.152	66	.4067	.9136	2.2460
.384	22	.9272	.3746	.4040	1.169	67	.3907	.9205	2.3559
.401	23	.9205	.3907	.4245	1.187	68	.3746	.9272	2.4751
.419	24	.9136	.4067	.4452	1.204	69	.3584	.9336	2.6051
.436	25	.9063	.4226	.4663	1.222	70	.3420	.9367	2.7475
.454	26	.8988	.4384	.4877	1.239	71	.3256	.9455	2.9042
.471	27	.8910	.4540	.5095	1.257	72	.3090	.9511	3.0777
.489	28	.8830	.4695	.5317	1.274	73	.2924	.9563	3.2709
.506	29	.8746	.4848	.5543	1.292	74	.2756	.9613	3.4874
.524	30	.8660	.5000	.5774	1.309	75	.2588	.9659	3.7312
.541	31	.8572	.5150	.6009	1.326	76	.2419	.9703	4.0108
.559	32	.8481	.5299	.6249	1.344	77	.2250	.9744	4.3315
.576	33	.8387	.5446	.6494	1.361	78	.2079	.9782	4.7046
.593	34	.8290	.5592	.6746	1.379	79	.1908	.9816	5.1446
.611	35	.8192	.5736	.7002	1.396	80	.1737	.9848	5.6713
.628	36	.8090	.5878	.7265	1.414	81	1.564	.9877	6.3138
.646	37	.7986	.6018	.7536	1.431	82	.1392	.9903	7.1154
.663	38	.7880	.6157	.7813	1.449	83	.1219	.9926	8.1443
.681	39	.7772	.6293	.8098	1.466	84	.1045	.9945	9.5144
.698	40	.7660	.6428	.8391	1.484	85	.0872	.9962	11.430
.716	41	.7547	.6561	.8693	1.501	86	.0698	.9976	14.301
.733	42	.7431	.6691	.9004	1.518	87	.0523	.9986	19.081
.750	43	.7314	.6820	.9325	1.536	88	.0349	.9994	28.636
.768	44	.7193	.6947	.9657	1.553	89	.0175	.9999	57.290

472

INDEX

105
+ 215.25
331.01
452.56

21.55

315.25